BEYOND HETEROCHRONY

BEYOND HETEROCHRONY

The Evolution of Development

EDITED BY

Miriam Leah Zelditch

Museum of Paleontology
University of Michigan

A JOHN WILEY & SONS, INC., PUBLICATION

New York • Chichester • Weinheim • Brisbane • Singapore • Toronto

This book is printed on acid-free paper. ♾

Published simultaneously in Canada.

For ordering and customer service, call 1-800-CALL-WILEY.

Library of Congress Cataloging-in-Publication Data:

Beyond heterochrony : the evolution of development / edited by Miriam Zelditch.
p. cm.
Includes bibliographical references (p.).
ISBN 0-471-37973-5 (cloth : alk. paper)
1. Heterochrony (Biology). 2. Developmental biology. 3. Morphogenesis. I. Zelditch, Miriam, 1952–
QH395 .B49 2001
576.8—dc21 2001023742

Printed in the United States of America.

10 9 8 7 6 5 4 3 2 1

CONTENTS

FOREWORD

The publication of *On The Origin of Species* by Charles Darwin initiated an intensive search for physical evidence of evolution. An important line of evidence emerged from embryos. The discovery in embryos of indications of relationships between organisms led to new groupings of animals and to a much greater understanding of the relationships, shared history, and ancestry of both invertebrates and vertebrates. For example, the discovery that vertebrates, cephalochordates, and urochordates shared a notochord led to the establishment of the chordates as a "natural" group. Along with such physical evidence came theories to explain how embryos illuminated evolution. What essential relationship(s) between embryos, adults, and ancestors allowed embryology to become a chief source of evidence for evolution?

An influential view, subsequently proven false, was Haeckel's theory that ontogeny recapitulates phylogeny (*die Ontogenie ist eine Rekapitulation der Phylogenie*), which specified (in later version but not as initially enunciated by Haeckel), that, in their development, animals repeat the adult stages of their ancestors.[1] Haeckel proposed heterochrony (change in time) and heterotopy (change in position) to explain how modification during ontogeny could influence evolution.

This book situates heterochrony and heterotopy in their appropriate phylogenetic contexts, evaluates the ways they have contributed to our understanding of links between ontogeny and phylogeny, and asks, Where do we go from here? What lies *Beyond Heterochrony*? Those who see less explanatory power in the concept might ask "what lies beyond the hegemony of heterochrony?"

Approaches to heterochrony have differed vastly, as has the importance given to heterochrony as an evolutionary mechanism. On the one hand

> We know of only one other example [other than] paedomorphosis in urodeles and the evolution of free-living vertebrates from sessile, tunicate-like ancestors ... in

[1] In its original form (which lacked any suggestion that adult stages were recapitulated), Haeckel's view reflected a recapitulation with which most would be quite comfortable: "Ontogeny is a brief and rapid recapitulation of Phylogeny, dependent on the physiological functions of Heredity (reproduction) and Adaptation (nutrition)" (Haeckel, 1866, vol. 2, p. 300). We tend to forget that Haeckel was a staunch Darwinian, not a radical anti-evolutionist; see the title of his book.

> which a heterochronic process has been implicated in the evolution of derived larval forms of marine invertebrates.—Hart and Wray, 1999, p. 161

On the other hand

> What I would argue is that it [heterochrony] permeates every nook and cranny of evolution. Indeed, without it evolution wouldn't have happened. For it explains everything, from the shape of a delphinium flower, to a horses' foot, to the song of a bird.—McNamara, 1997, p. 46

How can a concept (alteration in the timing of some aspect of development), which has a basis that is so simple and which acts on one of the most fundamental aspects of developmental processes (time) engender such divergent reactions? Reading this book will help you to answer this and many other questions pertaining to heterochrony, heterotopy, and indeed to the evolution of development. The reader will find most approaches to heterochrony that are based on testable hypotheses in this volume. For instance, is heterochrony a concept worth keeping? Does heterochrony contribute to the generation of evolutionary novelties (establishing new morphologies or even returning organisms to old trajectories) or merely move structures along an ancestral trajectory? Are time-keeping genes also genes for heterochrony?[2] Approaches that are not based on testable hypotheses are also discussed and found to be wanting.

What of heterotopy, that is, alteration in the position within an embryo (or larva or postnatally) where a structure forms? Heterotopy, a term also coined by Haeckel, but now the forgotten ugly sister to heterochrony, is a second developmental mechanism for evolutionary change. Some see vastly more evolutionary potential in heterotopy than in heterochrony. Some see even more potential in a combination of heterochrony and heterotopy and more still when heterotypy (change in type) and heterometry (change in amount) are included in the embryo's armamentarium of evolutionary developmental mechanisms.[3] These other "heteros" can take organisms beyond (outside) the ancestral ontogenetic trajectory perpetuated by heterochrony. This is where the origin of novelty lies.

This book deals with both theoretical and practical approaches to heterochrony and heterotopy, using examples from extant and extinct forms of animals and plants. Novel approaches are elaborated. Reflecting, in part, differences in how the time axis is interpreted, definitions of heterochrony abound, including

- change in developmental "timing" (where timing includes rate changes),
- change in the timing of a developmental process,

[2] See Adoutte (2000) and Pasquinelli et al. (2000) for a small (21 nucleotide) RNA that acts as a time-keeping gene in *C. elegans*, *Drosophila*, three mammals, two ascidians, a fish, a frog, an annelid, a mollusk, and an echinoderm.

[3] See Brylski and Hall (1988), Zelditch and Fink (1996), Hall (1984, 1999, 2001), Wake (1996), Rice (1997), Arthur (2000), and Li and Johnston (2000) for these positions. Li and Johnston also present the perspective that, in plants, heterotopy can be equated with homeosis.

- change in developmental rate or timing that produces parallelism between ontogeny and phylogeny, and
- evolutionary change in development.

The debate you will find in these pages is refreshing. In short, this book lives up to its title. It takes the evolution of development where it belongs, which is beyond heterochrony and into new and uncharted waters.

BRIAN K. HALL

Department of Biology, Dalhousie University, Halifax NS, Canada

REFERENCES

Adoutte A (2000): Small but mighty timekeepers. Nature 408: 37–38.

Arthur W (2000): The concept of developmental reprogramming and the quest for an inclusive theory of evolutionary mechanisms. Evol Dev 2: 49–57.

Brylski P, Hall BK (1988): Ontogeny of a macroevolutionary phenotype: the external cheek pouches of Geomyoid rodents. Evolution 42: 391–395.

Haeckel E (1866): *Generelle Morphologie der Organismen: Allgemeine GrundzŸge der organischen Formen-Wissenschaft, mechanisch begrŸndet durch die von* Charles Darwin *reformite Descendenz-Theorie*. Berlin: Georg Reimer, 2 vols.

Hall BK (1984): Developmental processes underlying heterochrony as an evolutionary mechanism. Can J Zoolol 62: 1–7.

Hall BK (1999): *Evolutionary Developmental Biology* (2nd ed). Dordrecht, The Netherlands: Kluwer Academic Publishers.

Hall BK (2001): Evolutionary developmental biology: where embryos and fossils meet. In: McNamara KJ, Minugh-Purves N (eds), *Human Evolution Through Developmental Change*. Baltimore, MD: Johns Hopkins University Press (in press).

Hart MW, Wray GA (1999): Heterochrony. In: Hall BK (ed), *The Origin and Evolution of Larval Forms*. San Diego, CA: Academic Press, p. 159–165.

Li P, Johnston MO (2000): Heterochrony in plant evolutionary studies through the Twentieth century. Bot Rev 66: 57–88.

McNamara KJ (1997): *Shapes of Time. The Evolution of Growth and Development*. Baltimore, MD: The Johns Hopkins University Press.

Pasquinelli AE, Reinhart BJ, Slack F et al. (2000): Conservation of the sequence and temporal expression of *let-7* heterochronic regulatory RNA. Nature 408: 86–89.

Rice SH (1997): The analysis of ontogenetic trajectories: When a change in size or shape is not heterochrony. Proc Natl Acad Sci USA 94: 907–912.

Wake DB (1996): Evolutionary developmental biology—prospects for an evolutionary synthesis at the developmental level. Mem Cal Acad Sci 20: 97–107.

Zelditch ML, Fink WL (1996): Heterochrony and heterotopy: innovation and stability in the evolution of form. Paleobiology 22: 241–254.

PREFACE

Waddington (1967) posed a fundamental question for evolutionary biology: "How do you come to have horses and tigers, and things?" Increasingly, the answer is being sought in developmental biology. Surprisingly often, heterochrony (evolutionary change in developmental rate or timing) seems to be the answer to that question. The concept of heterochrony is hardly new; it dominated evolutionary developmental biology in the late 19th century. However, it fell into disrepute for various reasons, to be rehabilitated in Gould's (1977) *Ontogeny and Phylogeny*. Soon thereafter, the seminal paper by Alberch and colleagues (Alberch et al., 1979) provided a conceptual framework (and semi-operational method) for empirical studies. Over a remarkably short time, heterochrony was documented in numerous groups, among them trilobites (McNamara, 1978; McNamara, 1981), salamanders (Alberch and Alberch, 1981; Wake, 1980), actinopteryigans (Fink, 1981), lung fish (Bemis, 1984), bryozoans (Anstey, 1987), and echinoderms (McKinney, 1984). Just over one decade later, hundreds of examples were compiled, providing what seems like compelling evidence for the frequency and evolutionary significance of heterochrony (McKinney, 1988). Further evidence continues to amass nearly weekly. Sometimes it seems that heterochrony *is* the proximate causes of morphological diversity.

The preeminence of heterochrony in the modern literature can be documented by a highly informal survey; searching Science Citation Index (the electronic version) for two key works, "heterochrony" and "heterotopy" (evolutionary changes in the location of development), another term coined by Haeckel, the search for heterochrony produced 333 papers, whereas the search for heterotopy produced only 21, of which 10 were about medical pathologies not evolution. Perhaps, this disparity accurately reflects the relative frequencies of heterochrony and heterotopy in nature. Perhaps, it reflects the recent broadening of the term "heterochrony" to the point that virtually any change in ontogeny can be interpreted as heterochrony (see Gould, 2000, on that recent broadening). Or perhaps it reflects the disproportionate attention lavished on heterochrony, at the expense of alternatives (such as heterotopy). In looking beyond heterochrony, our aim in this volume is to attend to a variety of explanations, not just heterochrony. In doing so, we do not dismiss the possibility that heterochrony might explain our data—indeed, this is one of the hypotheses tested by most authors. Certainly, we do not intend to disparage the scientific value of heterochrony. Rather, we aim to place it in a richer, broader context. The central premise of this book is that we cannot afford to single out one phenomenon at the expense of all others if we hope to understand how development evolves.

As evident from the chapters of this volume, there is more to evolution than heterochrony. There is also more than heterotopy, although heterotopy does appear to be a common phenomenon, based on studies in this book. I did not anticipate this emphasis on heterotopy when soliciting chapters for this book, but perhaps I should have. After all, the spatial patterning of development has been a major interest of developmental biologists for decades. Clearly, we cannot understand the evolution of pattern formation, including changes in the spatial organization of growth, until we examine both spatial and temporal aspects of development. Heterotopy deserves (even needs) as much attention as heterochrony; Hall (1994) predicted that heterotopy would come into its own as interest in heterochrony wanes and our knowledge of developmental mechanisms increases. Perhaps we do not need a waning of interest in heterochrony so much as more inclusive perspective on evolutionary developmental biology. This field includes more than heterochrony and more than heterotopy—development can evolve in more than its timing and spatial patterning, as documented by the chapters in this volume.

Each chapter contains one or more case studies exploring the developmental basis of morphological evolution. None simply presents a compilation of familiar examples. Rather, each chapter makes a substantive and original contribution to the literature. The issues addressed are varied and include the developmental basis for the loss of antipredator defenses in a lineage of gastropods (Nehm, Chapter 1), the cellular basis of the diversity of pigment patterns in salamanders (Parichy, Chapter 7), the origin of flowers (Frohlich, Chapter 3), the modularity of the axial skeleton in snakes (Polly, Head, and Cohn, Chapter 9), and the utility of ontogenetic sequences in phylogenetic reconstruction (Hufford, Chapter 2). Each chapter provides the data on which the conclusions rest, as well as the phylogenetic context of the evolutionary interpretations. Taken together, they show the value of looking beyond heterochrony, but they need not be viewed collectively—each chapter can stand on its own.

Half of the papers in this book analyze the ontogeny and phylogeny of shape, the data at the heart of the studies of heterochrony for decades. The traditional models for heterochrony were formulated in terms of size, shape, and age (Gould, 1977; Alberch et al., 1979), making the developmental basis of evolutionary changes in form of special interest. Using both traditional and novel geometric methods of shape analysis, these chapters explicitly test the hypothesis of heterochrony. It may seem that these studies test the hypothesis almost too rigorously, but Nehm's analysis of the loss of anti-predator defenses in marginellid gastropods shows that the hypothesis does not need to be rejected when subjected to stringent tests. For that case, heterochrony is a compelling explanation for the evolution of form. However, in some, it is not very important. Webster, Sheets, and Hughes (Chapter 4) examine the role of heterochrony in the evolution of cephalic form in Lower Cambrian olenellid trilobites, one of the paradigm cases of heterochrony. They document a complex ontogeny of form, containing distinguishable phases, which do not evolve by heterochrony. Guralnick and Kurpius (Chapter 6) analyze intraspecific variation in form and raise several methodological issues concerning the analysis of shape in bivalves. They show that variation is not constrained as expected under a hypothesis of heterochrony. Roopnarine (Chapter 8) tackles another difficult and important methodological issue—the phylogenetic interpretation of shape data—to examine the evolutionary changes in ontogeny in a genus of bivalves. Finally, my colleagues Sheets and Fink and I (Chapter 5) examine the spatial patterning of juvenile growth in piranhas, using a technique pioneered by Huxley (1932), the analysis of growth profiles, to explore the relationship between spatial complexity and evolutionary dynamics of growth. We also test and reject the hypothesis of conservatism.

The other five chapters exemplify novel approaches and sometimes startlingly original ideas. The questions and hypotheses are diverse, as are the methods of analysis. Each goes beyond heterochrony in a different way. Polly, Head and Cohn (Chapter 9) concentrate on one crucial (but often ignored premise) of studies of heterochrony—that it involves a dissociation in timing between two developmental modules; in their study, they ask whether the tail and trunk of snakes are indeed dissociable modules. Answering this question requires novel methods because the key question concerns modularity, a topic of considerable interest but one that has rarely been addressed so thoroughly, integrating development, morphology, and evolutionary analyses. Hufford (Chapter 2) develops novel methods to determine whether ontogenetic sequence data are useful in phylogenetic reconstructions and to ascertain what such reconstructions tell us about the evolution of morphology. He raises several important questions, especially about the units to which the concept of homology is applied. Parichy (Chapter 7) examines the evolution of pigment patterns in amphibians at the cellular level, focusing on their morphogenetic behavior, using experimental approaches to testing various hypotheses, not only about rates and timings but also about such features as the cues responsible for melanophore localization. Shapiro and Carl (Chapter 10) discuss a variety of factors that might affect limb development in two nontraditional model systems, the skink and direct-developing frog, focusing on cartilage condensation patterns in skinks and on limb outgrowth in direct-developing frogs lacking the apical ectodermal ridge. One clear message of their study is the importance of looking beyond model systems as well as beyond heterochrony. Frohlich (Chapter 3) offers a highly original and detailed scenario for the evolution of bisexual reproductive units in angiosperms, implicating a change in the position in which a structure develops as a key element. Although all these chapters discuss heterochrony, their major contribution may lie in the novelty of their questions and methods.

I should note that the concept of heterochrony has several meanings, both in the literature as a whole and in this book. This could prove disconcerting to readers who expect scientific terms to be unambiguous. Unfortunately, "heterochrony" may rival "homology" and "species" in its multiplicity of definitions. A full review of its semantics is beyond the scope of this introduction, but some clarification is important, if only to understand the tests of the hypothesis. Each test is framed in light of an author's understanding of the concept, and, because they understand it differently, they test it differently. At present, there are four widely used definitions of heterochrony in the literature. One is the traditional concept of heterochrony, the one stated by Gould (1977) and formalized by Alberch et al. (1979). According to this definition, heterochrony refers to changes in developmental rate or timing that result in parallelism between ontogeny and phylogeny. Given this definition, heterochrony is empirically documented by that parallelism or by finding that taxa share a common ontogenetic trajectory, differing only in its rate or timing. A second definition, stated by Raff and Wray (1989), focuses on dissociations in timing among individual developmental processes; according to this definition, heterochrony is an evolutionary change in the rate or timing of a developmental process relative to other processes. Given this definition, heterochrony is empirically documented when such temporal dissociations of processes occur and are responsible for the novelty of interest. A third concept of heterochrony refers to a permutation or change in rate/timing of ontogenetic sequences (Alberch and Alberch, 1981). According to this definition, heterochrony is documented by showing that conserved units of the sequence are altered in relative timings or rate. A fourth concept encompasses all of these, along with everything else. According to this very broad definition, heterochrony is virtually synonymous with evolutionary change of ontogeny (e.g., McKinney and McNamara, 1991; Klingenberg, 1998). This one requires no empirical documentation because there is no conceivable falsifier. As Klingenberg states, heterochrony (as defined in this broad sense) is uninformative.

As editor, I could have insisted that all authors adopt the same definition, but I chose not to for three reasons. The first is that authors use different definitions because they think about heterochrony differently. Asking them to adopt another definition for the sake of this book would amount to forcing a paradigm shift. The second is that the meaningful concepts are equally interesting and equally valid. Gould's definition has historical priority, and it is the one linking heterochrony to life-history theory and to the notion of intrinsic channels on the evolution of form. However, the mechanistic developmental concept has its own advantages because of its emphasis on process. Heterochrony is often defined simply as "evolutionary changes in developmental rate or timing," which does indeed imply that it is about process and attempts to explain why heterochrony is common often evoke theories of process. Taking Gould's definition out of its morphological context and placing it in the context of developmental process introduces some semantic confusion. Nevertheless, it would be perverse to object to evolutionary studies of developmental mechanisms. The third reason for not trying to enforce consensus is that the prospect of doing so brings to mind the image of herding cats. I doubt that a consensus will be achieved soon, but, if it is, it will not be by force. For these reasons, I did not strive for uniformity of definition but, rather, asked authors to define their terms precisely and clearly.

All of the meaningful definitions of heterochrony are about timing, rate, and sequence. There is no doubt that time, rate, and sequence are important, even fundamental, aspects of development. However, development involves more than that—it involves processes distributed in space as well as in time. There is no reason to think that the processes themselves are conserved in all their details nor that their spatial organization is conserved. Nor is there any reason to think that modifications of process and spatial patterning are less interesting (or common) than changes in timing. In looking beyond heterochrony, we can only enrich our theories of evolutionary developmental biology.

Miriam Leah Zelditch

Museum of Paleontology, University of Michigan, Ann Arbor, MI

ACKNOWLEDGMENTS

I thank Luna Han for her assistance, guidance, and occasional nudging. I also thank the authors for their thoughtful and novel contributions.

REFERENCES

Alberch P, Alberch J (1981): Heterochronic mechanisms of morphological diversification and evolutionary change in the Neotropical salamander *Bolitoglossa occidentalis* (Amphibia: Plethodontidae). J Morphol 167: 249–264.

Alberch P, Gould SJ, Oster GF, Wake DB (1979): Size and shape in ontogeny and phylogeny. Paleobiology 5: 296–317.

Anstey RL (1987): Astogeny and phylogeny: evolutionary heterochrony in Paleozoic bryozoans. Paleobiology 13: 20–43.

Bemis W (1984): Paedomorphosis and the evolution of the Dipnoi. Paleobiology 10: 293–307.

Fink WL (1981): Ontogeny and phylogeny of tooth attachment modes in actinotyperygian fishes. J Morphol 167: 167–184.

Gould SJ (1977): *Ontogeny and Phylogeny*. Cambridge, MA: Harvard University Press.

Gould SJ (1988): The uses of heterochrony. In: McKinney ML (ed), *Heterochrony in Evolution: A Multidisciplinary Approach*. New York: Plenum Press, p. 1–13.

Gould SJ (2000): Of coiled oysters and big brains: how to rescue the terminology of heterochrony, now gone astray. Evol Dev 2: 241–248.

Hall BK (1998): *Evolutionary Developmental Biology* (2nd ed). London: Kluwer Academic Publishers.

Huxley JS (1932): *Problems of Relative Growth*. London: MacVeigh.

Klingenberg CP (1998): Heterochrony and allometry: the analysis of evolutionary change in ontogeny. Biol Rev 73: 79–123.

McKinney ML (1984): Allometry and heterochrony in an Eocene echinoid lineage: morphological change as a by-product of size selection. Paleobiology 10: 407–419.

McKinney ML (1988): *Heterochrony in Evolution: A Multidisciplinary Approach*. New York: Plenum Press.

McKinney ML, McNamara KJ (1991): *Heterochrony: The Evolution of Ontogeny*. New York: Plenum Press.

McNamara KJ (1978): Paedomorphosis in Scottish olenelid trilobites (Early Cambrian). Palaeontology 21: 635–655.

McNamara KJ (1981): Paedomorphosis in Middle Cambrian xystridurine trilobites from northern Australia. Allcheringa 5: 209–224.

Raff RA, Wray GA (1989): Heterochrony: developmental mechanisms and evolutionary results. J Evol Biol 2: 409–434.

Waddington CH (1967): Discussion. In Moorehead PS, Kaplan MM (eds). *Mathematical Challenges to the Neo-Darwinian Interpretation of Evolution*. Philadelphia, PA: Wistar Institute of Anatomy and Biology.

Wake DB (1980): Evidence of heterochronic evolution: a nasal bone in the Olympic salamander, *Rhyacotriton olympicus*. J Herpetol 14: 292–295.

CONTRIBUTORS

Timothy F. Carl, Department of Organismic and Evolutionary Biology and Museum of Comparative Biology, Harvard University, Cambridge, Massachusetts

Martin J. Cohn, Division of Zoology, School of Animal and Microbial Sciences, University of Reading, Reading, United Kingdom

William L. Fink, Museum of Zoology and Department of Biology, University of Michigan, Ann Arbor, Michigan

Michael W. Frohlich, Herbarium and Department of Biology, University of Michigan, Ann Arbor, Michigan

Robert Guralnick, Colorado University Museum and Department of Environmental, Population and Organismal Biology, University of Colorado at Boulder, Boulder, Colorado

Jason J. Head, Department of Geological Sciences, Southern Methodist University, Dallas, Texas

Larry Hufford, School of Biological Sciences, Washington State University, Pullman, Washington

Nigel C. Hughes, Department of Earth Sciences, University of California, Riverside, Riverside, California

James Kurpius, Museum of Paleontology and Department of Integrative Biology, University of California, Berkeley, Berkeley, California

Ross H. Nehm, Barnard College, Columbia University, New York, New York

David M. Parichy, Sections of Integrative Biology and Molecular, Cellular and Developmental Biology, Institute for Cellular and Molecular Biology, University of Texas at Austin, Austin, Texas

P. David Polly, Molecular and Cellular Biology Section, Biomedical Science Division, Queen Mary & Westfield College, Department of Paleontology, The Natural History Museum, London, United Kingdom

Peter D. Roopnarine, Department of Invertebrate Zoology and Geology, California Academy of Sciences, San Francisco, California

Michael D. Shapiro, Department of Organismic and Evolutionary Biology, and Museum of Comparative Zoology, Harvard University, Cambridge, Massachusetts

H. David Sheets, Department of Physics, Canisius College, Buffalo, New York

Mark Webster, Department of Earth Sciences, University of California, Riverside, Riverside, California

Miriam Leah Zelditch, Museum of Paleontology, University of Michigan, Ann Arbor, Michigan

1

THE DEVELOPMENTAL BASIS OF MORPHOLOGICAL DISARMAMENT IN *PRUNUM* (NEOGASTROPODA: MARGINELLIDAE)

Ross H. Nehm
Barnard College, Columbia University, USA

INTRODUCTION

The rapid origin and loss of morphological structures are important components of the macroevolutionary history of marginellid gastropods (e.g., Nehm and Geary, 1994; Nehm, 1998, 2001a, b). Many of the shell features that display macroevolutionary trends, such as shell thickness, aperture shape, callus area, and lip width, are hypothesized to be defenses against predation by naticid gastropods and durophagous arthropods (Vermeij, 1987, 1993; Nehm, 1998). Since their diversification in the Upper Oligocene, marginellids of the genus *Prunum* have been subject to attack by these groups, as indicated by an abundance of well-preserved naticid borings and repair scars (Nehm, 1998; Fig. 1.1). Many of the features that display rapid rates of evolution may be tied to fluctuations in predation intensity (Kitchell, 1990; Nehm and Geary, 1994; Nehm, 1998, 2001).

Although geographic and temporal patterns of morphological escalation in

Beyond Heterochrony: The Evolution of Development, Edited by Miriam Leah Zelditch
ISBN 0-471-37973-5

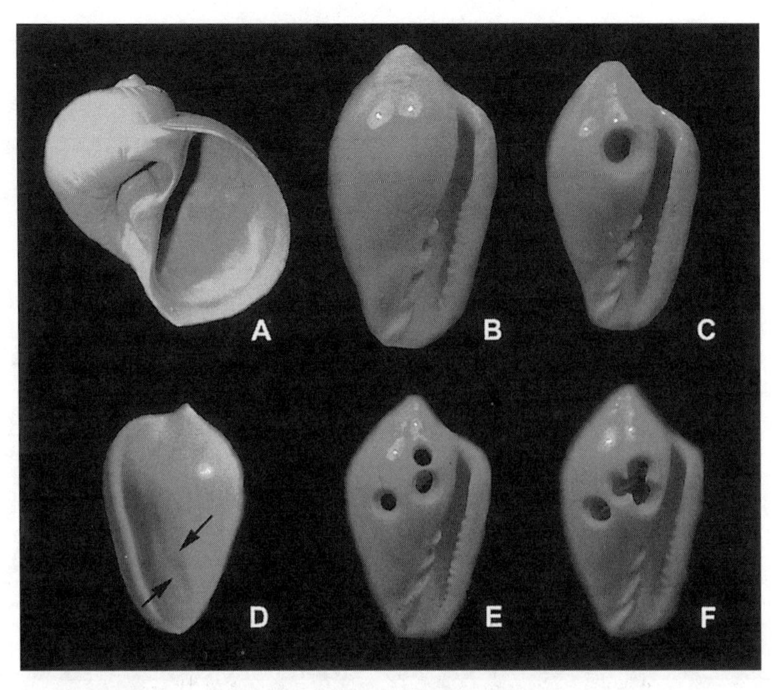

Figure 1.1. A. Neogene naticid gastropod predator. B. Neogene Adult *Prunum* shell. C, E, F. Neogene naticid borings on adult *Prunum*. D. Durophagous predator repair scar on an adult *Prunum* shell.

mollusks have been the focus of considerable attention (e.g., Vermeij, 1978, 1987, 1993), very little is known about how developmental processes generate or modify the defensive aptations that produce these macroevolutionary patterns (Gould and Vrba, 1982). The converse of morphological escalation, "morphological disarmament" (Nehm, 1998), is also poorly understood but has been hypothesized to be produced by paedomorphosis (Nehm, 1998). Morphological armament and disarmament often involve the evolution of multiple morphological features, such as shell thickness, callus area, and varix thickness (Vermeij, 1978; Nehm, 2001a; Fig. 1.1). Because many anatomical structures may be involved in defense, studies of morphological armament and disarmament require the study of multiple shell features and their integration through ontogenetic and geologic time.

In this study, I investigate the role of developmental processes in the loss of defensive shell features in a clade of marginellid gastropods (the *Prunum maoense* group). I compare juveniles and adults of ancestors and descendants using traditional and geometric morphometric analyses, partition ontogenetic change into its spatial and temporal components, and document the magnitudes of size, shape, and age dissociation. These approaches are used to test the hypotheses that morphological evolution is constrained along ancestral developmental pathways and that heterochronic change alone accounts for observed changes in antipredatory morphologies in marginellid gastropods.

MARGINELLID GASTROPODS AS AN EVOLUTIONARY-DEVELOPMENTAL RESEARCH SYSTEM

Marginellid gastropods of the genus *Prunum* are well suited for morphological and developmental studies for several reasons: First, all species have determinate growth, which permits the recognition of juvenile and adult shells (Vermeij and Signor, 1992). Second, adult shells preserve a complete compositional and morphological record of ontogeny, which is easily examined using many different techniques (e.g., direct imaging of the shell exterior, hard-tissue histology, X-radiographic images, and so forth). Third, unlike many gastropod groups, *Prunum* does not remodel the shell interior during or after sexual maturity, permitting accurate X-radiographic, morphometric, and microstructural studies of ontogenetic change within and among species. Fourth, a large amount of information has been assembled about the geographic, stratigraphic, and bathymetric distributions of *Prunum* species and their relationship to ontogenetic and morphologic variation (Nehm and Geary, 1994; Nehm, 1998, 2001a, b). This research system is used to investigate developmental evolution in a *Prunum* clade referred to as the "*Prunum maoense* group."

The *P. maoense* group consists of three extinct species [*P. maoense* (Maury, 1917), *P. latissimum* (Dall, 1896), and *P. dasum* (Gardner, 1928)], which were abundant members of highly diverse shallow-marine benthic communities (Saunders et al., 1986). These communities contain naticid and durophagous predators, as indicated by a rich record of predator fossils and their marks on prey (Allmon et al., 1993; Vermeij, 1978; Nehm, 2001b). The *P. maoense* clade ranges temporally from the early-middle Miocene Chipola Formation of Florida to the lower Pliocene Mao Formation of the Dominican Republic. The clade was endemic to the Caloosahatchian tropical marine province of the Caribbean Basin (Woodring, 1974; Petuch, 1981, 1982). More detailed stratigraphic, geographic, paleoecological, and phylogenetic information may be found in Nehm (2001b).

Prunum latissimum and *P. maoense* display pronounced morphological differences at adulthood (Figs. 1.2 and 1.3). *P. latissimum* is large, obovate, and heavily callused, whereas *P. maoense* is small, cylindrical, and poorly callused. In *P. maoense*, the callus covers less than 40% of the dorsal face of the body whorl and lacks lobes or processes on the aperture margin. In contrast, callus covers most of the dorsal body whorl of *P. latissimum*, and thick but localized callus processes and ridges occur along the aperture margin. These processes are often thicker than the layers comprising the body whorl. The most prominent callus process occurs near the posterior aperture margin and expands into the aperture and onto the body whorl. In some individuals, the callus covering the posterior aperture and lip forms a process as prominent as the spire.

Despite pronounced *adult* morphological differences, the ontogenies of *P. latissimum* and *P. maoense* are very similar: The juvenile shells of both species are proportionally narrower than the adult shells, and the outer lips are very thin until adulthood, when varices are added. A small callus deposit surrounds the columellar plications on the anterior margin of the shell in the youngest

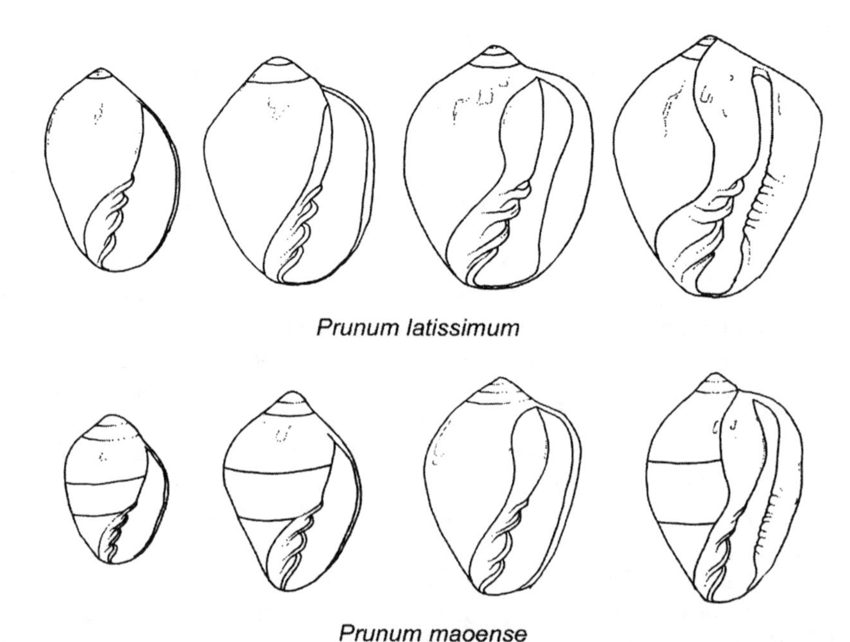

Figure 1.2. Illustrations of ontogenetic change in *P. latissimum* (top) and *P. maoense* (bottom). The first three shells in each row are sub-adults whereas the last shells in each row are adults.

shells of both species. The thickening of the inner and outer lips, in combination with the expansion of the aperture margin calluses, produces a proportional decrease in the aperture area through ontogeny in both species. In both *P. latissimum* and *P. maoense*, four very thin and sharp plications occur in juveniles and thicken through ontogeny. In addition, the height of the aperture becomes proportionally higher through ontogeny. Finally, adulthood in both species is marked by the addition of five shell features: (1) an external varix, (2) a terminal inflection of the body whorl, (3) lip denticulations and/or crenulations, (4) inner lip thickening, and (5) a posterior lip callus (Fig. 1.4).

The *P. maoense* clade experienced high rates of naticid gastropod predation, as indicated by an abundance of well-preserved parabolic drill holes that are uniquely characteristic of naticids (Kitchell et al., 1990; Anderson et al., 1991; Nehm, 1998). Attacks by durophagous predators probably occurred infrequently, as suggested by the paucity of shell repair scars in the clade. Naticid predation on species in the *P. maoense* clade is highly stereotyped, as indicated by the preponderance of drill holes in a very localized region of the shell; more than 95% of all naticid drill holes occur along the posterior aperture margin (Fig. 1.1). Interestingly, this region also contains the thickest and most localized concentration of shell callus. The presence of callus in this region doubles the shell's thickness. The absence of callus in other regions of the shell is associated with the absence of naticid boreholes. From the Miocene to the Pliocene, naticid predation on these *Prunum* species decreased from approximately

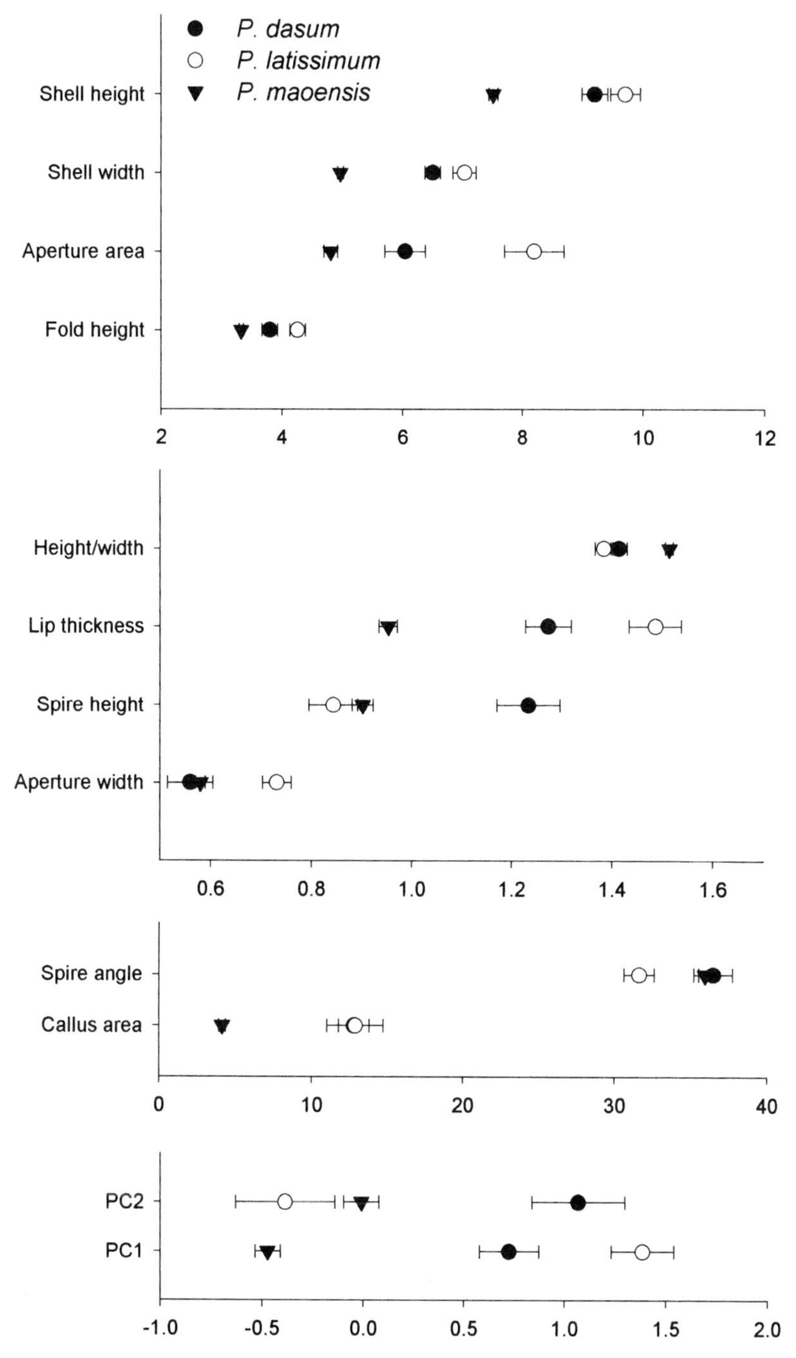

Figure 1.3. Adult morphological differences between species in the *P. maoense* group. *Prunum dasum* is the outgroup. Means and two standard errors are indicated for shell height, shell width, aperture area, columellar fold height, lip thickness, spire height, aperture width, spire angle, and callus area on adult specimens of each species. Additionally, mean Principal Components Analysis factor scores are compared between species.

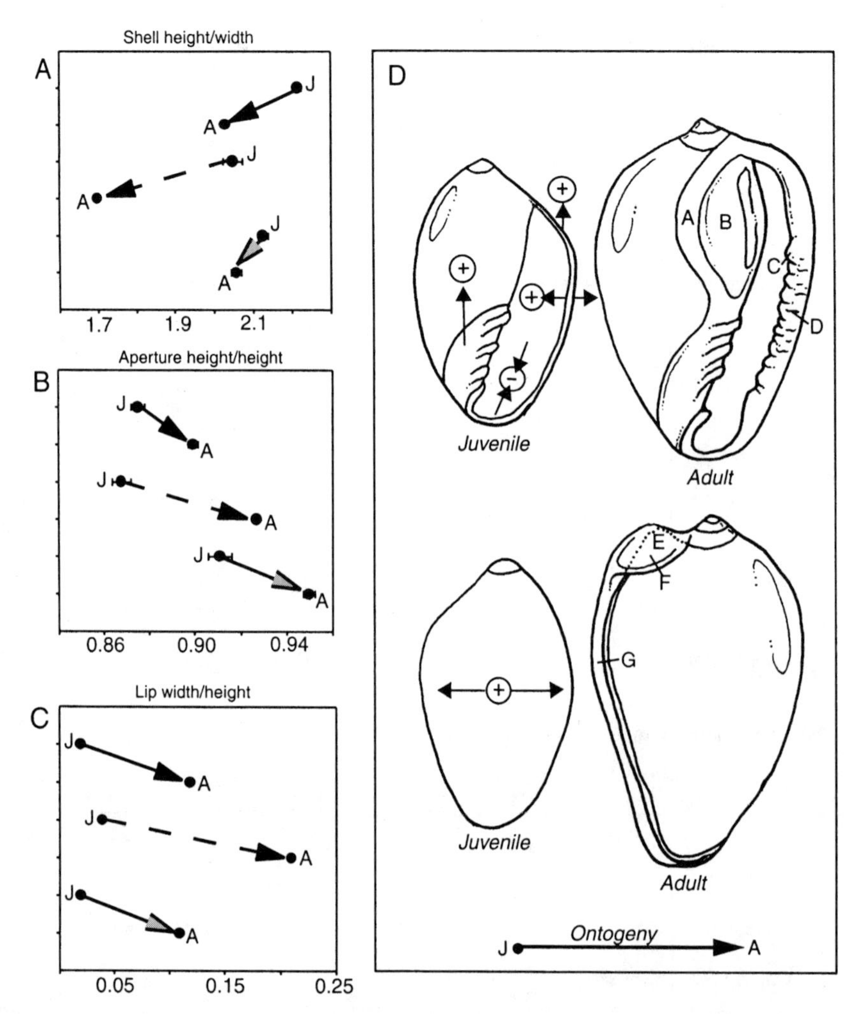

Figure 1.4. Ontogenetic change in *Prunum*. A–C Morphometric change from juvenile (J) to adult (A) specimens of three *Prunum* species for: A. shell height/width, B. aperture height/height, and C. lip width/height. All *Prunum* species studied to date display these patterns. D. Morphological and morphometric differences between juvenile and adult *Prunum* shells. "+" positively allometric, "–" negative allometric. Adult shells display the following morphological features: A = aperture margin callus, B = posterior callus process, C = denticle, D = crenulation, E = dorsal lip callus, F = terminal lip inflection, G = external varix. Juvenile shells lack these features.

20% to less than 5% (Nehm, 1998). Predation reduction is correlated with morphological reduction in callus area and shell thickness. The developmental processes that produced the loss of these morphological structures are the focus of this study.

HYPOTHESES OF DEVELOPMENTAL EVOLUTION

Developmentally, how was the ontogeny of *P. latissimum* modified to produce the morphological reduction in *P. maoense*? Were these changes produced by the retention of an ancestral ontogenetic state into adulthood, the introduction of novel developmental changes, or combinations thereof? In this study, evolutionary changes in development are considered to be the product of two classes of change: heterochrony and "novelty" (sensu Gould, 1977, p. 4, and 1988, p. 4; also see preface of this book). Cases in which morphological change is channeled along ancestral ontogenetic trajectories (that is, ontogeny parallels phylogeny) are considered to represent heterochrony (Gould, 1977, p. 4, and 1988, p. 4; also see preface of this book). Alternatively, cases in which shapes or structures that are not present in, or extrapolated from, ancestral ontogenies are introduced in the descendant's ontogeny are referred to as "novelty" (sensu Gould, 1977, p. 4, and 1988, p. 4; also see preface of this book). Heterochrony and novelty may be considered to be extremes along a continuum of possible change; combinations of heterochrony and novelty are possible and expected. The null hypothesis for this study is that the developmental evolution from *P. latissimum* to *P. maoense* is a product of pure heterochrony.

Methodologically, geometric morphometric analysis (such as relative warp analysis or RWA) may be used to identify pure novelty, pure heterochrony, or combinations of the two patterns. Ancestral and descendant ontogenies may be compared in RWA shape space. Ontogenetic evolution may be visualized in shape space in two different ways. First, ontogenetic *vectors* (from juveniles to adults) may be compared between ancestors and descendants in shape space. Heterochronic change is recognized as no change in the direction but rather a change in length between the ancestral and descendant ontogenetic vectors. Pure paedomorphosis, for example, may be recognized in shape space by a truncated descendant ontogenetic vector that overlaps with the ancestral ontogenetic vector. Alternatively, novelty may be recognized in shape space when descendent ontogenetic trajectories deviate into regions of shape space not extrapolated from those of the ancestor. In such cases, ontogenetic vectors will display differences in their angles; thus, heterochrony alone does not explain developmental evolution.

Paedomorphosis and novelty may also be visualized and identified by comparing the *distributions* of ancestor and descendant ontogenies in shape space. Pure paedomorphosis, for example, is characterized by the complete overlap of the descendant within the shape space of the ancestral ontogeny. In such cases, no novel shapes have been introduced into the descendent's ontogeny, and heterochrony alone explains developmental change. Pure paedomorphosis is falsified in cases in which shape space occupation differs significantly between ancestors and descendants. In such cases, peramorphosis, novelty, or a combination thereof explains developmental evolution; ontogenetic *vectors* may be used to determine the relative contributions of peramorphosis and novelty (see

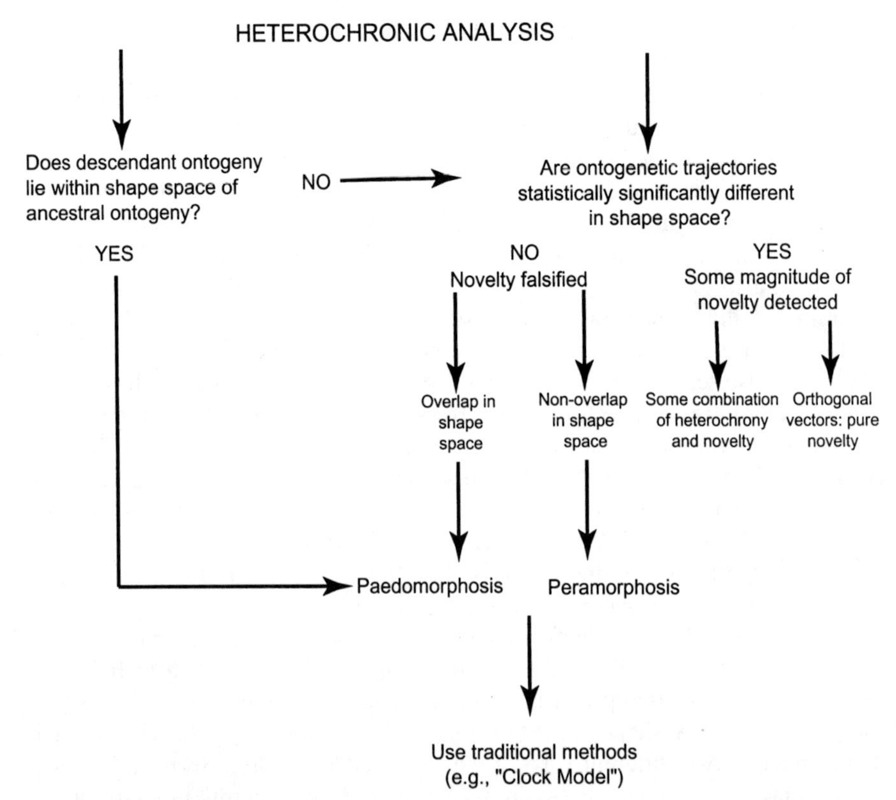

Figure 1.5. Flowchart illustrating the methodological approach used to distinguish heterochrony and novelty in marginellid gastropods.

above). Comparing ontogenetic trajectories and distributions in shape spaces provides a rigorous methodology for differentiating heterochrony and novelty in developmental evolution (Fig. 1.5; Zelditch and Fink, 1996).

If pure heterochrony cannot be falsified using the methodologies outlined above, traditional methods of heterochronic analysis such as Gould's "clock model" may be used to infer heterochronic modes (Gould, 1977). However, if the null hypothesis of pure heterochrony is falsified, the use of traditional methods is not appropriate because they assume heterochronic change exclusively. On the basis of the above conceptual framework, developmental evolution from *P. latissimum* to *P. maoense* is characterized as heterochrony, novelty, or a combination thereof using geometric morphometric (RWA) shape spaces. If the null hypothesis is not falsified, traditional morphometric methods (bivariate regression and multivariate analysis) will be used to place ontogenetic patterns within the theoretical "clock models" of Gould (1977).

CASE STUDY

Phylogenetic Framework

Robust phylogenetic hypotheses are prerequisite to investigations of the polarity of morphological and developmental change (Harvey and Pagel, 1991). Several different approaches contribute to a very good understanding of marginellid gastropod phylogeny. Phylogenetic analyses based on morphological data (Coovert and Coovert, 1995; Nehm, 1996) and molecular data (Nehm and Tran, 1997; Nehm and Simison, unpublished observations) establish the monophyly of marginellid gastropods and several clades therein. Within the Marginellidae, the monophyletic tribe Prunini is diagnosed by the type 6 radula (Coovert, 1989) and the presence of an esophogeal caecum (Coovert and Coovert, 1995; Nehm, 1996). Molecular data also support the monophyly of this clade (Nehm and Simison, unpublished observations). The Prunini include the genera *Prunum*, *Volvarina*, *Bullata*, *Rivomarginella*, *Hyalina*, and *Cryptospira*.

Within the Prunini, shell morphology, radular morphology, internal anatomy, stratophenetic data, and molecular data provide very strong support for the monophyly of Prunum + *Volvarina* (Coovert and Coovert, 1995; Nehm, 1996; Nehm and Simison, unpublished observations). This study focuses on a small but temporally long-ranging clade within the *Prunum* + *Volvarina* clade referred to here as the *Prunum maoense* group. The *P. maoense* group is a well-supported monophyletic group that is distinguished morphologically from other *Prunum* and *Volvarina* species by the presence of (1) a large posterior aperture margin callus that is thicker than the body whorl shell layers, (2) a posterior lip indentation, (3) anterior lip margin denticles, and (4) three thin body whorl stripes. These character states do not collectively occur in any other living or fossil marginellids (Nehm, 1998, 2001b). Morphological and developmental comparisons are made between an extensively studied and sampled ancestral-descendant stratigraphic sequence of *P. latissimum* and *P. maoense* in the Dominican Republic Neogene (Saunders et al., 1986; Nehm, 2001b).

Morphometric Analyses: Distances and Landmarks

Distance measurements and landmark placements on the shell were made using several different shell orientations (aperture and apical views) and visualizations (images of the shell exterior and X-radiographs). For distance measurements, juvenile and adult shells from each species were mounted on cardboard trays in apertural view with the columellar axis oriented horizontally to the surface (orientation shown in Fig. 1.6). Images of the shells were captured using a video camera and manipulated in Scion Image to increase the clarity of the morphological features before measurement. Nine variables were measured on 520 adult specimens: six distances, two areas, and one angle (Fig. 1.6A). X-radiographic films were made of shells in aperture view, and their images were captured using

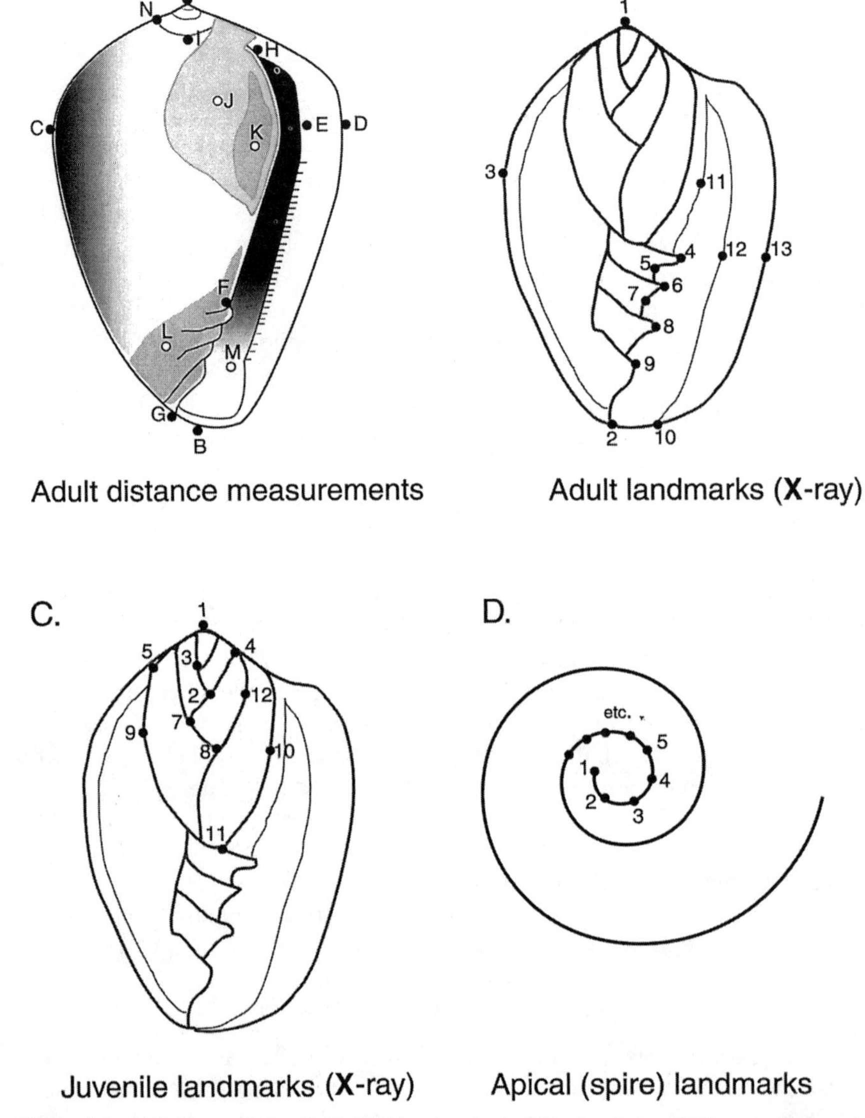

Figure 1.6. Distance measurements and landmark placements on the shell were made using several different orientations (aperture and apical views) and visualizations (images of the shell exterior and X-radiographs). A. Distance measurements on the exterior shell. B. Adult landmarks on X-radiograph. C. Juvenile landmarks on X-radiograph. D. Apical landmarks on the exterior shell for studies of the complete shell ontogeny.

a video camera and manipulated in Scion Image. Eight distance measurements were made on X-radiographs of 75 adult specimens. In addition, 12 landmarks were identified on X-radiographs for RWA of juveniles (Fig. 1.6B). For RWA of adults, 13 landmarks were identified on X-radiographs (Fig. 1.6C). For RWA of the spire (and each whorl), images were captured of shells in apical view. Landmarks in apical view were spaced every 15 degrees (Fig. 1.6D). Collectively, these different orientations and visualizations provide a comprehensive characterization of ontogenetic and morphologic variation in the marginellid shell.

Statistical Analyses

Several morphometric analyses were performed to document patterns of ontogenetic and morphologic change in the *P. maoense* group. RWA was used to compare ancestral and descendent ontogenetic trajectories in shape space. Two shape spaces were constructed: one using apical landmarks and one using ventral landmarks. In addition, reduced major axis (RMA) regression was used to compare juvenile allometric patterns between species. Finally, tests for covariance homogeneity and common principal components (CPC) were used to study morphological differences between juveniles and adults of *P. latissimum* and *P. maoense*.

RWA was specifically used to (1) separate size and shape variation in ancestors and descendants and (2) compare shape space occupation between ancestors and descendants (Zelditch and Fink, 1996). Ontogenetic trajectories are difficult to construct using aperture-view RWA because adults develop a suite of morphological features that are not present in juveniles. These differences prohibit comparisons of the same morphological landmarks and thus preclude shape space comparisons. However, comparisons of aperture-view RWA can be and were made between juvenile ancestors and juvenile descendants and between adult ancestors and adult descendants. Apical-view RWA was also used to compare juvenile ancestors to juvenile descendants and adult ancestors to adult descendants. Both approaches permit equivalent landmark comparisons.

RWA is a geometric analogue of principal components analysis. RWA characterizes shape change as a deformation of landmarks from a "reference" form for each specimen under analysis (Rohlf, 1993a). The reference form or tangent configuration used here was generated by calculating a generalized least squares consensus of all specimens in TPSRELW version 1.13 (Rohlf, 1993b). RWA was used to compare shape variation between ancestors and descendants using TPSRELW (alpha set to 0). Variation in uniform + nonuniform shape space is presented as a series of bivariate plots comparing the relative warps (RWs) that explain the most variation in each RWA. Five RWA comparisons were performed: (1) juvenile ancestors vs. juvenile descendants in *aperture* view, (2) adult ancestors vs. adult descendants in *aperture* view, (3) juvenile ancestors vs. juvenile descendants in *apical* view, (4) adult ancestors vs. adult descendants in *apical* view, and (5) juvenile + adult descendants vs. juvenile ancestors in *apical* view.

Tests for significant differences in shape space occupation between species were performed using a Mann-Whitney *U* test on the RWs that collectively explain >80% of the variation in shape space. This nonparametric test was used because the distributions of RW scores were not distributed normally. Alpha was set to 0.05 in each test.

RMA regression was used to compare juvenile allometries between species and to help interpret (and test) geometric morphometic results. Model II regression (such as RMA) is appropriate in cases in which both of the morphometric variables under study are not under the control of the investigator or are measured with error (such as shell height and shell width) (Sokal and Rohlf, 1995, p. 558; McKinney and McNamara, 1991, p. 34–35). RMA tests established bivariate allometric relationships of shell width, aperture width, shell thickness, and aperture height with shell height for each species. In addition, RMA slopes and intercepts were calculated for aperture area in relation to total shell area for each species. These separate analyses were necessary because RMA calculations require that the two variables under study have the same measurement units. RMA slopes and intercepts (and their confidence intervals) were calculated on log-transformed variables for juveniles of each species using BIOMstat 3.0.

Quantitative Change Through Ontogeny

Juvenile Ancestors vs. Juvenile Descendants in Aperture View. RWA of juvenile ancestors and juvenile descendants produced four RW axes that collectively explained more than 75% of the variation in the data set. RW1 explained the most variation (40.4%) but produced poor separation of specimens in shape space. A Mann-Whitney U test indicated that the distributions of the RW1 scores were not significantly different between species (test statistic = 93.0, 1 df, $P > 0.17$). RW2 explained 18.4% of the variation in the data set and, like RW1, produced poor separation between species (test statistic = 100, 1 df, $P > 0.07$). Variation in *P. latissimum* was noticeably greater along RW2 relative to *P. maoense*. RW3 and RW4 also produced poor separation of species and were not significantly different (RW3: test statistic = 48, 1 df, $P > 0.19$; RW4: test statistic = 76, 1 df, $P > 0.72$). In summary, juveniles of ancestors and descendants were not significantly different in shape space (Fig. 1.7).

Adult Ancestors vs. Adult Descendants in Aperture View. RWA of adult ancestors and adult descendants produced four RW axes that collectively explained about 80% of the variation in the data set. RW1 explained the most variation (36.2%) and produced some separation of specimens in shape space. A Mann-Whitney U test indicated that the distributions of the RW1 scores were significantly different between species (test statistic = 37, 1 df, $P < 0.05$). RW2 explained 28.4% of the variation in the data set but, unlike RW1, produced excellent separation of species (test statistic = 15, 1 df, $P < 0.01$). As was observed in juvenile shape space, variation in adult *P. latissimum* was noticeably greater along RW2 relative to *P. maoense*. RW3 produced poor separation of species, and the distributions are not significantly different (RW3:

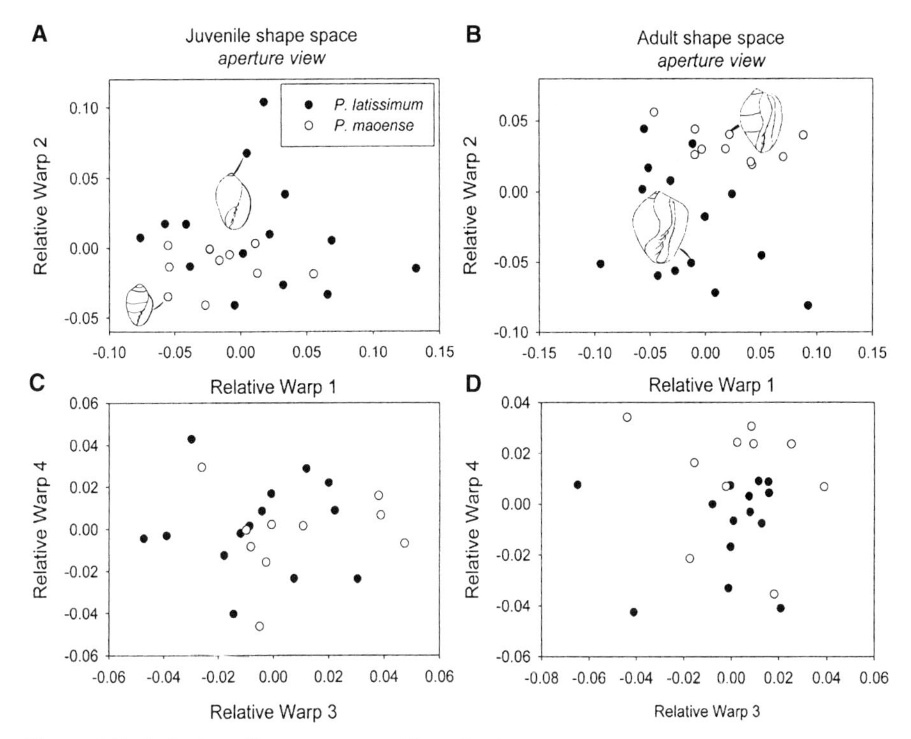

Figure 1.7. A, C. Juvenile ancestors and juvenile descendants (aperture view analysis) in uniform + non-uniform shape space using Relative Warp Analysis of landmarks shown in Figure 1.6C. B, D. Adult ancestors and adult descendants (aperture view analysis) in uniform + non-uniform shape space using Relative Warp Analysis of landmarks shown in Figure 1.6B.

test statistic = 66, 1 df, $P > 0.80$). RW4 produced good separation of species, and the distributions are significantly different (U-test statistic = 76, 1 df, $P < 0.030$). In summary, unlike juvenile-juvenile comparisons, adults of ancestors and descendants were significantly different in shape space comparisons (Fig. 1.7).

Apical View Shape Spaces. RWA of juvenile ancestor and juvenile descendant shells in *apical* view produced similar patterns to RWA of shells in *aperture* view: specimens of both species displayed considerable overlap and were not significantly different. RWA of apical landmarks in *P. latissimum* and *P. maoense* captured both uniform and nonuniform shape differences among specimens. RW1 explained 49% and RW2 explained 28% of the shape variation in the data set. A Mann-Whitney U test indicated that the distributions of the RW1 and RW2 scores were not significantly different between species (RW1: test statistic = 1,407.5, 1 df, $P > 0.05$; RW2: test statistic = 1,206.5, 1 df, $P > 0.60$). In summary, juvenile-juvenile comparisons in apical view indicated that ancestors and descendants were not significantly different in shape space (Fig. 1.8).

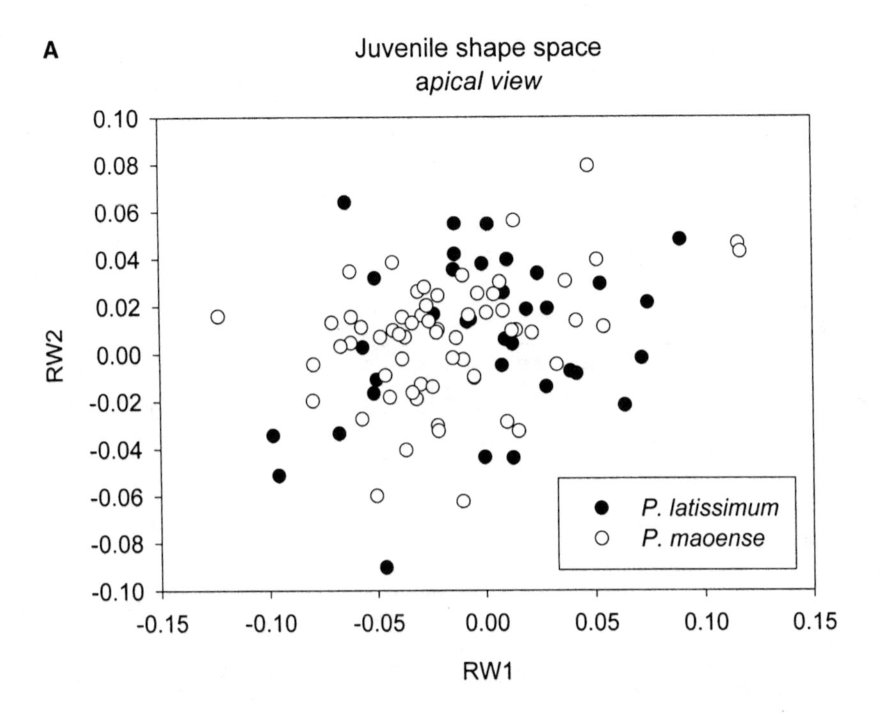

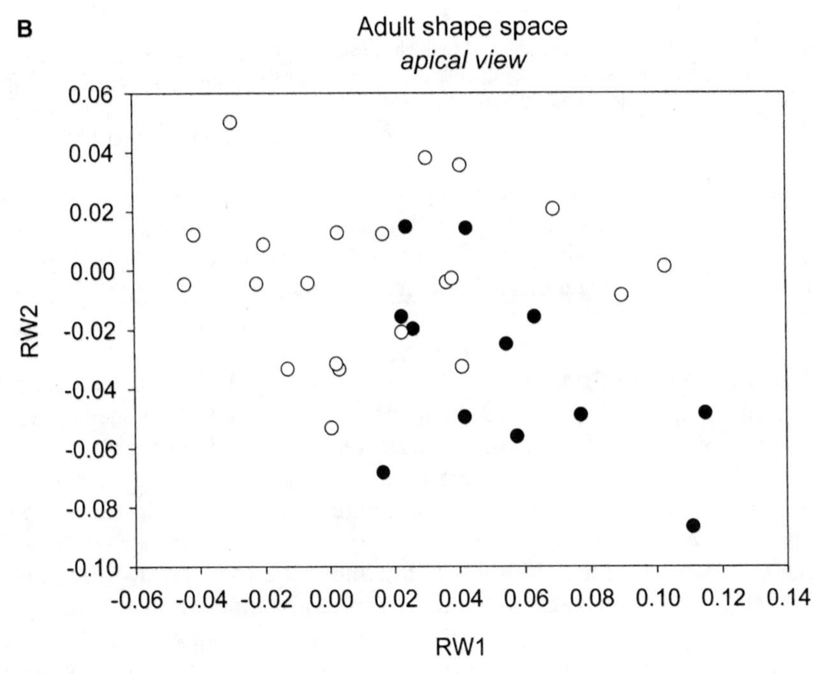

Figure 1.8. A. Juvenile ancestors and juvenile descendants (spiral view analysis) in uniform + non-uniform shape space using Relative Warp Analysis of landmarks shown in Figure 1.6D. B. Adult ancestors and adult descendants (spiral view analysis) in uniform + non-uniform shape space using RWA of landmarks shown in Figure 1.6D.

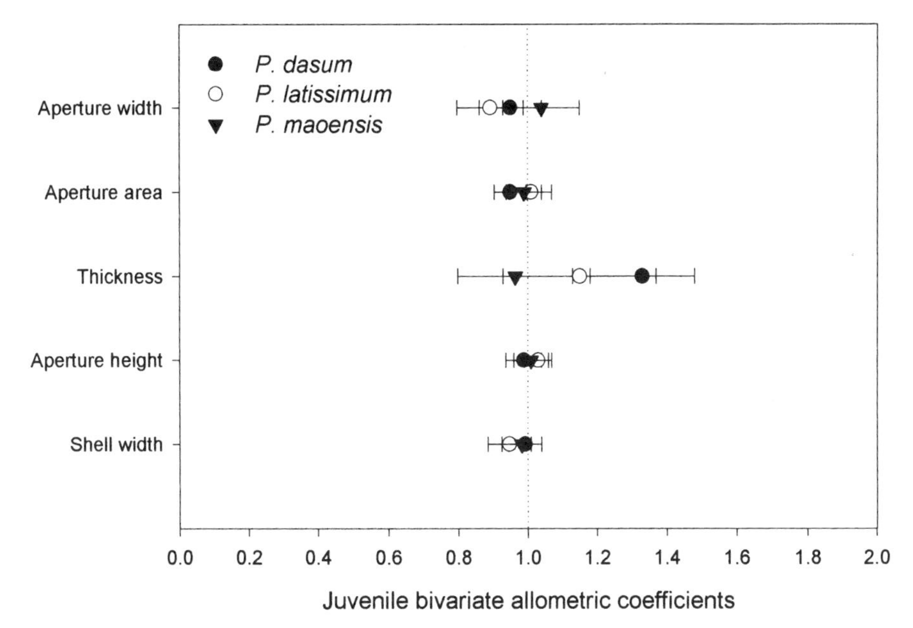

Figure 1.9. Comparisons of juvenile allometries using Reduced Major Axis "regression". See text for details of the analysis. *Prunum dasum* is the outgroup.

RWA of adult ancestor and adult descendant shells in *apical* view also produced similar patterns as RWA of shells in *aperture* view: specimens of both species separated well in shape space. A Mann-Whitney U test indicated that the distributions of the RW1 and RW2 were significantly different between adults of both species (RW1: test statistic = 200.5, 1 df, $P < 0.01$; RW2: test statistic = 60, 1 df, $P < 0.01$).

Comparisons of juveniles + adults of *P. maoense* with juveniles of *P. latissimum* produced overlap in shape space; adult *P. maoense* are similar in shape to juveniles of *P. latissimum*. The two distributions were also not significantly different using a Mann-Whitney U test (RW1: test statistic = 1,714.5, 1 df, $P > 0.25$; RW2: test statistic = 1,659, 1 df, $P > 0.40$). In summary, descendant adults overlap in shape space with ancestral juveniles. No new shapes appear to have been introduced in the evolution of *P. maoense*.

RMA regression of juveniles of *P. latissimum* and *P. maoense* for aperture width, aperture area, shell thickness, aperture height, and shell width are shown in Figure 1.9; 95% confidence intervals of RMA slopes indicate that shell width, aperture height, and aperture area are generally isometric, and aperture width and shell thickness exhibit the most variation among species. Nevertheless, allometric patterns in juveniles of all three species are not significantly different, corroborating the qualitative observation that early ontogenies are very similar and the RWA shape spaces overlap.

Tests for CPC and covariation homogeneity also suggest that juveniles of both species are very similar. CPCs characterize juveniles of *P. latissimum* and *P. maoense* ($\chi^2 = 14.26$, df $= 10$, $P = 0.161$). In addition, analyses of homogeneity of covariation between juveniles of each species indicate that covariation matrices are not significantly different ($\chi^2 = 27.97$, df $= 15$, $P = 0.02$). However, the addition of adults produces significant differences in both CPC and covariance homogeneity tests ($\chi^2 = 51.89$, df $= 15$, $P < 0.001$). These results corroborate the conclusions from RWA: the morphological differences between *P. latissimum* and *P. maoense* originate in late ontogeny.

INTERPRETATION

Heterochrony and Novelty

The null hypothesis that developmental evolution from *P. latissimum* to *P. maoense* is a product of pure heterochrony cannot be rejected based on the results of geometric morphometric analysis of shells in aperture and apical view: (1) juveniles of *P. latissimum* and *P. maoense* do not occupy significantly different regions of shape space and (2) the adults of *P. maoense* do not occupy significantly different regions of shape space from those of juvenile *P. latissimum*. No novel shapes appear to have been introduced in evolution of *P. maoense* from *P. latissimum*. Therefore, it appears that developmental evolution is a product of pure heterochrony. Ontogeny parallels phylogeny in the *P. maoense* clade: adult descendants resemble juvenile ancestors.

Traditional morphometric approaches corroborate the RWA analyses and indicate that juveniles are not significantly different between ancestors and descendants but adults are different: (1) RMA regression indicates that juvenile allometries are not significantly different between *P. latissimum* and *P. maoense*, (2) covariance matrices indicate that juvenile covariance structures are not significantly different between *P. latissimum* and *P. maoense*, and (3) CPCs are shared by juveniles of *P. latissimum* and *P. maoense*. The addition of adults to these analyses produced significant differences in *all* cases. Because ancestral and descendant adults differ significantly in all traditional and geometric morphometric analyses, but juveniles do not, evolutionary change must have occurred in late ontogeny in this clade.

The presence of descendant adults within the ancestral juvenile shape space suggests that pure heterochrony via paedomorphosis characterizes developmental evolution in *P. maoense*. This conclusion is corroborated by the observations that (1) *P. maoense* is morphologically "reduced" (e.g., smaller, less callused, thinner) relative to *P. latissimum*, (2) adult *P. maoense* resemble late-stage juveniles of *P. latissimum*, and (3) adult *P. maoense* have slightly fewer shell whorls and are smaller than adult *P. latissimum*, suggesting that *P. maoense* may attain adulthood at a younger age than *P. latissimum*. Because heterochrony was not falsified as an exclusive contributor to developmental evolution, traditional methods may be used to infer heterochronic mode(s). Gould's (1977) "clock model" is used to determine which mode of paedomorphosis best characterizes evolutionary change in *P. maoense*. To use this model, estimates of time change are needed, in addition to information on size and shape change.

Temporal change is a crucial component of the study of heterochrony (Gould, 1977; McKinney and McNamara, 1991). Traditional models of paedomorphic change from ancestor to descendent involve comparisons of three temporal guideposts: the onset of growth (onset time), the termination of growth (offset time), and the rate of growth (from onset to offset). Three paedomorphic shifts are possible by altering these three parameters: deceleration, hypomorphosis, or postdisplacement. *Deceleration* is defined as no change in onset or offset time and a decrease in the rate of growth. *Hypomorphosis* is defined as no change in onset time or growth rate and a truncation of offset time. *Postdisplacement* is defined as no change in the rate of growth or offset time and a delay in onset time. To place patterns of paedomorphosis in the *P. maoense* clade in a heterochronic framework (e.g., Gould, 1977; Alberch et al., 1979) it is necessary to answer the following questions: (1) Is onset time the same in *P. latissimum* and *P. maoense*? (2) Is offset time the same in these species? (3) Does growth rate differ between the two species?

Unfortunately, estimates of absolute time of growth are currently lacking for species in the *P. maoense* clade. Therefore, it is necessary to speculate on the magnitude of growth rate change in the evolutionary history of these species until stable isotopic work is complete (Nehm, unpublished observations). Speculation on the relationship between morphological and temporal change in the *P. maoense* clade is based on observations from closely related living species.

Onset Time, Offset Time, and the Rate of Change

Several lines of evidence suggest that the onset of shell growth is the same in *P. latissimum* and *P. maoense*. The first protoconch whorl of *Prunum* species generally occurs at or near the protoconch-teleoconch boundary. This boundary is correlated with the transition from intracapsular juvenile to hatched juvenile. Shell size, area, aperture area, width, aperture height, and shell thickness are not significantly different between species near the protoconch-teleoconch boundry, suggesting that no differences appear to have occurred by this developmental stage. This suggests that hatching took place at a similar growth stage in both species. Although this does not rule out temporal changes, it suggests that postdisplacement is not a cause of heterochronic change in *P. maoense*.

Two different "clocks" may be used to estimate offset time (the timing of growth termination) in *P. latissimum* and *P. maoense*. The first "clock" is the number of shell whorls. Each consecutive whorl in one species may be considered to be temporally equivalent to the respective whorl in the other species. The second "clock" is animal size (shell height). Similar size increases between both species may approximate equivalent amounts of time. Interestingly, both "clocks" produce similar estimates of offset time in *P. maoense*. Whorl number suggests that a reduction in offset time occurred between *P. latissimum* and *P. maoense* because *P. maoense* has 0.25–1.0 less whorls than *P. latissimum*. Animal size indicates that offset time occured earlier in *P. maoense* because adult *P. maoense* are significantly smaller than adult *P. latissimum* (Fig. 1.3). Therefore, assuming that the size and whorl "clocks" tell time accurately, hypomorphosis is a likely explanation of patterns of heterochronic change in *P. maoense*.

The rates of change of *P. latissimum* and *P. maoense* appear to be the same in early ontogeny. Juveniles of both species share similar allometric relationships (determined using RMA analysis). Comparisons of shell height, width, thickness, aperture area, and shell area between juveniles of *P. latissimum* and *P. maoense* produced no significant differences (Fig. 1.9). The similarities of size and allometric relationships in juveniles of both species suggest that no changes in growth rate occured between *P. latissimum* and *P. maoense*. However, significant differences in size and shape occur between adult *P. latissimum* and *P. maoense*. Therefore, it appears that the morphological differences between *P. maoense* and *P. latissimum* were produced in late ontogeny.

In summary, no change in growth onset time appears to have occurred between species, based on the similarities of size and whorl number at the protoconch-teleoconch boundary. A change in offset time is suggested by the smaller size and whorl reduction in *P. maoense* relative to *P. latissimum*. No change in the rate of growth is suggested by the absence of significant differences in bivariate allometric relationships between juveniles of both species. Finally, adults of *P. maoense* overlap with the shape space of juveniles of *P. latissimum*. All of this evidence suggests that paedomorphosis via hypomorphosis (progenesis) produced the morphological reduction of shell features in *P. maoense*.

Using the Clock Model to Distinguish Heterochronic Modes

Gould (1977) developed a model for comparing evolutionary changes in size, shape, and age between two species, thereby facilitating the recognition of different heterochronic modes. Comparisons between species are made using a semicircular "clock" with three scales: size, shape, and age. In cases of paedomorphosis, which are the focus of this study, the *scales* on the clock represent measurements from the ontogeny of the ancestor (from the onset of growth to adulthood). The two *hands* of the clock represent the size and shape of the descendant. Comparisons between ancestor and descendant are made at a common time or developmental stage (such as adulthood). For example, in cases of hypomorphosis (progenesis), in which the descendant ontogeny is truncated by early sexual maturity, the correlations of size and shape are unchanged during evolution: the descendant is a sexually mature ancestral juvenile. This pattern may be visualized using the clock model (Fig. 1.10). The hands of size and shape move together because no dissociation between size and shape has occurred.

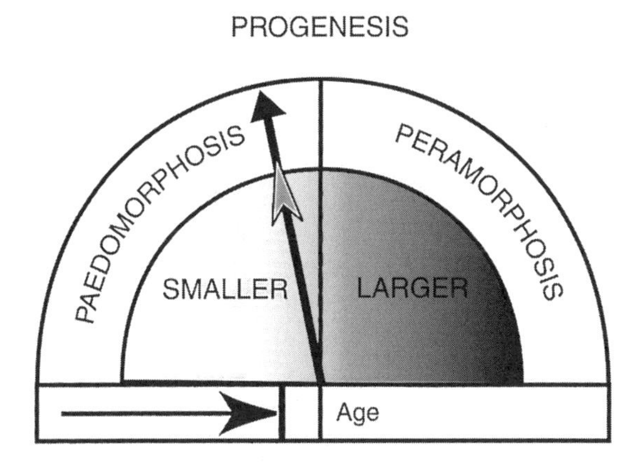

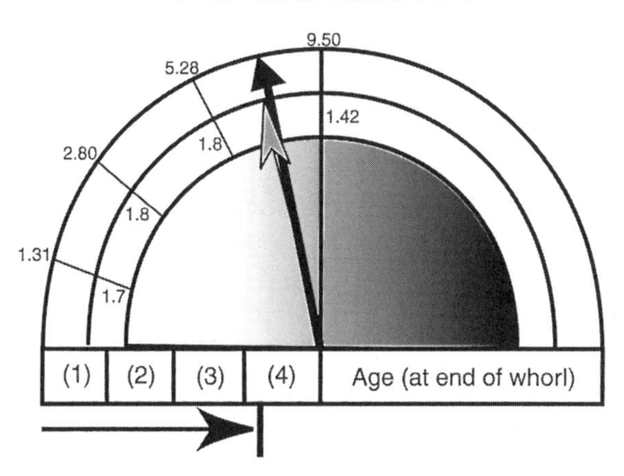

Figure 1.10. Top. Gould's clock model for comparing evolutionary changes in size, shape, and age between two species. Comparisons between species are made using a semicircular "clock" with three scales: size, shape, and age. The two *hands* of the clock represent the size and shape of the descendant. Comparisons between ancestor and descendant are made at a common time or developmental stage (such as adulthood). A theoretical case of progenesis is illustrated above and the empirical results of comparisons between *P. latissimum* and *P. maoense* are illustrated below.

Both hands move from the "12 o'clock" comparison point (adulthood) to the left because the descendant adults fall within the size and shape scales of ancestral juveniles (that is, they are smaller and have a juvenilized shape). Time to adulthood has also decreased because, like size and shape, it has not become dissociated in evolution. Gould (1977, p. 257–260) illustrates other heterochronic modes using the clock model.

The clock model may be used to infer the mode of paedomorphosis that occurred in the evolution of *P. maoense*. The ancestral size scale is estimated using shell height in *P. latissimum*, and the ancestral shape scale is estimated using the degree of shell globosity in *P. latissimum* (measured as the ratio of shell width relative to size). The hands of the clock represent the size and shape of *P. maoense* (the descendant) at adulthood. The clock is now ready to run: the scales of size and shape are calibrated using the ontogeny of *P. latissimum*, and the size and shape hands of *P. maoense* are set to adulthood in order to compare the two species at a common developmental stage. When the clock is started, the hands of size and shape move together to the left and stop at a point during the late ontogeny of *P. latissimum* (Fig. 1.10). Temporal reduction is suggested by fewer whorls in *P. maoense* relative to *P. latissimum*. Assuming the scales of the clock are accurate, the model indicates that hypomorphosis (progenesis) produced paedomorphic change in the *P. maoense* group.

The clock model conclusion of hypomorphosis (progenesis) agrees with all of the traditional and geometric morphometric analyses as well as the speculative conclusions of onset stability and offset truncation. RMA regression analyses indicated no significant differences between juvenile allometric relationships but significant differences in all morphometric variables between adults. This indicates that developmental change occurred during the late ontogeny of *P. maoense*. RWA indicated that the descendant adults fell within the shape space of ancestral juveniles, in concordance with analyses using shell globosity as a measure of shape. Adult descendants are smaller and have fewer whorls than the ancestors, suggesting a slight truncation of ontogeny. Therefore, paedomorphosis via hypomorphosis appears to explain the reduction of morphological features in evolutionary history of the *P. maoense* clade.

CONCLUSIONS

Summary

Paedomorphosis via hypomorphosis (progenesis) explains the morphological disarmament in the evolutionary history of the *P. maoense* clade as suggested by (1) no change in growth onset time based on the similarities of size and whorl number at the protoconch-teleoconch boundary, (2) a change in offset time based on the smaller size and fewer number of whorls in *P. maoense* relative to *P. latissimum*, (3) no rate change in bivariate allometric relationships between juveniles of both species, (4) significant differences between all morphological features of adults in traditional and geometric morphometric analyses, and (5) the overlap of adult *P. maoense* in the shape space of juveniles of *P. latissimum*. All of the analyses in this study support the conclusion that pure paedomorphosis via hypmorphosis (progenesis) produced the morphological reduction of shell features in *P. maoense*.

Implications for the Study of Evolution and Development

An understanding of the tension between selection and constraint is essential for a full understanding of macroevolutionary change. Developmental constraints are considered to play a large role in patterns of macroevolutionary change by limiting what selection can produce (Gould, 1977). Studies of developmental evolution during episodes of morphological armament and disarmament may be one of the most promising arenas for exploring the interplay between intrinsic and extrinsic processes in macroevolution. In such cases, selection intensity may be inferred by the magnitude of predation (or other biotic or abiotic parameters) and constraint may be inferred in cases of pure heterochrony in which size, shape, and age are not dissociated.

This study demonstrates that large and prominent (i.e., metabolically expensive) ancestral morphological features (such as the margin callus) may be reduced in a descendant by maintaining an ancestral ontogenetic stage that lacked these structures into adulthood. In this study, evolutionary change was highly conserved and did not involve dissociations between size, shape, or age. In addition, the absence of novelty and the presence of these tight correlations suggest that evolutionary change may have been constrained or channeled along the ancestral ontogenetic pathway. More studies are clearly needed to determine whether this pattern typifies developmental changes that produce morphological disarmament in marginellids. Nevertheless, it appears that selection was limited in the changes it could produce. The reduction of the callus in the descendant was not produced by dissociating callus development from overall development in the ancestor; rather, the callus was reduced by limiting the overall duration of development in the descendant.

Studies of the developmental basis of the addition or accentuation of morphological features (i.e., morphological armament) could also provide important evidence on the frequency or magnitude of developmental constraint in marginellids. For example, if evolutionary additions were also produced by pure heterochrony and a lack of size, shape, and age dissociation (i.e., hypermorphosis), ontogenetic constraint would contribute to patterns of morphological armament. Determining the cause(s) of one-time evolutionary events is difficult; more studies of the developmental basis of morphological armament and disarmament are clearly needed to determine the role of developmental constraint in marginellid gastropod evolution.

Future Directions

Sclerochronology. Time is essential for studies of heterochrony, but it is difficult (and expensive) to determine the ontogenetic age of fossil organisms. Nevertheless, to understand how development evolves through macroevolutionary time, we must establish absolute timescales for comparisons between ancestors and descendants (e.g., Jones and Gould, 1998). This study exemplifies the complexities of telling time without an absolute timescale and how the absence of time weakens studies of heterochrony. Although the conclusion from this study that paedomorphosis occurred in *P. maoense* is not dependent on the accuracy of the whorl and size "clocks," the hypothesis of the mode of paedomorphosis (hypomorphosis) *is* dependent on the estimates of absolute time. Therefore, although general interpretations of developmental evolution can be determined with a fair degree of certainty without absolute time (e.g., novelty and heterochrony or types of heterochrony such as peramorphosis and paedomorphosis), many of the particulars of developmental evolution we seek to address cannot be answered.

Constructional Morphology. Morphometric comparisons of size and shape do not elucidate patterns and processes by which morphological structures are constructed. Although structures may be similar in size or shape, they may be constructed of different materials or the same materials in different spatial arrangements (e.g., shell crystallography; Nehm, 2001a). For example, the whorls of the marginellid shell are constructed using three to four shell layers deposited in at least two different ways (marginally and surficially). Similarity in size and shape does not necessarily indicate constructional or compositional similarity (although see Nehm, 2001a, for an example in which they do). Further research is needed to determine whether shape space overlap coincides with structural and constructional homogeneity. This is important because, although pure heterochrony may characterize morphological change at a gross level, microstructural studies could indicate rearrangements of the spatial ordering of shell layering (i.e., novelty). For these reasons, constructional morphology has great potential in expanding our understanding of both patterns and processes of developmental evolution.

Methodological Comparisons of Morphometric Approaches. The ways in which shape spaces are constructed and differences in how shape space occupation is quantified may influence our perceptions of heterochrony and novelty. Our conceptual understandings of ontogenetic and evolutionary variation are dependent on the methodological approaches that we use to characterize or quantify variation, the units that are compared in the analysis of variation, and the sampling necessary to capture patterns of variation. All three factors have been poorly investigated in gastropods and other lineages. For example, both geometric morphometric and eigenshape approaches can be used in analyses of developmental evolution of gastropods using shape spaces, but the appropriateness of these different methods for different research questions has not been investigated (MacLeod, 1999). In addition, although geometric morphometric analysis permits quantitative studies of the organism as an integrated whole, little is known about how analyses of different developmental or morphological units (e.g., shell whorls) alter interpretations of evolutionary and developmental variation and change. Finally, although inexpensive visualization hardware and software now permits rapid and extensive landmark-gathering capabilities, little is known of how different magnitudes of landmark sampling influence constructions and comparisons of shape spaces. All of these factors may bias investigations of evolutionary-developmental change and distort our perceptions of heterochrony and novelty.

Theoretical Pluralism. Theoretically, progress in our understanding of heterochrony requires (1) rejection of the assumption that all developmental evolution is a product of heterochrony, (2) the adoption of alternative hypotheses such as novelty or heterotopy (Gould, 1977; Zelditch and Fink, 1996; also see preface of this book), and (3) consideration of multiple theoretical frameworks in which to conceptualize heterochrony. Heterochrony has been used by various authors to refer to a wide variety of patterns and processes (see preface). None of these heterochronic frameworks (e.g., event-based heterochrony) has been demonstrated to be less fruitful than the heterochrony/novelty model adopted in this study. For example, are conceptualizations of developmental evolution using the heterochrony/novelty model more appropriate than conceptualizations based on changes in the timing of developmental events? Unfortunately, few studies have applied different theoretical frameworks to the same empirical data. Certain research systems, such as fossil organisms, prevent the investigation of some conceptual models of developmental evolution (e.g., mechanistic models, Alberch, 1985; Alberch and Blanco, 1996). Regardless, it is paramount that we interpret our data using as many conceptual frameworks as our data permit in order to evaluate the robustness of evolutionary-developmental conclusions.

In summary, progress in heterochronic research requires the rejection of panheterochrony, the adoption of alternative hypotheses to heterochrony (such as novelty or heterotopy), and an exploration of how different conceptual models differentially inform us of developmental evolution. In addition, sclerochronological studies, combined with constructional morphological approaches and methodological pluralism, provide promising avenues for future studies of evolution and development. Finally, more studies of morphological and developmental evolution within the *same* clades are necessary to determine the relative importance of developmental constraints in macroevolution.

ACKNOWLEDGMENTS

The material for this study was provided by Peter Jung and Rene Panchaud of the Naturhistorisches Museum, Basel, Switzerland; Jack and Winifred Gibson-Smith of Surrey, England; Emily Vokes of Tulane University; Tom Waller and Warren Blow of the Smithsonian Institution; James McLean, Edward Wilson, and Linsey Groves of the Los Angeles Museum of Natural History; Roger Portell and Kurt Auffenburg of the Florida Museum of Natural History at Gainesville; Robert Van Syk of the California Academy of Sciences; Gary Rosenberg of the Academy of Natural Sciences, Philadelphia; and David Lindberg of the University of California Museum of Paleontology. I am grateful for the access to these collections and the generous hospitality and assistance provided by these individuals. I thank Miriam Zelditch for expanding my understanding of developmental evolution and providing an unpublished manuscript. Mark Groves provided a helpful review of the manuscript. Field assistance in the Dominican Republic by Bryan Bemis is also appreciated. Financial support was provided by a National Science Foundation International Fellowship.

REFERENCES

Alberch P (1985): Problems with the interpretation of developmental sequences. Syst Zool 34: 46–58.

Alberch P, Blanco MJ (1996): Evolutionary patterns in ontogenetic transformation: From laws to regularities. Int J Dev Biol 40: 845–858.

Alberch P, Gould SJ, Oster GF, Wake DB (1979): Size and shape in ontogeny and phylogeny. Paleobiology 5: 296–317.

Allmon WD, Rosenberg GA, Portell R, Schindler K (1993): Diversity of Atlantic coastal plain mollusks since the Pliocene. Science 260: 1626–1629.

Anderson LC, Geary DH, Nehm RH, Allmon WD (1991): A comparative study of gastropod predation on *Varicorbula caloosae* and *Chione cancellata*, Plio-Pleistocene of Florida, U.S.A. Palaeogeogr Palaeoclimatol Palaeoecol 85: 29–46.

BIOMstat. Exeter Software, Inc. 1996.

Bookstein FL (1996): Combining the tools of geometric morphometrics. In: Marcus LF et al. (eds), *Advances in Morphometrics*. New York: Plenum Press, p. 131–152.

Coovert GA (1989): A literature review and summary of published marginellid radulae. Marginella Marginalia (Dayton Museum of Natural History) 7: 1–37.

Coovert GA, Coovert HK (1995): Revision of the suprageneric classification of marginelliform gastropods. Nautilus 109: 43–110.

Dall WH (1896): Diagnoses of new Tertiary fossils from the southern United States. Proc US Natl Mus 18: 21–46.

Gardner JA (1928): Molluscan fauna of the Allum Bluff group of Florida. United States Geological Survey Professional Paper 142 A-I, p. 1–709.

Gould SJ (1977): *Ontogeny and Phylogeny*. Cambridge, MA: Harvard University Press.

Gould SJ (1988): The uses of heterochrony. In: McKinney M (ed), *Heterochrony in Evolution. A Multidisciplinary Approach*. New York: Plenum Press, p. 1–13.

Gould SJ, Vrba ES (1982): Exaptation—a missing term in the science of form. Paleobiology 8: 4–15.

Harvey PH, Pagel MD (1991): *The Comparative Method*. Oxford, UK: Oxford University Press.

Jones D, Gould SJ (1998): Direct measurement of age in fossil Gryphaea: the solution to a classic problem in paleobiology. Paleobiology 25: 158–187.

Kitchell JA (1990): The reciprocal interaction of organism and effective environment: Learning more about "and." In: Ross RM, Allmon, WD (eds), *Causes of Evolution, a Paleontological Perspective*, Chicago, IL: University of Chicago Press, p. 151–169.

MacLeod N (1999): Generalizing and extending the eigenshape method of shape space visualization and analysis. Paleobiology 25: 197–138.

Maury CJ (1917): Santo Domingo type sections and fossils. Part 1. Bull Am Paleontol 5: 1–251.

McKinney ML, McNamara KJ (1991): *Heterochrony: The Evolution of Ontogeny*. New York: Plenum Press.

Nehm RH (1996): Estimating phylogenetic relationships in marginelliform gastropods: revisiting replicability and testability in molluscan systematics. American Malacological Union Programs and Abstracts.

Nehm RH (1998): *Macroevolution and Development in Marginellid Gastropods from the Neogene of the Caribbean Basin*. PhD dissertation, University of California-Berkeley.

Nehm RH (2001a): Linking macroevolutionary pattern and developmental process in marginellid gastropods. In: Jackson JBC, McKinney FK, Lidgard S (eds), *Process from Pattern in the Fossil Record*. Chicago, IL: University of Chicago Press.

Nehm RH (2001b): Neogene Paleontology in the northern Dominican Republic: the genus *Prunum*. Bull Am Paleontol.

Nehm RH, Geary DH (1994): A gradual morphologic transition during a rapid speciation event in marginellid gastropods (Neogene: Dominican Republic). J Paleontol 68: 787–795.

Nehm RH, Tran C (1997): Molecular phylogeny of marginelliform gastropods: a progress report. *American Malacological Union Program with Abstracts*.

NTSYS 2.0 (1998): Exeter Software.

Petuch EJ (1981): A relict Neogene Caenogastropod fauna from northern South America. Malacologia 20: 307–347.

Petuch EJ (1982): Geographical heterochrony: contemporaneous coexistence of Neogene and Recent molluscan faunas in the Americas. Palaeogeogr Palaeoclimatol Palaeoecol 37: 277–312.

Raff RA (1996): The Shape of Life. Genes, Development, and Evolution. Chicago, IL: University of Chicago Press.

Reilly SM, Wiley EO, Meinhardt DJ (1997): An integrative appoach to heterochrony: the distinction between intraspecific and interspecific phenomena. Biol J Linn Soc 60: 119–143.

Rohlf FJ, Marcus LF (1993): A revolution in morphometrics. Trends Ecol Evol 8: 129–132.

Rohlf FJ (1993a): Relative warp analysis and an example of its application to mosquito wings. In: Marcus LF, Bello E, Garcia-Valdecasas A (eds), *Contributions to Morphometrics*. Madrid: Monografias del Museo Nacional de Ciencias Naturales 8, p. 131–159.

Rohlf FJ (1993b). TPS programs distributed by the author. SUNY-Stonybrook.

Saunders JB, Jung P, Biju-Duval B (1986): Neogene paleontology of the northern Dominican Republic 1. Field surveys, lithology, environment, and age. Bull Am Paleontol 89: 1–79.

Sokal RR, Rohlf FJ (1995): *Biometry* (2nd ed.). New York: Freeman.

Vermeij GJ (1978): *Biogeography and Adaptation*. Cambridge: Belknap Press.

Vermeij GJ (1987): *Evolution and Escalation: An Ecological History of Life*: Princeton, NJ: Princeton University Press.

Vermeij GJ, Signor PW (1992): The geographic, taxonomic and temporal distribution of determinate growth in marine gastropods. Biol J Linn Soc 47: 233–247.

Vermeij GJ (1993): *A Natural History of Shells*. Princeton, NJ: Princeton University Press.

Woodring WP (1974): The Miocene Caribbean faunal province and its subprovinces, Verh Naturforsch Ges Basel 84: 209–213.

Zelditch ML, Fink WL (1996): Heterochrony and heterotopy: stability and innovation in the evolution of form. Paleobiology 22: 241–254.

2

ONTOGENETIC SEQUENCES: HOMOLOGY, EVOLUTION, AND THE PATTERNING OF CLADE DIVERSITY

Larry Hufford

School of Biological Sciences, Washington State University, Pullman, Washington

INTRODUCTION

Development is the deployment of phenotype, and diversification depends on alterations in that deployment. Comparative studies of development, especially those that apply robust reconstructions of organismal phylogeny, can elucidate not only the ontogenetic changes that create diversity but also how the patterning of diversity in clades emanates from the succession of ontogenies that compose lineages.

Since the publication of Gould's (1977) *Ontogeny and Phylogeny*, developmental studies have had a central role in our understanding of clade diversification. Most of those studies have emphasized evolutionary differences in developmental timing, which is known as heterochrony. Heterochrony has been modelled in various organismal systems, especially through the characterization of different rates of morphological development (whether expressed as relative or absolute rates) and of heterogeneity in initiation and cessation times of de-

Beyond Heterochrony: The Evolution of Development, Edited by Miriam Leah Zelditch
ISBN 0-471-37973-5

velopmental events. Following the formalized heterochrony models of Gould (1977) and, especially, Alberch et al. (1979), most studies of heterochrony have used either allometry or comparative time courses of developmental events to demonstrate the effects of changes in developmental timing on morphological diversification. Various studies of the diversification of angiosperm flowers, the primary focus of this chapter, have used heterochrony models, and many of the earliest were reviewed by Diggle (1992).

Various approaches, in addition to those centered on developmental timing and heterochrony, can be used to describe development and its evolution. Prominent among these alternatives are (1) developmental-genetic, (2) biophysical, and (3) sequence approaches. These options along with developmental timing are best seen as complementary ways to understand the development of organismal form and its diversification. Each approach has strengths and weaknesses, and each can provide unique insights. Among these approaches, biophysical aspects of floral development have been investigated little. The biophysical approach used by Green has involved modelling the floral apical meristem as a hoop-reinforced dome in which growth and cell division positions create stresses that cause buckling (organogenesis), and his experiments have shown that changes in the physical parameters of the meristem can modify organography in ways that mimic evolutionary diversity (Green, 1988, 1992, 1996; Selker et al., 1992; Hernández and Green, 1993). In contrast, the intensive recent research on the developmental genetics of flowers has stimulated a plethora of models for the molecular basis of floral development (e.g., Coen and Meyerowitz, 1991) and evolution (e.g., Irish and Yamamoto, 1995; Doebley and Lukens, 1998; Kramer and Irish, 2000). Most investigators interested in morphological evolution, however, have focused primarily on developmental timing or ontogenetic sequences as ways of describing organismal development and its diversification.

Alberch (1985) criticized the use of ontogenetic sequences in evolutionary studies. He discussed particularly the need for sequences to consist of causally related stages for them to have value in evolutionary studies. The analysis that follows suggests that causality is not critical; however, the later states of ontogenetic sequences should be contingent on those that occurred earlier. Alberch (1985) also criticized attempts to hypothesize homologies between stages of different organismal ontogenies. In contrast, I argue below that homology assessment is fundamental to the comparative study of ontogenies. Indeed, we should recognize from the outset that all homology assessments in comparative biology select and compare ontogenetic stages among organisms even if they are from reproductive phases or more terminal states of development.

I discuss the application of ontogenetic sequence models for understanding the patterning of morphological diversity in a clade of flowering plants. This analysis focuses particularly on unique vs. iterative novelties in the patterning morphological diversity in clades. The origin of novelties has been one of the great concerns of evolutionary biology (e.g., Mayr, 1960; Bock, 1965; Müller and Wagner, 1991), and, ultimately, diversification depends on the invention of new structures. However, one of the most prominent outcomes of over two decades of concerted emphasis on cladistic analysis is the recognition that homoplastic evolution—the reiteration of novelties—is pervasive across all data sets and all kinds of organisms (Sanderson and Donoghue, 1989). This raises questions not only about the relative importance of unique novelties vs. those that are iterated and homoplastic in the patterning of phenotypic diversity but also how evolutionary transformations of ontogenies are involved in each.

In the sections that follow, I first characterize ontogenetic sequence models and then apply them to the description of floral development in the angiosperm family Hydrangeaceae. The Hydrangeaceae example is used to examine criticisms of ontogenetic sequences made by Alberch (1985). Ultimately, on the basis of the homology assessments and a phylogenetic hypothesis for the family, inferences are drawn about the relationship between ontogenetic evolution and the patterning of morphological diversity in Hydrangeaceae.

ONTOGENETIC SEQUENCES AS MODELS OF ONTOGENY

Sequence models are formalizations of traditional narrative descriptions of development. They emphasize the sequence of forms and events that modify forms in the course of development. An ontogenetic sequence can be defined explicitly as a model of an ontogeny composed of a series of instantaneous phenotypic states (that is, ontogenetic states). Later ontogenetic states are developmental transformations of earlier states. This approach corresponds to what Velhagen (1997) called a transformation sequence, which he contrasted with an event sequence in which the model consists of a succession of distinct developmental events.

The ontogenetic sequence for even the simplest organism would be a complex, highly branched chart; thus, we typically encounter sequences for characters. This approach is used in the example below that focuses on flowers of the Hydrangeaceae. I have discussed previously how floral morphological development can be conceived as a hierarchy of ontogenetic sequences (Hufford, 1996a). Important aspects of floral diversity among the angiosperms can be investigated by focusing on the evolution of ontogenetic sequences at different levels in the floral developmental hierarchy. Various floral characters can be described as discrete ontogenetic sequences that proceed from development on an apical meristem, and they are found at numerous levels in the floral developmental hierarchy.

Ontogenetic sequences can undergo various evolutionary transformations, and these have been discussed by Hufford (1996a, 1997b). Classifications of the evolutionary transformations that occur in sequences (e.g., O'Grady, 1985; Mabee, 1993; Hufford, 1996a, 1997b) focus specifically on patterns of phenotypic change and not on underlying developmental processes. The use of sequence models in comparative biology can focus attention on those aspects of development that require further study, especially at lower levels of investigation (such as biophysical or genetic analysis). The application of sequence models in studies of diversification is valuable primarily for showing how phenotypic diversity arises as a consequence of alterations in the course of morphological development.

CASE STUDY: ANDROECIAL ONTOGENY AND DIVERSITY IN HYDRANGEACEAE

Hydrangeaceae

The flowering plant family Hydrangeaceae is part of the order Cornales of the asterids (Angiosperm Phylogeny Group, 1998). Phylogenetic analyses of Cornales showed that Hydrangeaceae is the sister of Loasaceae (Soltis et al., 1995; Xiang et al., 1998; Xiang, 1999; Olmstead et al., 2000). Hydrangeaceae consists of 17 genera and approximately 220 species. Phylogenetic relationships in Hydrangeaceae have been examined using both morphological data (Hufford, 1997a) and DNA sequences of the plastid genes *rbcL* (Soltis et al., 1995) and *matK* (Hufford et al., in press). The most robust phylogenetic estimate for the family (Fig. 2.1) is that based on cladograms derived from maximum parsimony analyses of DNA sequences of *rbcL* and *matK* (Hufford et al., in press).

The evolution of reproductive characters in Hydrangeaceae involves tradeoffs between flowers and inflorescences that include flower number per inflorescence, presence/absence of sterile display flowers, and flower size (Hufford, unpublished observations). Fertile flowers differ across clades of Hydrangeaceae in calyx, corolla, androecium (region of male reproductive structures), and gynoecium (region of female reproductive structures) states. In this chapter, I examine the patterning of diversity in the evolution of the androecium of Hydrangeaceae. The androecium of a flower consists of male reproductive appendages, that is, the stamens. Each stamen in the Hydrangeaceae has a largely filamentous basal region (the filament) and is terminated distally by an anther, in which pollen grains are produced in four pollen sacs. The number of stamens in the androecium has a direct impact on reproductive ecology and male fitness through factors that include the total amount of pollen produced, the ratio of pollen to ovules in populations, floral display to potential biotic pollinators, and the efficiency and effectiveness of pollen removal (e.g., Queller, 1984; Harder and Thomson, 1989; Willson, 1991; Cruden, 1997).

Androecial Forms

The classification of androecia uses the number of stamens relative to the number of perianth appendages, the positions of stamens relative to perianth appendages, and the number of stamen whorls as criteria. The following three general forms of androecia are recognized.

Haplostemony. The number of stamens equals the number of appendages in either the calyx or corolla (i.e., usually half of the total number of sepals and petals) in haplostemonous androecia. One stamen is positioned opposite each

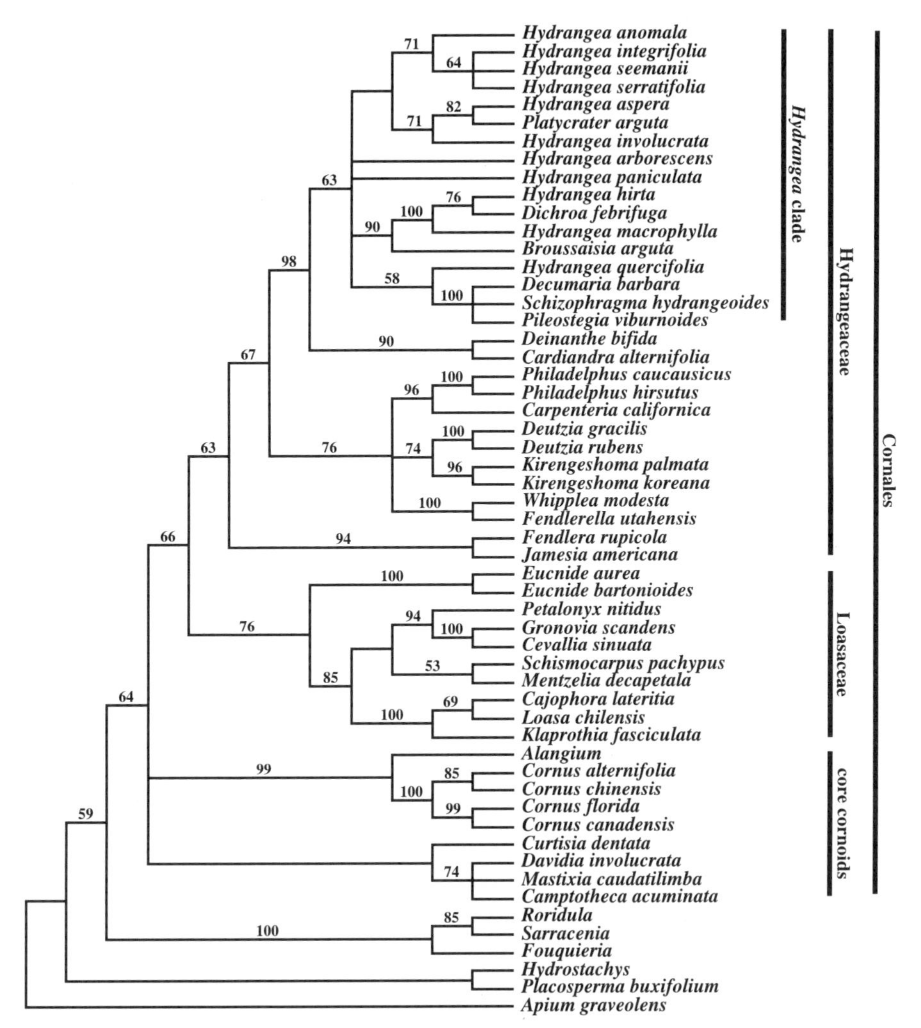

Figure 2.1. Strict consensus cladogram of 504 equally parsimonious trees of Hydrangeaceae based on DNA sequences from the plastid genes *matK* and *rbcL* (data from Hempel et al., 1995; Soltis et al., 1995; Xiang et al., 1998; Hufford et al., in press; Moody et al., 2001; and GenBank). The maximum parsimony analysis of 558 informative characters used heuristic search procedures in PAUP (Swofford, 1998), including 100 replicated searches with random taxon addition to search for islands of equally parsimonious cladograms. The analysis found 504 trees of 2,466 steps (consistency index = 0.46, retention index = 0.61, rescaled consistency index = 0.36). Numbers above the branches are bootstrap values (indicated when above 50%). The bootstrap analysis was conducted using PAUP (Swofford, 1998) and included 100 replicates.

appendage of one perianth series (e.g., either the sepals or petals; if stamens are opposite petals, then the condition is called obhaplostemony). The androecium has a single whorl of stamens.

Diplostemony. The number of stamens equals the total number of sepals and petals in diplostemonous androecia. One stamen is positioned opposite each appendage of the perianth (i.e., opposite each sepal and petal). The androecium has two whorls of stamens. Among flowering plants that have diplostemony, it is most common for the outer whorl of stamens to be positioned opposite the sepals and the inner whorl opposite the petals (when this is reversed, the condition is called obdiplostemony).

Polystemony. The number of stamens is greater than the total number of perianth appendages in polystemonous androecia. There are various morphological expressions of polystemonous androecia, and they can differ in stamen numbers and positions as well as in number of whorls of stamens (Hirmer, 1918; Ronse Decraene and Smets, 1992, 1993). A traditional means of distinguishing among the many forms of polystemonous androecia is by their pattern of stamen initiation (Payer, 1857). Polystemonous androecia that have (1) helical, (2) centrifugal, or (3) centripetal initiation of stamen primordia on the floral apical meristem have been recognized.

Hydrangeaceae Androecia

Pattern of Evolution. Most Hydrangeaceae have diplostemonous or polystemonous androecia. Haplostemony is characteristic only of *Dichroa pentandra* Schl. and *D. platyphylla* Merrill and is clearly derived in *Dichroa*. A first step toward understanding the evolution of these different androecial forms is to map their distribution on a phylogenetic hypothesis. A phylogenetic hypothesis for Cornales (Fig. 2.2) is used here because the optimization of androecial states on internal nodes of Hydrangeaceae depends on outgroup states. This problem is heightened because the sister clade of Cornales is uncertain (Fig. 2.2a), and the androecial state of this sister clade influences the optimization of states on the internal nodes of Hydrangeaceae. It is possible that the androecial state of the sister clade of Cornales is haplostemony, and, if true, this offers no resolution of the states of the equivocal internal nodes of Hydrangeaceae (Fig. 2.2b). In contrast, if the sister clade of Cornales is hypothesized to be diplostemonous (Fig. 2.2c), then the androecial states of most internal nodes of Hydrangeaceae are resolved and the plesiomorphic state for the family is shown to be diplostemony. However, if the sister clade of the Cornales is hypothesized to be polystemonous (Fig. 2.2d), then the androecial state at the lowermost node and various internal nodes of Hydrangeaceae can be either diplostemony or polystemony. For each of these alternative scenarios, the plesiomorphic androecial state for Hydrangeaceae will be either diplostemony or polystemony. Under the possible phylogenetic scenarios suggested by recent analyses, haplostemony could not be the plesiomorphic androecial state in Hydrangeaceae.

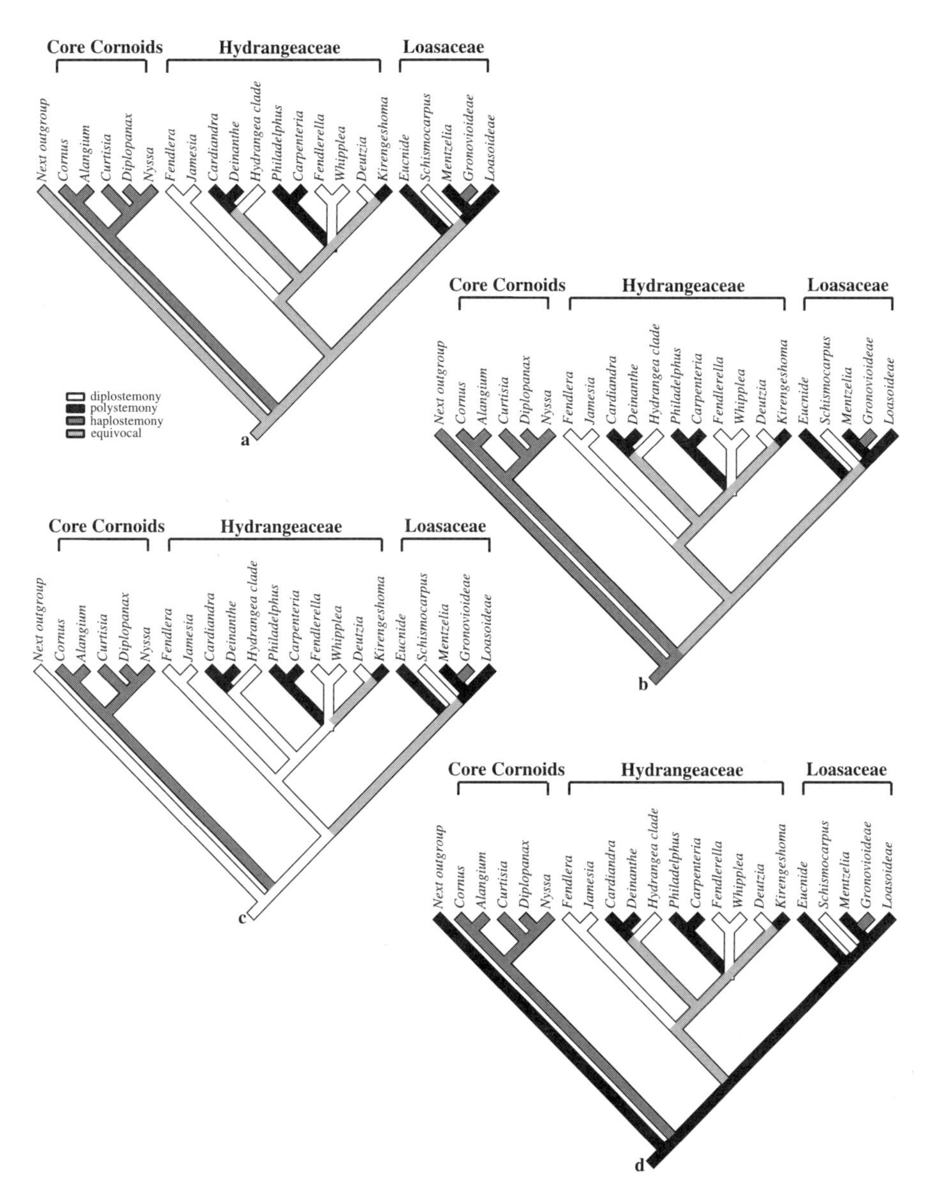

Figure 2.2. Phylogenetic supertree constructed to show androecial diversity in Cornales. Relationships among clades of Cornales based on Xiang et al. (1998, for relationships of core cornoids), Moody et al. (in press, for relationships in Loasaceae), and this chapter (for relationships in Hydrangeaceae). The closest relatives of the Cornales are uncertain, and this impacts the resolution of androecial states at internal nodes of Hydrangeaceae. a: If the androecial state of the sister clade of Cornales is coded as equivocal, then the lowermost node and various internal nodes of Hydrangeaceae are also equivocal. b: If the sister clade of the Cornales is hypothesized to be haplostemonous, then the androecial state at the lowermost node and various internal nodes of Hydrangeaceae is equivocal. c: If the sister clade of Cornales is hypothesized to be diplostemonous, then the androecial state of most internal nodes of Hydrangeaceae are resolved. The plesiomorphic condition of the androecium is shown to be diplostemonous. d: If the sister clade of the Cornales is hypothesized to be polystemonous, then the androecial state at the lowermost node and various internal nodes of Hydrangeaceae is equivocal.

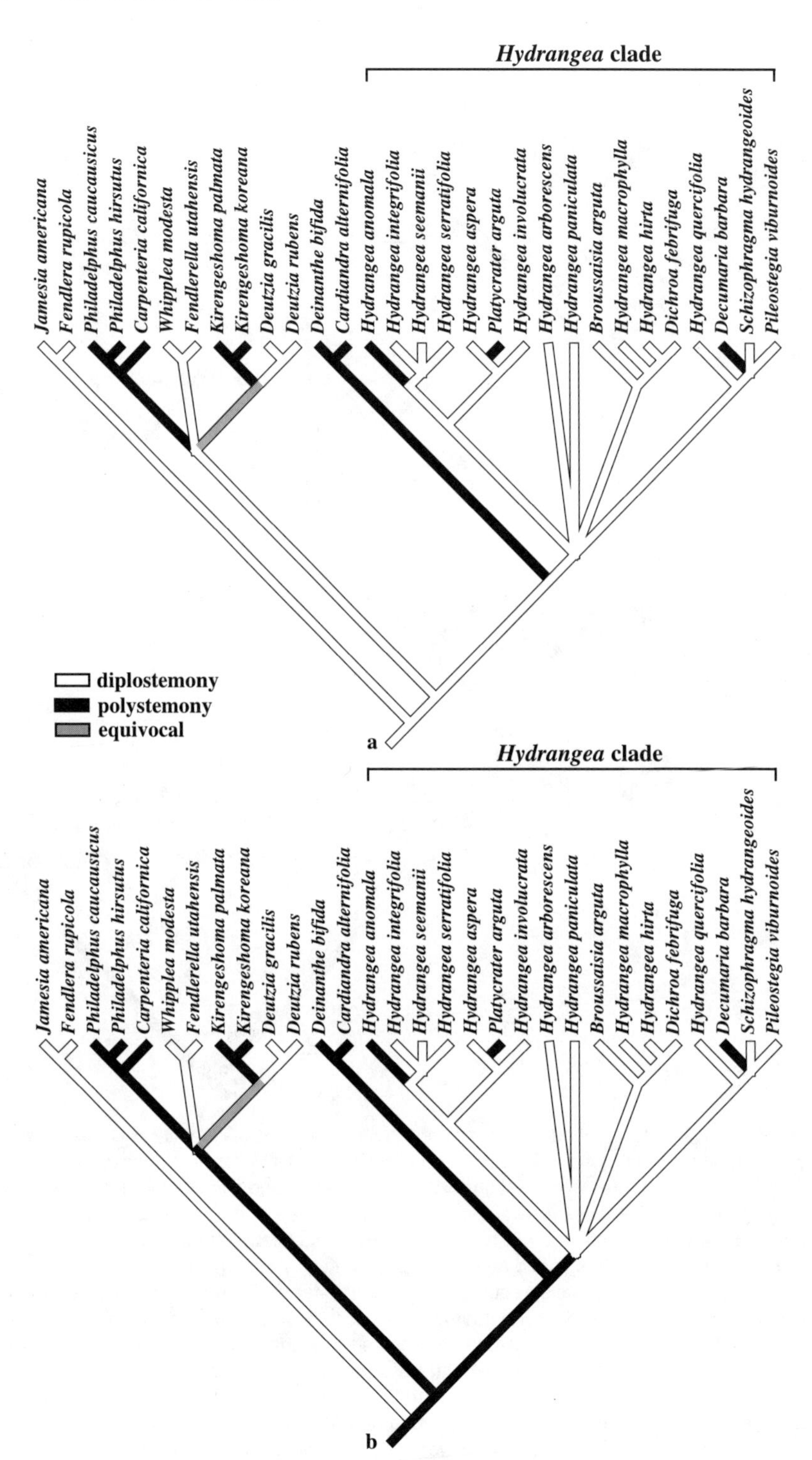
Hydrangea clade
Jamesia americana
Fendlera rupicola
Philadelphus caucausicus
Philadelphus hirsutus
Carpenteria californica
Whipplea modesta
Fendlerella utahensis
Kirengeshoma palmata
Kirengeshoma koreana
Deutzia gracilis
Deutzia rubens
Deinanthe bifida
Cardiandra alternifolia
Hydrangea anomala
Hydrangea integrifolia
Hydrangea seemanii
Hydrangea serratifolia
Hydrangea aspera
Platycrater arguta
Hydrangea involucrata
Hydrangea arborescens
Hydrangea paniculata
Broussaisia arguta
Hydrangea macrophylla
Hydrangea hirta
Dichroa febrifuga
Hydrangea quercifolia
Decumaria barbara
Schizophragma hydrangeoides
Pileostegia viburnoides
diplostemony
polystemony
equivocal
a
Hydrangea clade
Jamesia americana
Fendlera rupicola
Philadelphus caucausicus
Philadelphus hirsutus
Carpenteria californica
Whipplea modesta
Fendlerella utahensis
Kirengeshoma palmata
Kirengeshoma koreana
Deutzia gracilis
Deutzia rubens
Deinanthe bifida
Cardiandra alternifolia
Hydrangea anomala
Hydrangea integrifolia
Hydrangea seemanii
Hydrangea serratifolia
Hydrangea aspera
Platycrater arguta
Hydrangea involucrata
Hydrangea arborescens
Hydrangea paniculata
Broussaisia arguta
Hydrangea macrophylla
Hydrangea hirta
Dichroa febrifuga
Hydrangea quercifolia
Decumaria barbara
Schizophragma hydrangeoides
Pileostegia viburnoides
b

We can reconstruct androecial evolution in the Hydrangeaceae using the results of the phylogenetic analysis of the plastid *matK* and *rbcL* sequences (Fig. 2.1) and assume that diplostemony and polystemony are equally possible as the plesiomorphic condition for the family (Fig. 2.3). If we assume that diplostemony is plesiomorphic (Fig. 2.3a), then most of the internal nodes are resolved as diplostemonous and several independent origins of polystemony are indicated. If polystemony is plesiomorphic for the family (Fig. 2.3b), then we find various independent shifts to diplostemony as well as reversals to polystemony. The *Hydrangea* clade notably has three independent shifts to polystemony under either scenario (Fig. 2.3b). The optimization in which diplostemony is plesiomorphic is slightly more parsimonious than the scenario in which polystemony is plesiomorphic based on the strict consensus cladogram. It seems reasonable at present to consider the scenarios for either diplostemony or polystemony as the plesiomorphic state in the family to be equally possible.

Androecial Development. Androecial development in all investigated Hydrangeaceae begins after the sepal primordia of the calyx and petal primordia of the corolla have been established on the periphery of the floral apical meristem. As both the calyx and corolla primordia are initiated, tissue subjacent to them extends upward, causing the conformation of the floral apex to shift from flat to concave. The funnel of tissue between the central apical dome and the bases of the individual sepal and petal primordia is the hypanthium. Individual stamen primordia are initiated centripetal to the perianth on the inner slope of the hypanthium. Subsequently, gynoecial structures (styles and ovarian septa) are also initiated on the inner flank of the concavity. The gynoecium forms lower on the hypanthium than the androecium, giving rise to an inferior ovary. The androecium of most Hydrangeaceae occupies a relatively short region of the hypanthium between the corolla and the gynoecium.

The early development of the androecium in all investigated diplostemonous members of the family is the same (Figs. 2.4 and 2.5–2.8). It involves the nearly simultaneous initiation of a whorl of stamens (Figs. 2.5 and 2.7). Each stamen of this whorl is positioned opposite a sepal (antesepalous positions). Subsequently, another whorl of stamens is formed (Figs. 2.6 and 2.8), and each of these stamens is positioned opposite a petal (antepetalous positions). Thus, the ontogenetic sequence for all of the diplostemonous androecia in the family has two ontogenetic states: (1) median antesepalous stamens and (2) median antesepalous stamens and median antepetalous stamens (shown in Fig. 2.4 as the trajectory A → B).

◄

Figure 2.3. Strict consensus cladogram for Hydrangeaceae based on the maximum parsimony analysis of DNA sequences from the plastid genes *matK* and *rbcL*. Different evolutionary scenarios for androecial diversification are required if diplostemony rather than polystemony is hypothesized to be the plesiomorphic state for the family. a: Diplostemony plesiomorphic. b: Polystemony plesiomorphic.

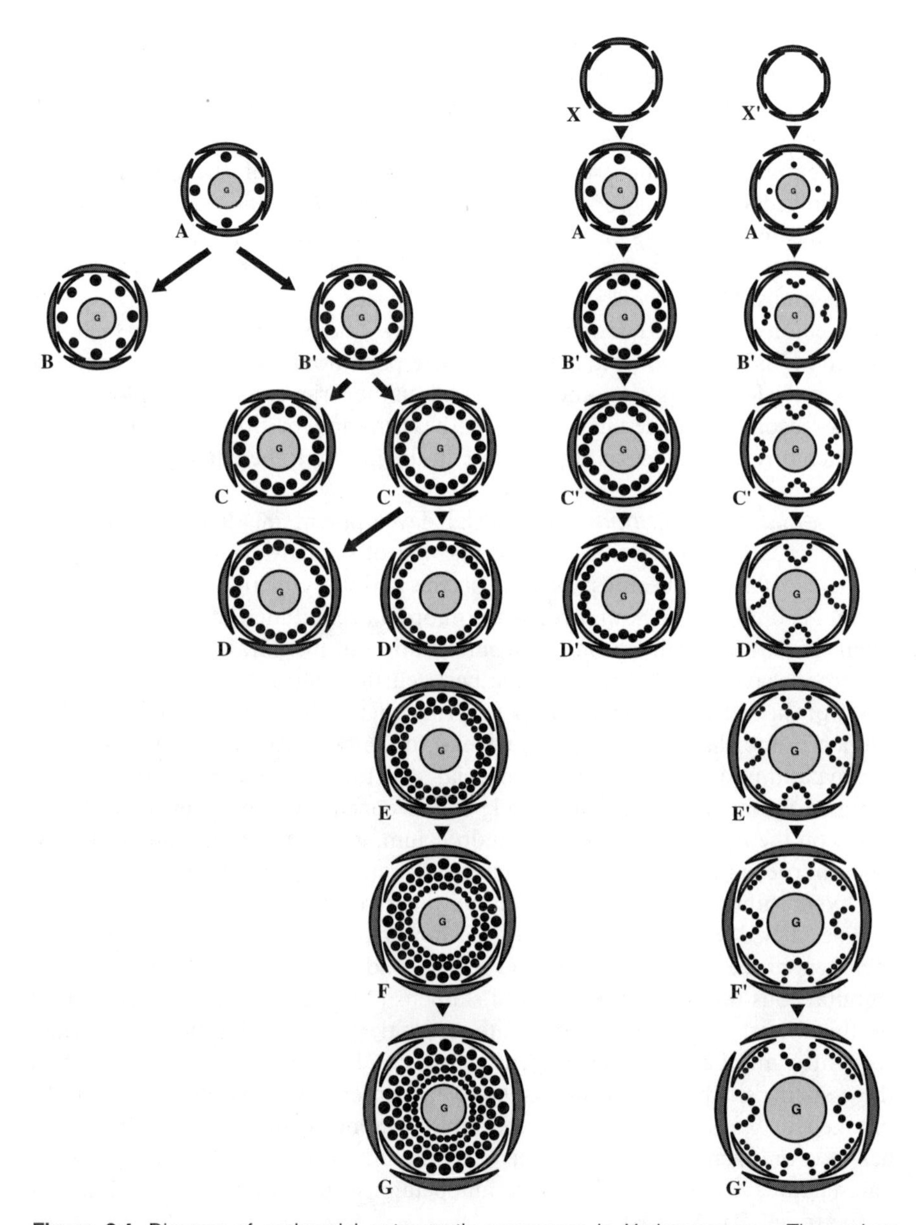

Figure 2.4. Diagram of androecial ontogenetic sequences in Hydrangeaceae. The various androecial ontogenetic sequences reconstructed for the members of the family proceed from the top of the page toward the bottom. Ontogenetic states are identified by letters. The developmental procession of ontogenetic states is indicated by arrows. Stamens of the androecium are indicated by solid circles. The thin lines around antesepalous stamen groups in the ontogenetic sequences $X \rightarrow A \rightarrow B' \rightarrow C' \rightarrow D'$ and $X' \rightarrow A \rightarrow B' \rightarrow C' \rightarrow D' \rightarrow E' \rightarrow F' \rightarrow G'$ are common primordia. The gynoecium (G) is lightly shaded. The sepals (the outer, most darkly shaded whorl of the perianth) and the petals (the inner, more lightly shaded whorl of the perianth) are indicated by crescent shapes.

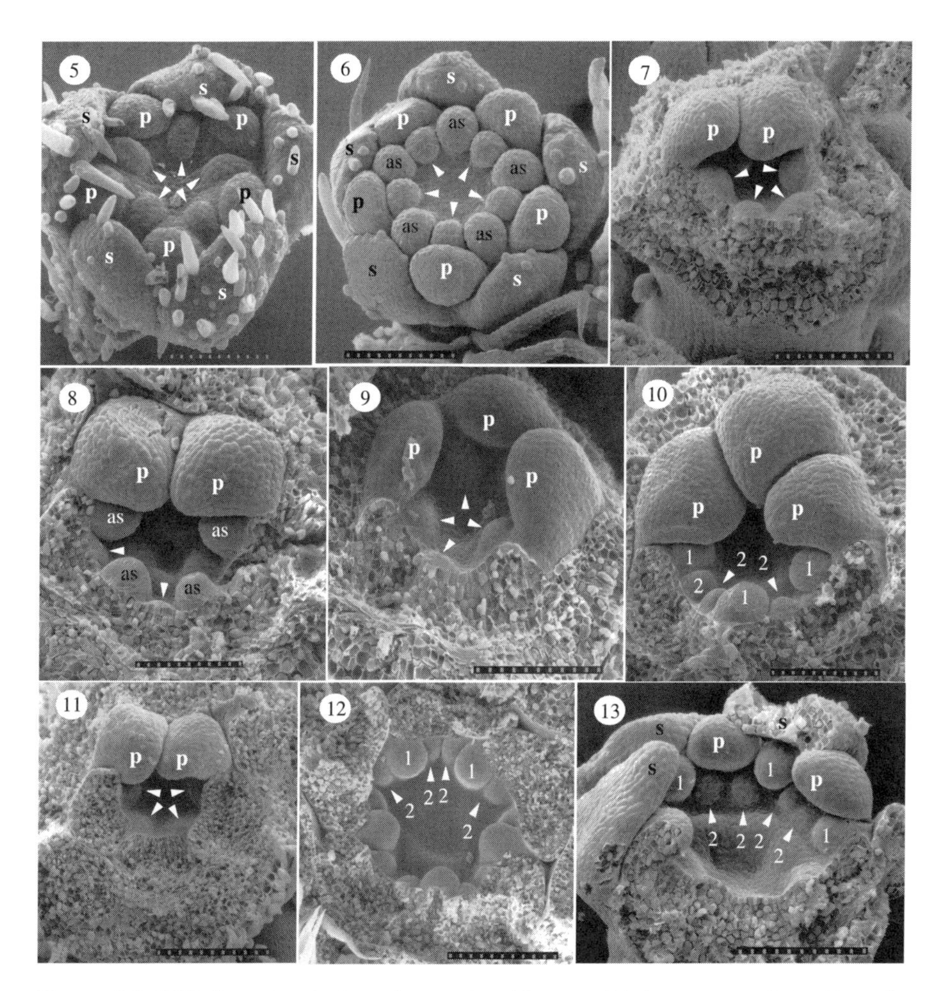

Figures 2.5–2.13. Scanning electron micrographs of flowers of Hydrangeaceae during the early development of the androecium. 2.5 and 2.6: *Jamesia americana* Torr. and A. Gray. 2.5: After the initiation of antesepalous stamens (arrowheads); scale = 100 μm. 2.6: Androecium after the initiation of both antesepalous and antepetalous (arrowheads) stamens; scale = 100 μm. 2.7 and 2.8: *Hydrangea macrophylla* (Thunb.) Ser. 2.7: After the initiation of antesepalous stamens (arrowheads); scale = 176 μm. 2.8: Androecium after the formation of antesepalous stamens and during the initiation of antepetalous (arrowheads) stamens; scale = 136 μm. 2.9 and 2.10: *Hydrangea anomala* subspecies *petiolaris*. 2.9: After the initiation of the whorl of median antesepalous stamens (arrowheads); scale = 120 μm. 2.10: Androecium after the initiation of median antesepalous stamens (1) and a pair of stamens (2) on the lateral flanks of each, forming antesepalous triplets; scale = 136 μm. 2.11 and 2.12: *Kirengeshoma palmata* Yatabe. 2.11: After the initiation of the whorl of median antesepalous stamens (arrowheads); scale = 270 μm. 2.12: Androecium after the initiation of median antesepalous stamens (1) and a pair of stamens (2) on the lateral flanks of each, forming antesepalous triplets; scale = 231 μm. 2.13: *Decumaria barbara* L. after antesepalous triplets have formed (1 = median antesepalous stamen; 2 = lateral flanking stamen); scale = 200 μm. as = antesepalous stamen; p = petal; s = sepal.

All of the polystemonous members of the family also form their first stamens in median positions opposite sepals (Figs. 2.9, 2.11, 2.14, 2.16, 2.21, 2.25). Thus, the development of polystemony begins with the formation of a whorl of antesepalous stamens just as occurs in diplostemonous flowers. Unlike diplostemonous flowers, however, the second step in the development of polystemonous androecia is the formation of a stamen on each lateral flank of each median antesepalous stamen (Figs. 2.10, 2.12, 2.13, 2.15, 2.16, 2.21, 2.25). Thus, all polystemonous taxa have an ontogenetic state in which three stamens are positioned opposite each sepal (Fig. 2.4: state B′). These three stamens are called "antesepalous triplets" (Hufford, 1998). Among the polystemonous Hydrangeaceae, *Decumaria, Hydrangea anomala* D. Don subspecies *petiolaris* (Siebold & Zucc.) McClintock, and *Kirengeshoma* have androecial development that is generally limited only to the formation of antesepalous triplets (Figs. 2.10, 2.12, and 2.13). This ontogenetic sequence is shown in Fig. 2.4 as A → B′. In *Decumaria*, antesepalous triplets are sometimes incomplete because of irregularities in the merosity (appendage numbers per floral whorl) and positioning of primordia in the perianth (Hufford, 1998). *Decumaria*, *Hydrangea anomala* subspecies *petiolaris*, and *Kirengeshoma* have each been observed periodically to form a set of antepetalous stamens after the establishment of the antesepalous triplets (Fig. 2.4: A → B′ → C). *Cardiandra* has an androecium in which the formation of antesepalous triplets and a set of median antepetalous stamens is common (Fig. 2.4: A → B′ → C). Alternatively, flowers of *Cardiandra* have also been observed to form an additional pair of lateral flanking stamens outside of each antesepalous triplet (resulting in five stamens opposite each sepal instead of just an antesepalous triplet) and, ultimately, give rise to stamens in median antepetalous positions (Fig. 2.4: A → B′ → C′ → D). Other polystemonous members of the family have more numerous stamens in their androecia, and this is facilitated by specific developmental modifications.

Deinanthe, the sister of *Cardiandra* (Fig. 2.1), has among the highest stamen numbers (~300 per flower) in the family. To accomodate a high number of stamens, flowers must typically undergo evolutionary modifications that create additional space. In *Deinanthe*, the portion of the hypanthium between the top part of the gynoecium and the insertion of the petals (the epigynoecial zone of the hypanthium) continues to extend upward after the initiation of the whorl of antesepalous stamens. In most other members of the family, the epigynoecial zone of the hypanthium extends little after the initiation of the whorl of antesepalous stamens. The hyperextension of the epigynoecial zone of the hypanthium in *Deinanthe* creates space that is adopted by the androecium for the formation of numerous stamens. Its androecial ontogeny proceeds through the formation of antesepalous triplets to the formation of additional pairs of stamens on their flanks (Fig. 2.16). This results in the formation of a single ring of stamen primordia around the upper part of the inner surface of the hypanthium (Fig. 2.17). Subsequently, additional whorls of stamens form downward over the length of the hypanthium (Figs. 2.18 and 2.19). Thus, the ontogenetic sequence of the androecium in *Deinanthe* begins as in the preceding taxa

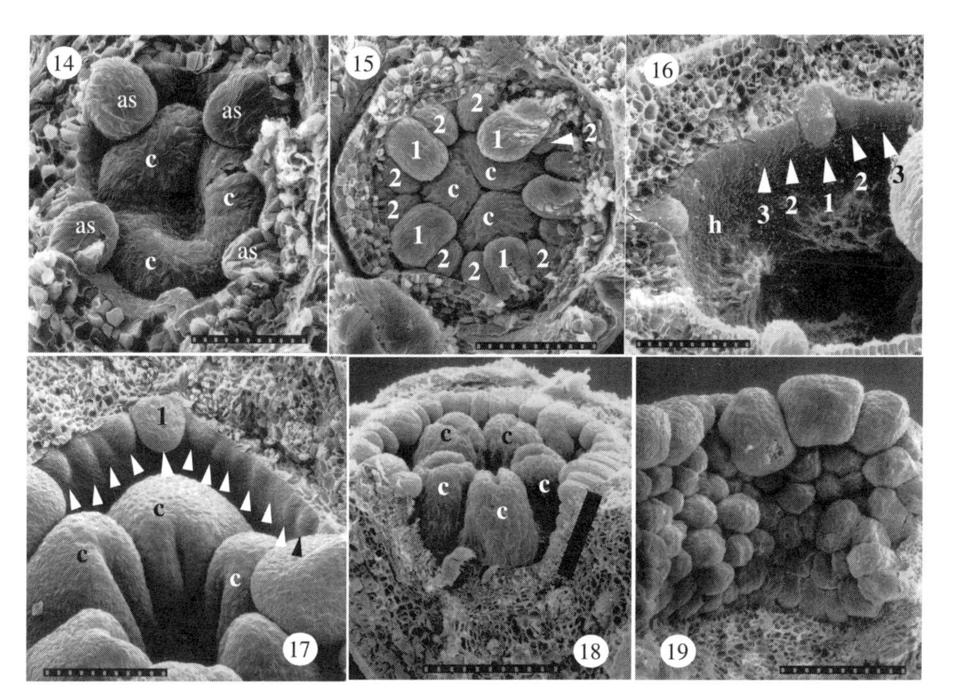

Figures 2.14–2.19. Scanning electron micrographs of flowers of *Cardiandra alternifolia* Siebold & Zucc. and *Deinanthe bifida* Maxim during the early development of the androecium. 2.14 and 2.15: *Cardiandra alternifolia*. 2.14: Androecium after the initiation of median antesepalous stamens; scale = 60 μm. 2.15: Androecium after the antesepalous triplets have formed and at the beginning of the formation of median antepetalous stamens (1 = median antesepalous stamen; 2 = lateral flanking stamen); scale = 120 μm. 2.16–2.19: *Deinanthe bifida*. 2.16: Initial median antesepalous stamens (1) have formed and pairs of lateral flanking stamens (2–3) are being initiated; scale = 86 μm. 2.17: Androecium in which a single ring of stamens has formed around the upper part of the hypanthium (1 = median antesepalous stamen; other stamen primordia are indicated by arrowheads); scale = 150 μm. 2.18: Flower from which a portion of the hypanthium as well as the perianth has been removed to show that several whorls of stamens have begun to form along the length of the hypanthium (solid black bar indicates the extent of epigynoecial hypanthium, and stamen primordia line the face to its left); scale = 200 μm. 2.19: Portion of the hypanthium removed from a flower to show the numerous whorls of stamens that have been formed; scale = 176 μm. as = antesepalous stamen; c = style of carpel; h = epigynoecial hypanthium; p = petal; s= sepal.

but has additional ontogenetic states and the formation of further whorls of stamens on the epigynoecial zone of the hypanthium (Fig. 2.4: $A \rightarrow B' \rightarrow C' \rightarrow D' \rightarrow E \rightarrow F \rightarrow G$). Hypanthial extension, including the hyperextension of the epigynoecial zone, in *Deinanthe* is strictly not part of androecial development and, thus, is not included in the ontogenetic sequence. However, it is an example of a novelty that involves heterochrony, changing growth proportions in the hypanthium, in a synorganized zone of floral appendages that is not strictly part of the androecium, which has a significant impact on ontogenetic evolution and innovation in the androecium.

Philadelphus and *Carpenteria* share the novelty of having common primordia in their androecia. Common primordia are initiated and have the form of typical appendage primordia, although they are sometimes much larger than other appendage primordia. Common primordia do not develop into a single appendage; instead, more than one appendage primordium will form on the surface of a common primordium. *Philadelphus* and *Carpenteria* form common primordia in antesepalous positions at the beginning of androecial development (Figs. 2.20 and 2.23). The antesepalous triplets form on the surface of the common primordia (Fig. 2.4: $X \rightarrow A \rightarrow B'$ and $X' \rightarrow A \rightarrow B'$; Figs. 2.21 and 2.25). Both genera, however, form stamens in addition to those in antesepalous triplets.

In *Philadelphus*, androecial development continues with the formation of additional stamens on the lateral flanks of the antesepalous triplets (Fig. 2.22). The common primordia expand laterally, and these additional stamens form on them. Ultimately, the androecium consists of a single ring of stamens (Fig. 2.4: $X \rightarrow A \rightarrow B' \rightarrow C' \rightarrow D'$). No investigated species of *Philadelphus* have been found to have more than one whorl of stamens, although species such as *P. hirsutus* that have relatively high numbers of stamens may likely have additional whorls.

The common primordia of *Carpenteria* are U-shaped and continuous with the petals (Figs. 2.23 and 2.24). The individual stamen primordia form first on the rim of the common primordia (Fig. 2.25). The initial median antesepalous stamens form on the common primordia at the base of the U (Fig. 2.25). Lateral flanking stamens form adjacent to the median antesepalous stamens, and, subsequently, additional lateral flanking stamens are initiated centrifugally along the outward projecting arms of the U-shaped common primordia (Fig. 2.4: $X' \rightarrow A \rightarrow B' \rightarrow C' \rightarrow D'$; Fig. 2.26). In addition to the stamens that form on the rim of the common primordia in antesepalous positions, other stamens also form in antepetalous positions (beginning at state E' in Fig. 2.4; Figs. 2.27 and 2.28). These antepetalous stamens arise along the margins of the common primordia where they meet the petals. The stamens opposite any particular petal form on the margins of the two adjacent common primordia. The antepetalous stamen primordia are initiated first at the base of the common primordia (near the base of the petals), and additional primordia arise successively in lines toward the rims of the common primordia. The ontogenetic sequence of the *Carpenteria* androecium can be modeled as shown in Fig. 2.4 as $X' \rightarrow A \rightarrow B' \rightarrow C' \rightarrow D' \rightarrow E' \rightarrow F' \rightarrow G'$.

Figures 2.20–2.28. Scanning electron micrographs of flowers of *Philadelphus satsumanus* Siebold ex Miq. and *Carpenteria californica* during the early development of the androecium. 2.20–2.22: *Philadelphus satsumanus*. 2.20: Androecium after the initiation of common primordia (arrowheads) in antesepalous positions; scale = 136 μm. 2.21: Androecium in which the common primordia are at different stages of development. Only a median antesepalous stamen has formed on the uppermost common primordium. Lateral flanking stamens have begun to form at the sides of the median antesepalous stamens of the two common primordia on the sides of the

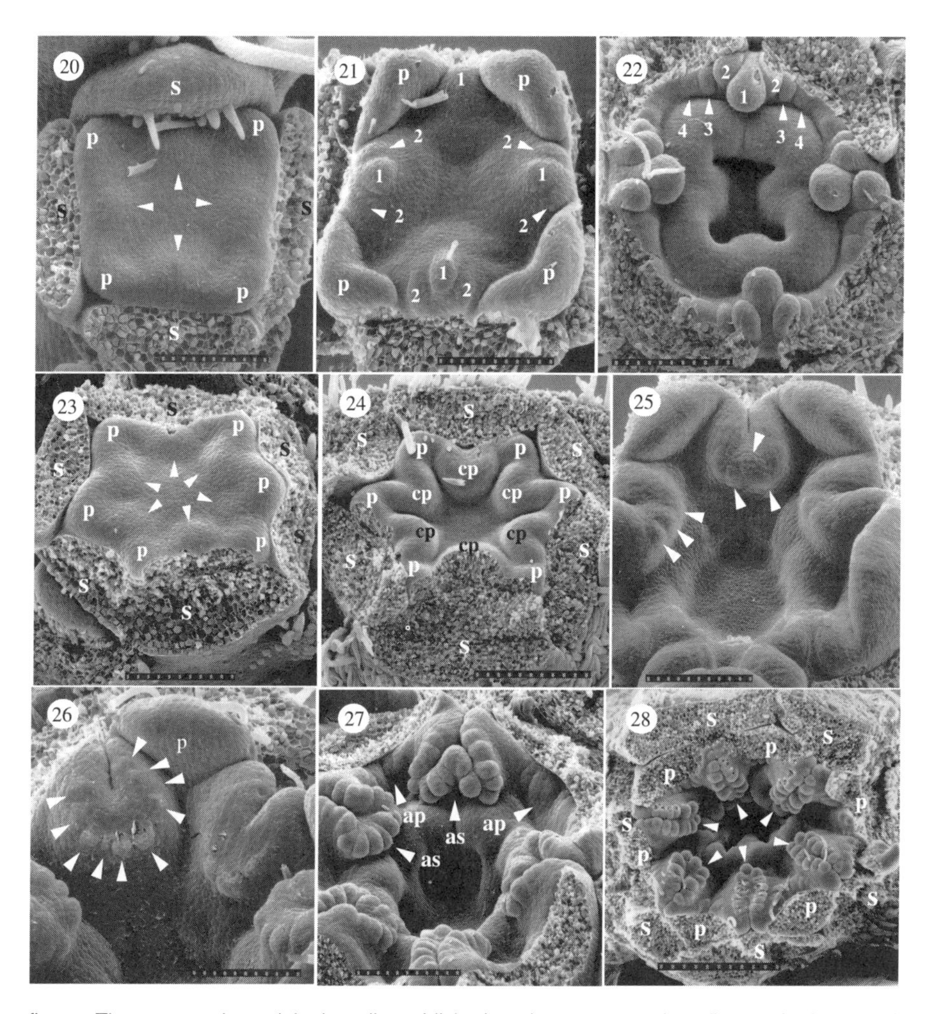

flower. The antesepalous triplet is well established on the common primordium at the bottom of the flower; 1 = median antesepalous stamen; 2 = lateral flanking stamen; scale =151 μm. 2.22: Androecium in which successive pairs of stamens have formed on the flanks of the antesepalous triplets on each common primordium; 1 = median antesepalous stamen; 2 = lateral flanking stamen; 3 and 4 = successive pairs of lateral flanking stamens in antesepalous groups; scale = 200 μm. 2.23–2.28. *Carpenteria californica*. 2.23: Androecium after the initiation of common primordia (arrows) in antesepalous positions; scale = 176 μm. 2.24: Enlargement of the common primordia of the androecium is associated with their heightened plication; scale = 270 μm. 2.25: Median antesepalous stamens and lateral flanking stamens form on the rim at the base of each U-shaped common primordium. Arrows indicate the antesepalous triplets on two common primordia; scale = 176 μm. 2.26: Successive stamens (arrows) form on the rim of the common primordia; scale = 136 μm. 2.27: After stamens have formed on the rim of the common primordia, additional stamen primordia form opposite petals along the margins where each common primordium meets a petal; scale = 231 μm. 2.28: Distal view of a flower from which the perianth has been removed to show an androecium after both antesepalous (arrows) and antepetalous stamens have formed; scale = 430 μm. ap = antepetalous groups of stamens; as = antesepalous groups of stamens; cp = common primordium; p = petal; s = sepal.

A polystemonous androecium is also found in *Platycrater*, which is characterized by a high number of stamens similar to that found in *Carpenteria* and *Deinanthe*. Unfortunately, specimens of *Platycrater* have not been available for developmental study, and, thus, it is unclear how it compares with other members of Hydrangeaceae. *Deutzia*, a predominantly diplostemonous taxon, also includes a few polystemonous species. The polystemonous species of *Deutzia* also have not been available for investigation.

CONTINGENCY OF STATES IN ONTOGENETIC SEQUENCES

Alberch (1985) cautioned about the use of developmental sequences in systematics (cf Langille and Hall, 1989). He suggested that it was important to distinguish between causal and temporal sequences. Alberch (1985, p. 49) suggested that temporal sequences "lack causality between the antecedent and the descendant stages" and that this "renders the sequence useless for systematic and heterochronic purposes." In causal sequences, one ontogenetic state is a prerequisite for the next. Alberch (1985) valued causal sequences because they were constrained. For example, he suggested that a "temporal sequence is not constrained and it can be altered. In general, it will not be conserved through phylogeny (a fundamental requirement in systematic and heterochronic analysis)" (Alberch, 1985, p. 49).

Although caution is warranted in the application of ontogenetic sequences in systematics and diversification analyses, Alberch's assertions are questionable (Kluge, 1988). Clearly, ontogenetic sequences must be variable and cannot be conserved if they are to have systematic value (without variability there will be no synapomorphies). Indeed, Weston (1988) and Velhagen (1997) have proposed strategies for coding ontogenetic sequences for phylogenetic analyses. Alberch's assertion that only causal sequences have value is also questionable. In the Hydrangeaceae androecium data (Fig. 2.4), I would suggest that the ontogenetic sequence $X \rightarrow A \rightarrow B' \rightarrow C' \rightarrow D'$ for *Philadelphus* is not causal in the sense of Alberch (1985). Alberch's sense of causality implies that state X be required for the expression of state A, which is not true in Hydrangeaceae. State X of *Philadelphus* is a novelty that appears to impact the extent to which polystemony can develop. The extent and course of development taken in polystemonous androecia appear to be contingent on earlier events, such as the formation of common primordia, which are characteristic of state X, or epigynoecial hypanthial elongation as occurs in *Deinanthe*. Contingency rather than causality may be the key element of useful ontogenetic sequences. Thus, states B and B′ (Fig. 2.4) are not caused by state A (or the developmental event that led to state A) in the sense of Alberch (1985), but they are contingent on state A. Similarly, in the ontogenetic sequence $X \rightarrow A \rightarrow B' \rightarrow C' \rightarrow D'$ for *Philadelphus* (Fig. 2.4), states A and B′ may not be caused by or even contingent on state X, but the later states C′ and D′ of polystemony are likely to be contingent on X.

Sequences that have the property of contingeny of later states on some earlier state or states can be valuably used in comparative biology. We can readily see the correspondence of the A and B′ states of the X → A → B′ → C′ → D′ for *Philadelphus* with the same states that occur in other polystemonous taxa (Fig. 2.4). Such a comparison calls attention to the novel state X, demonstrating that ontogenies can be modified at their initiation as well as terminally and that initial additions need not disrupt the entire course of ontogeny.

ONTOGENETIC SEQUENCES AND HOMOLOGY

The concept of homology exists because comparative biologists see hypotheses of correspondences among different organisms as being central to their goals. It is thus intriguing that Alberch (1985) suggested that acceptance of von Baer's principle that ontogenies of different species diverge continuously over the course of ontogeny implies that we cannot establish homologies between ontogenetic sequences. In contrast, we might better see the von Baerian concept of organismal ontogenies as a driving force behind attempts to homologize the ontogenetic states of different organisms. We need to hypothesize homologies because comparable structures in different organisms are not identical (if there was no divergence, then a homology concept would be unnecessary). As framed by Collazo (2000, p. 5), when organismal ontogenetic sequences are compared, the "more [developmental] character states that are shared, the more likely that the structures formed by these processes are homologous." Investigations of closely related taxa, such as the genera of Hydrangeaceae, that focus on the patterning of evolutionary diversity may be the best examples of cases in which ontogenetic sequence models can be applied and states homologized among the ontogenetic sequences of different organisms.

It is important to distinguish between structural and phylogenetic homology in discussions of morphology (Hufford, 1996b). Structural homologies are nonphylogenetic hypotheses of morphological correspondence (also the primary homology of de Pinna, 1991). They are best inferred on the basis of explicit criteria, and those of Remane (1952)—similarity of position, presence of intermediates, and special attributes—are applied here. Phylogenetic homologies are correspondences between organisms deduced to be derived from the same character state in a common ancestor on the basis of a phylogenetic hypothesis. Structural homologues are generally the states tested for phylogenetic homology using cladograms. Such tests may be consistent with the hypothesis that structural homologues are derived from the same ancestral state or they may be inconsistent, in which case we would recognize that strutural homologues are homoplasious (the congruence test of Patterson, 1982). Thus, structural homologues can be either phylogenetically homologous or homoplasious (Donoghue, 1992; Hufford, 1996b).

As a first step in formulating hypotheses of homology in the Hydrangeaceae example, we need to consider whether the androecia are homologous. I suggest they are both structurally and phylogenetically homologous. The androecia of the investigated genera have the same position between the corolla and gynoecium, and they all share the special property of consisting of stamens. The groups most closely related to Hydrangeaceae—Loasaceae and other Cornales—also have androecia in similar positions and have stamens. Thus, it seems reasonable to suggest that androecia among Hydrangeaceae are structurally and phylogenetically homologous.

A critical concern for the problem of establishing homologies among ontogenies and, especially, among ontogenetic states is whether it is possible to homologize stamen positions among the different genera over the course of ontogeny. I suggest that the regularity of stamen positions, especially during early androecial development, is consistent with the goal of hypothesizing homologies. For example, all examined androecia, whether diplostemonous or polystemonous, begin with the formation of a stamen in a median position opposite each sepal. The regular positioning of a second whorl of stamens in median antepetalous positions in diplostemonous taxa and a second set of stamens on the lateral flanks of initial antesepalous stamens in polystemonous taxa also provides the kind of consistency that can be applied to using position as a criterion for inferring structural homology. That is, median antesepalous stamens are structurally homologous among all of the diplostemonous and polystemonous taxa, median antepetalous stamens are structurally homologous among diplostemonous taxa, and the initial pair of lateral flanking stamens is structurally homologous among polystemonous taxa. The hypothesis of structural homology for median antepetalous stamens among diplostemonous taxa and for the initial pair of lateral flanking stamens among polystemonous taxa does not imply that they are phylogenetically homologous. Given the alternative scenarios for the optimization of diplostemony and polystemony on internal nodes of the most parsimonious cladogram that is being used in the example (Fig. 2.3), it is most reasonable to suggest that at least some of the structurally homologous states of polystemonous ontogenies are evolutionary homoplasies and thus not phylogenetically homologous. Indeed, the identical ontogenetic trajectories (Fig. 2.4: A → B′) of *Kirengeshoma* (Figs. 2.11 and 2.12) and *Hydrangea anomala* subspecies *petiolaris* (Figs. 2.9 and 2.10) are unequivocally homoplasious (Fig. 2.3) but just as clearly structurally homologous. Similarly, structurally homologous states among the diplostemonous ontogenies of some taxa may not be phylogenetically homologous for the same reason (Fig. 2.3b).

Hydrangeaceae present interesting dilemmas for formulating hypotheses of homology. For example, are the B′ states (Fig. 2.4) homologous or does the presence of common primordia in *Philadelphus* and *Carpenteria* at B′ render them nonhomologous to the other polystemonous taxa at B′? They have been regarded as homologous here because they share the special characteristic of having stamens in antesepalous triplets and the stamens in those triplets share positional correspondences. Thus, I suggest that the androecial ontogenies of *Philadelphus* and *Carpenteria* that begin uniquely with common primordia do not present features inconsistent with a hypothesis that the B′ states are structurally homologous. The same question asked about the D′ states of the polystemonous taxa seems more complicated, partly because the shapes of the common primordia of *Philadelphus* and *Carpenteria* have diverged during development and this has impacted the arrangement of the stamens. Despite this difference in stamen arrangement caused by special attributes of development in the common primordia, it seems reasonable to suggest that antesepalous groups of stamens are structurally and phylogenetically homologous between *Philadelphus* and *Carpenteria* at this ontogenetic state. If we go further along the ontogenies, it seems unreasonable to suggest that state F of *Deinanthe* is structually homologous to state F′ of *Carpenteria*. These two genera have diverged into distinctive ontogenetic trajectories by this state. We understand this because of the process of homologizing earlier states between the ontogenetic sequences of these two genera.

It has been suggested that attempts to homologize states between two ontogenies assume that their evolution was conservative (i.e., reversal, deletions, and insertions would be rare, and ontogeny would be modified mainly by terminal addition) (Alberch, 1985; Weston, 1988). The assumption that ontogenies are modified mainly by terminal addition has been implicit in suggestions that ontogenies can be used to root cladograms (e.g., Nelson, 1978; Weston, 1988; Meier, 1997). Those assumptions are probably not reasonable (Mabee, 1993; Hufford, 1996a; Meier, 1997), and concerns about the specific application of ontogenies to root cladograms have been elegantly discussed by de Queiroz (1985). To formulate hypotheses of homology does not necessarily imply that ontogenies are conservative or that specific modes of change, such as terminal addition, are particularly common. For example, Hufford (1996a, 1997b) compared and homologized among sequences in different clades to show the prevalence of homoplasies and the variety of evolutionary transformations in ontogenies that created them. One of the values of attempting to homologize states among a set of ontogenies is that it can provide insights about the alterations in morphological development that underlie phenotypic diversity. Indeed, if one attempted to homologize a series of states between two ontogenies and found no correspondences, it would certainly suggest that a major phenotypic discontinuity existed between the two taxa, and one might infer that since the two taxa diverged from their most recent common ancestor that one or both of their ontogenies had been substantially repatterned (sensu Roth and Wake, 1985).

Alberch (1985) wants us to recognize that ontogenies can be homologous if they share initial and terminal states but not intermediate, and this seems acceptable; however, we must ask about intermediate states. I have previously discussed the evolution of androecia of Piperales in which some taxa have similar initial and terminal states but have different intermediate states, and the implications for structural and phylogenetic homology have been enlightening (Hufford, 1997b). For example, if different clades of species differ in intermediate states but share a similar terminal state, then it may be reasonable to infer that the terminal states are homoplasies and not phylogenetic homologies. If intermediate states differ within a species, then an inference of homoplasy may still be reasonable (e.g., if separate populations have become canalized for different developmental trajectories despite strong selection on the functional terminal state). Alternatively, if intermediate states differ within a species, then we may be seeing alternative developmental strategies permitted by the genetic and epigenetic milieu (Schlichting and Pigliucci, 1998).

ONTOGENETIC EVOLUTION AND THE PATTERNING OF DIVERSITY

Evolutionary Transformations of Ontogenetic Sequences

To explore the roles of ontogenetic sequence evolution in the patterning of phenotypic diversity, including the relative importance of unique vs. iterated novelties, it is necessary to optimize ontogenetic transformations on phylogenetic hypotheses. As discussed above, the Hydrangeaceae example presents the dilemma of having equivocal internal nodes (Figs. 2.2 and 2.3) that can be considered to have either diplostemony or polystemony present. This equivocation is, however, a problem faced in studies of diversification based on existing cladograms for most monophyletic groups. Although the problem is not resolved here, I offer alternative scenarios for androecial evolution (Fig. 2.3) and for ontogenetic sequence evolution (Figs. 2.29 and 2.30).

Figure 2.29. Evolutionary transformations of ontogenetic sequences optimized on the strict consensus cladogram (same as in Fig. 2.3) for Hydrangeaceae based on the maximum parsimony analysis of DNA sequences from the plastid genes *matK* and *rbcL*. This scenario is based on the hypothesis that a diplostemonous androecium is plesiomorphic for the family. The plesiomorphic diplostemonous ontogenetic sequence is shown at the base of the cladogram. For each evolutionary transformation, the hypothesized plesiomorphic ontogenetic sequence is shown to the left of the arrow and the derived to the right. For each ontogenetic sequence, the states correspond to those shown in Fig. 2.4. a: Unequivocal transformations. The evolutionary transformation for *Platycrater* is shown as a question mark because data on its androecial development are not available. **Deutzia* is hypothesized to be plesiomorphically diplostemonous; however, polystemony is present in the small number of North American species. Thus, at least one additional androecial change would have occurred in *Deutzia*. ***Dichroa* includes two haplostemonous species, and these are considered to be derived within the genus. b and c: Alternative evolutionary transformations for *Philadelphus* and *Carpenteria*. b: Shift to an androecial ontogenetic sequence like that of *Philadelphus* is hypothesized to be plesiomorphic for the *Philadelphus* + *Carpenteria* clade. c: Shift to an androecial ontogenetic sequence like that of *Carpenteria* is hypothesized to be plesiomorphic for the *Philadelphus* + *Carpenteria* clade. d–f: Alternative evolutionary transformations for *Cardiandra* and *Deinanthe*. The alternative ontogenetic sequences of *Cardiandra* are shown. d: Shift to an androecial ontogenetic sequence like that of *Cardiandra* is hypothesized to be plesiomorphic for the *Cardiandra* + *Deinanthe* clade. e: Shift to an androecial ontogenetic sequence like that of *Deinanthe* is hypothesized to be plesiomorphic for the *Cardiandra* + *Deinanthe* clade. f: Slightly less parsimonious alternative that emphasizes that *Cardiandra* and *Deinanthe* share regularly only the A and B′ states of the ontogenetic sequences shown in the Fig. 2.4. In this evolutionary scenario, the novel androecial ontogenetic sequences of both *Cardiandra* and *Deinanthe* would have been derived from a common ancestor in which the evolutionary transformation from diplostemony to polystemony had already occurred.

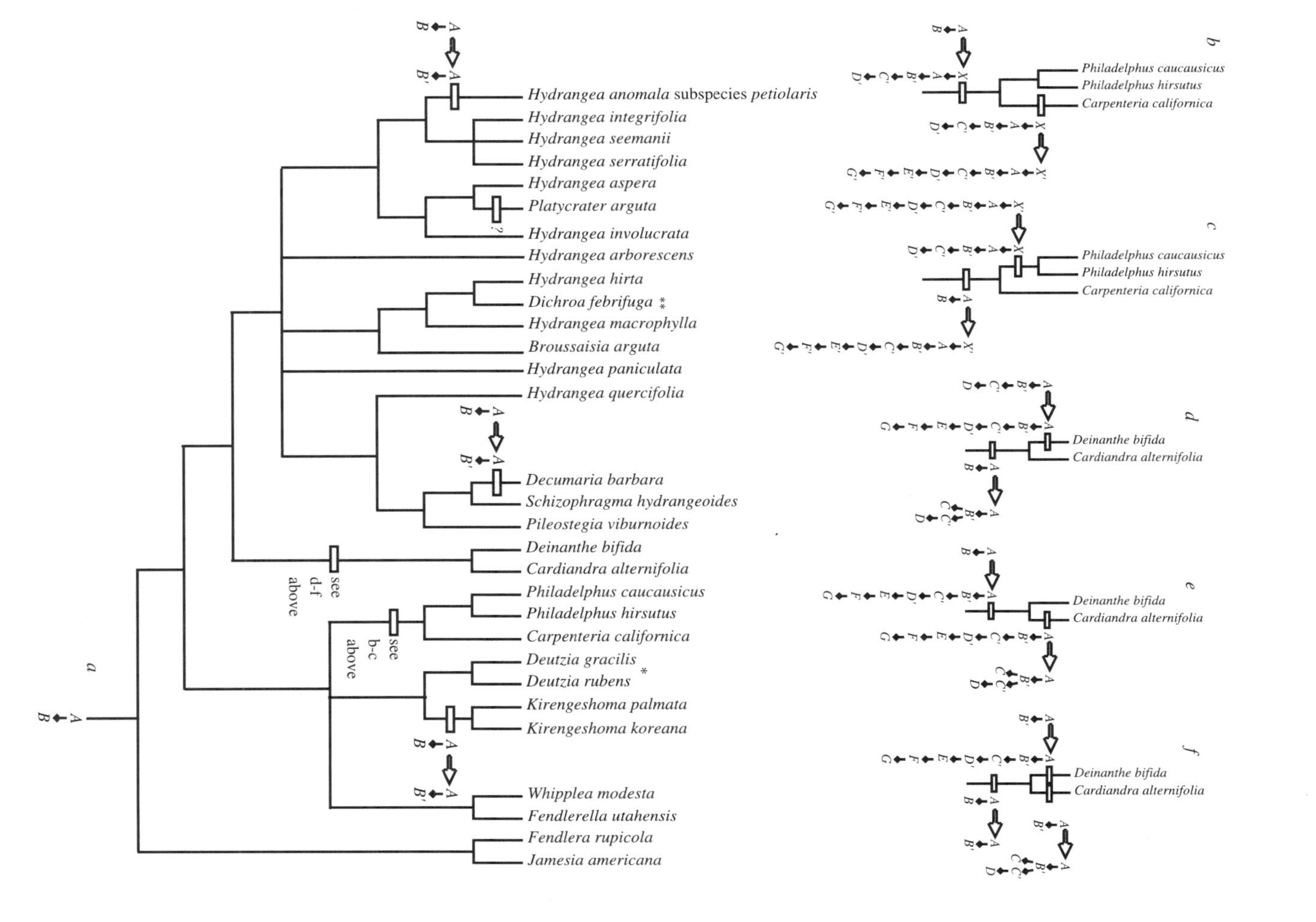
a
Hydrangea anomala subspecies petiolaris
Hydrangea integrifolia
Hydrangea seemanii
Hydrangea serratifolia
Hydrangea aspera
Platycrater arguta
Hydrangea involucrata
Hydrangea arborescens
Hydrangea hirta
Dichroa febrifuga **
Hydrangea macrophylla
Broussaisia arguta
Hydrangea paniculata
Hydrangea quercifolia
Decumaria barbara
Schizophragma hydrangeoides
Pileostegia viburnoides
Deinanthe bifida
Cardiandra alternifolia
Philadelphus caucausicus
Philadelphus hirsutus
Carpenteria californica
Deutzia gracilis
Deutzia rubens *
Kirengeshoma palmata
Kirengeshoma koreana
Whipplea modesta
Fendlerella utahensis
Fendlera rupicola
Jamesia americana
see d-f above
see b-c above
b
Philadelphus caucausicus
Philadelphus hirsutus
Carpenteria californica
c
Philadelphus caucausicus
Philadelphus hirsutus
Carpenteria californica
d
Deinanthe bifida
Cardiandra alternifolia
e
Deinanthe bifida
Cardiandra alternifolia
f
Deinanthe bifida
Cardiandra alternifolia

The optimization of ontogenetic transformations onto cladograms involves determining which ontogenetic sequence is derived from the more plesiomorphic sequence at each character state transition noted in Fig. 2.3. This is relatively unequivocal when the transformations involve changes such as those from diplostemony to the form of polystemony in which only antesepalous triplets are present (Fig. 2.29a). This transformation involves only a substitution of terminal ontogenetic states (Fig. 2.29a), and the clades of Hydrangeaceae in which these transformations occurred are unambiguous given the assumption that diplostemony is the plesiomorphic androecial state for the family. The transformations of ontogenetic sequences in the *Cardiandra* + *Deinanthe* and *Carpenteria* + *Philadelphus* clades are less certain under the assumption that diplostemony is plesiomorphic. In both of these clades, the evolution of polystemony has involved at least two ontogenetic changes (the character resolution provided in the analysis of ontogenies is greater than that in Fig. 2.3, in which all of the polystemonous taxa are coded as having the same state). For both clades, it is uncertain which of these changes occurred first, and, thus, multiple scenarios for ontogenetic evolution in the *Carpenteria* + *Philadelphus* (Fig. 2.29, b and c) and *Cardiandra* + *Deinanthe* (Fig. 2.29, d–f) clades are provided.

To assume that polystemony is plesiomorphic for Hydrangeaceae presents numerous difficulties for attempts to reconstruct ontogenetic evolution. Given the various expressions of polystemony in the family as well as the fact that the sister clade Loasaceae also has various expressions of polystemony (Hufford, 1988, 1990), the plesiomorphic ontogenetic sequence of polystemony for the family is equivocal. The option selected for discussion here (Fig. 2.30) is conservative because it takes only the common ontogenetic states found among all Hydrangeaceae and those shared with various Loasaceae [particularly *Eucnide* (Hufford, 1988) and *Mentzelia* (Leins and Winhard, 1973; Hufford, 1990)] as

Figure 2.30. Evolutionary transformations of ontogenetic sequences optimized on the strict consensus cladogram (same as in Fig. 2.3) for Hydrangeaceae based on the maximum parsimony analysis of DNA sequences from the plastid genes *matK* and *rbcL*. This scenario is based on the hypothesis that a polystemonous androecium characterized by the formation of antesepalous triplets is plesiomorphic for the family. The plesiomorphic polystemonous ontogenetic sequence is shown at the base of the cladogram. For each evolutionary transformation, the hypothesized plesiomorphic ontogenetic sequence is shown to the left of the arrow and the derived to the right. For each ontogenetic sequence, the states correspond to those shown in Fig. 2.4. a: Unequivocal transformations. The evolutionary transformation for *Platycrater* is shown as a question mark because data on its androecial development are not available. **Deutzia* is hypothesized to be plesiomorphically diplostemonous; however, polystemony is present in the small number of North American species. Thus, at least one additional androecial change would have occurred in *Deutzia*. ***Dichroa* includes two haplostemonous species, and these are considered to be derived within the genus. b and c: Alternative evolutionary transformations for *Philadelphus* and *Carpenteria*. b: Shift to an androecial ontogenetic sequence like that of *Philadelphus* is hypothesized to be plesiomorphic for the *Philadelphus* + *Carpenteria* clade. c: Shift to an androecial ontogenetic sequence like that of *Carpenteria* is hypothesized to be plesiomorphic for the *Philadelphus* + *Carpenteria* clade. d: Independent elaboration of ontogenetic sequences in both *Cardiandra* and *Deinanthe* is required.

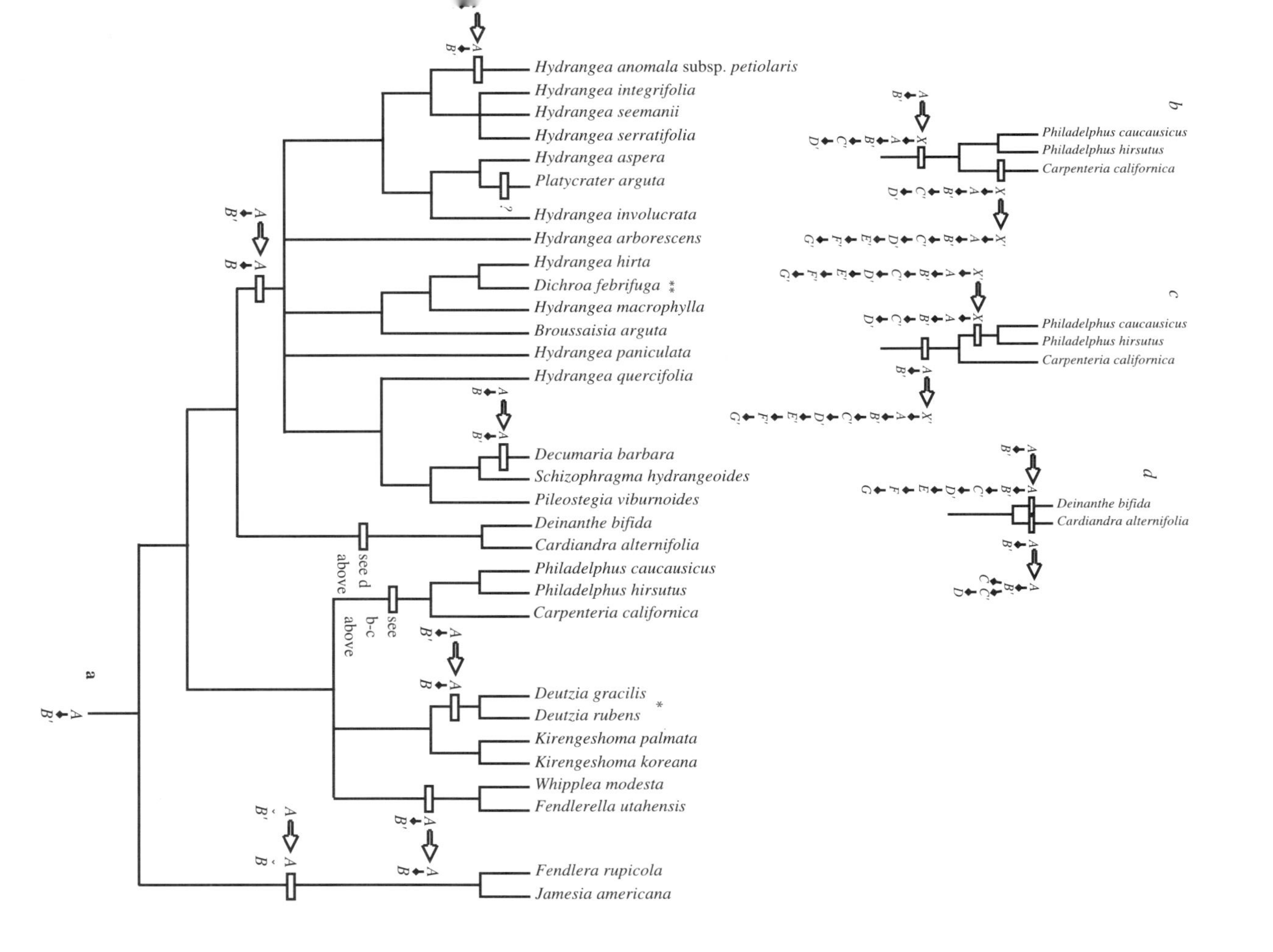

a
Hydrangea anomala subsp. petiolaris
Hydrangea integrifolia
Hydrangea seemanii
Hydrangea serratifolia
Hydrangea aspera
Platycrater arguta
Hydrangea involucrata
Hydrangea arborescens
Hydrangea hirta
Dichroa febrifuga **
Hydrangea macrophylla
Broussaisia arguta
Hydrangea paniculata
Hydrangea quercifolia
Decumaria barbara
Schizophragma hydrangeoides
Pileostegia viburnoides
Deinanthe bifida
Cardiandra alternifolia
Philadelphus caucasicus
Philadelphus hirsutus
Carpenteria californica
Deutzia gracilis *
Deutzia rubens
Kirengeshoma palmata
Kirengeshoma koreana
Whipplea modesta
Fendlerella utahensis
Fendlera rupicola
Jamesia americana
see d above
see b-c above
b
Philadelphus caucasicus
Philadelphus hirsutus
Carpenteria californica
c
Philadelphus caucasicus
Philadelphus hirsutus
Carpenteria californica
d
Deinanthe bifida
Cardiandra alternifolia

composing the plesiomorphic ontogeny. That is, under an assumption that polystemony is plesiomorphic, Figure 2.30 presents the plesiomorphic ontogeny for Hydrangeaceae as the sequence A → B′ (using states shown in Fig. 2.4). If a more complex polystemonous ontogenetic sequence is truly plesiomorphic for Hydrangeaceae, then the avenues of ontogenetic change presented in Figure 2.30 will be incorrect. For example, if the plesiomorphic ontogeny for Hydrangeaceae involves not only the formation of antesepalous triplets but also the inception of successive pairs of lateral flanking stamens to each antesepalous group of stamens and also involves the formation of additional whorls of stamens (a pattern similar to that in *Deinanthe*), then the transformations to the polystemony of *Decumaria*, *Hydrangea anomala* subspecies *petiolaris*, and *Kirengeshoma* would have involved juvenilizations, resulting from the loss of the more terminal ontogenetic states. It is critical to recognize that the interpretations of diversification drawn from reconstructions of ontogenetic evolution are dependent on the ability to reconstruct accurately both plesiomorphic and derived states as well as to identify the clade where evolutionary transformations occur. The scenario presented here in which polystemony is plesiomorphic (Fig. 2.30) has an entirely hypothetical ontogenetic sequence presented as the plesiomorphic condition for the family.

Diplostemony/Polystemony Transformations

The most prominent aspects of the patterning of androecial diversity in Hydrangeaceae are the homoplastic transitions between diplostemony and polystemony. Regardless of the true optimization of androecial states for internal nodes (e.g., whether diplostemony or polystemony is plesiomorphic for the family), numerous transitions between diplostemony and polystemony have occurred in the evolution of the family.

The diplostemony/polystemony transformation is interesting because both forms of androecia begin with the initiation of antesepalous stamens and the evolutionary transformation (e.g., A → B to/from A → B′ in Fig. 2.4) is centered in alternatives at the second developmental event. The transformation between diplostemony and polystemony does not involve additional androecial or floral innovations, which may be one reason for its evolutionary lability. Comparisons of scanning electron micrographs of diplostemonous and polystemonous androecia at ontogenetic states that correspond to A, B, and B′ in Fig. 2.4 show that they differ in the relative sizes of stamens to petals and sepals. This difference in size proportions between diplostemonous and polystemonous androecia is indicative of heterochronic effects that are also associated with the evolutionary transformation.

Antesepalous Triplets of Polystemonous Androecia

All polystemonous Hydrangeaceae share the first two steps of stamen initiation, leading to an ontogenetic state characterized by the presence of antesepalous triplets. We can infer from the recurrence of the antesepalous triplet state in clades in which polystemony has evolved independently (e.g., *Decumaria* and *Hydrangea anomala* subsp. *petiolaris*) that it must be a key feature of polystemonous androecia in the family. The development of an ontogenetic state characterized by antesepalous triplets also occurs in other families of polystemonous angiosperms, especially among the eudicots (sensu Angiosperm Phylogeny Group, 1998) and may be limited to them. For example, antesepalous triplets are not characteristic of polystemonous androecia of the Magnoliales, Laurales, Piperales, or monocot clades (Ronse Decraene and Smets, 1993). These noneudicot groups tend to be clades in which flowers are trimerous and/or have more or less helical (rather than whorled) arrangements of perianth appendages and often have perianth appendage numbers that have higher coefficients of variation than found among the eudicots (Endress, 1994). Thus, phyllotactic and meristic factors as well as those arising from floral meristem size and shape must prevent groups outside of the eudicots from forming polystemonous androecia that have antesepalous triplets. Polystemonous androecia that form antesepalous triplets have clearly evolved independently among the eudicots. They are found in Loasaceae (Hufford, 1988, 1990), the sister family of Hydrangeaceae (and this is a possible symplesiomorphy if polystemony is plesiomorphic for Hydrangeaceae). Outside of these two families of the Cornales, antesepalous triplets are not commonly found in other asterid taxa, although this may be largely a consequence of the relative rarity of polystemony among asterids. Polystemonous androecia that have an ontogenetic state characterized by antesepalous triplets are found in the rosid clade in families such as Rosaceae (Dickson, 1866) and Zygophyllaceae (Ronse Decraene and Smets, 1991).

Novelties of Polystemonous Taxa

Although various polystemonous clades of Hydrangeaceae are characterized by the formation of androecia that generally have only antesepalous triplets (including *Decumaria*, *Hydrangea anomala* subspecies *petiolaris*, and *Kirengeshoma*), others have greater meristic complexity. In terms of the patterning of polystemonous diversity, increased meristic complexity is associated with novelties that diversify ontogenetic sequences.

For understanding the elaboration of polystemonous ontogenies, *Cardiandra* is very interesting. As noted above, the two possible ontogenetic trajectories observed in *Cardiandra* androecia differ only in the number of pairs of lateral flanking stamens that form before the initiation of median antepetalous stamens. What is most interesting about *Cardiandra* compared with the taxa with higher stamen numbers is that it does not display any innovations in an ontogenetic sequence that extends beyond the formation of antesepalous triplets. In contrast, *Deinanthe*, *Carpenteria*, and *Philadelphus* have ontogenetic novelties and much higher stamen numbers (Figs. 2.29, b–f, and 2.30, b–d). Thus, ontogenetic novelties appear not to be associated simply with transcending the antesepalous triplet threshold but with facilitating high numbers of stamens.

High stamen numbers in *Deinanthe* are facilitated by the novel extension of the epigynoecial zone of the hypanthium described above. This extension involves zonal growth subjacent to the perianth appendages, and, thus, its association with heightened androecial complexity is interesting because it demonstrates how changes in other appendage whorls can impact androecial ontogenies. It also shows that ontogenetic sequences or alternative descriptions of other aspects of floral development can be critical for understanding the elaboration of androecial ontogenetic sequences.

The modes of ontogenetic sequence modification encountered in diplostemony/polystemony transformations, the extension of the polystemonous sequence in *Cardiandra* beyond the antesepalous triplet state, and increased meristic complexity of *Deinanthe* all involved terminal changes to androecial ontogenetic sequences. The elaboration of polystemony in *Philadelphus* and *Carpenteria* contrasts with those ontogenetic sequence transformations discussed so far because it involves the addition of a novelty at the beginning of androecial development. Both *Philadelphus* and *Carpenteria* begin androecial development with the formation of common primordia. After common primordia are established, both genera then form antesepalous triplets, just as all other polystemonous members of the family. Although the formation of common primordia is synapomorphic for these two taxa, the early shapes and developmental courses of the common primordia differ between them (Hufford, 1998). *Philadelphus* has relatively simple common primordia on which a single ring of stamens forms. This contrasts with what we observe in *Carpenteria*, in which the shape of the common primordium is elaborated (becoming plicate and having a broad base below its U-shaped rim), and they enlarge to protrude into the center of the flower. These shape and size novelties in the common primordia of *Carpenteria* are associated with the unique (at the level of Hydrangeaceae) formation of stamens not only in antesepalous groups on the rim of the common primordia but also in antepetalous positions along the margins of the common primordia.

If we turn to the issue of the relative importance of unique vs. iterated novelties in the patterning of morphological diversity in the androecial diversity of Hydrangeaceae, we find they occur with nearly equal frequency. Unique novelties, however, are limited to the most complex ontogenies. These unique ontogenetic novelties are associated particularly with the development of high numbers of stamens. In contrast, the homoplastic iterated novelties are centered in the common ontogenetic transformation between polystemony and diplostemony. The homoplasies are centered in options at the second developmental event in the androecial ontogenetic sequences in which either antepetalous stamens form to give rise to diplostemony or lateral flanking stamens form to establish polystemony.

CONCLUSIONS

Ontogenetic sequence models consist of a series of instantaneous ontogenetic states. Later ontogenetic states are developmental transformations of earlier states. Sequence models in which later states are contingent on some earlier state or states can be valuably used in comparative biology. Ontogenetic sequence models provide a means of characterizing development and its evolution that is complementary to approaches centered in developmental timing, biophysics, and developmental genetics. Ontogenetic sequence models can be especially complementary to heterochrony models in studies of evolutionary diversification. For example, heterochrony models based in allometry assume that whole-character ontogenies are homologous. These allometry studies focus on shifts in proportions among character states that must be present in all of the compared organisms. For this reason, heterochrony models that emphasize allometric comparisons can have limited value for understanding the origin of novelties. Ontogenetic sequences can be applied to those systems to understand elements of character ontogenies that may be novel.

All homology assessment involves ontogenetic states, whether or not the states come from late ontogeny or represent "terminal" stages. Ontogenetic sequences can have a particular value in formulating hypotheses of homology. Whole ontogenies can be compared, and structurally homologous states shared by ontogenies can be identified. If structural homologies are unclear in compared ontogenies, then alternatives such as the following need to be considered. (1) The point of difference between two compared ontogenies may be the focus of evolutionary developmental change, and other approaches to the study of developmental dynamics could concentrate on the formation of the nonhomologous states in the compared ontogenies in an attempt to understand how novel states have originated. (2) More intensive taxon sampling in the clade that includes the compared ontogenies or among outgroups may be required to clarify the structural homologies that are present. When we can identify correspondences between ontogenetic states of different organisms, it is possible to reconstruct how phenotypes have diversified by modifications at particular points in ontogenetic sequences.

The optimization of ontogenetic sequences on phylogenetic hypotheses can help us to understand the evolutionary patterning of phenotypic diversity. For example, the analysis of androecia in the Hydrangeaceae has shown that unique and iterated novelties occur largely with equal frequency. Unique novelties are associated with the development of more complex androecial states than are the iterated homoplastic novelties. For example, the homoplastic shifts between diplostemony and polystemony in Hydrangeaceae center on a single ontogenetic state subject to labile shifts in developmental evolution. The case study also demonstrated that many of the weaknesses of analyzing the evolution of ontogenies arise from problems inherent to phylogenetic systematics, particularly the equivocal optimization of states at internal nodes of cladograms and the difficulty in reconstructing plesiomorphic states.

ACKNOWLEDGMENTS

I thank Miriam Zelditch for the invitation to prepare a chapter for this book, Michelle McMahon for assistance at many turns and discussions of morphological homology and evolution, and Janice Newlon, Teresa Sanford, Trina Wiedenbach, and Andrea Winbauer for technical assistance. This research was supported in part by National Science Foundation Grant DEB-9496127.

REFERENCES

Alberch P (1985): Problems with the interpretation of developmental sequences. Syst Zool 34: 46–58.

Alberch P, Gould SJ, Oster GF, Wake DB (1979): Size and shape in ontogeny and phylogeny. Paleobiology 5: 296–317.

Angiosperm Phylogeny Group (1998): An ordinal classification for the families of flowering plants. Ann Missouri Bot Gard 85: 531–553.

Bock WJ (1965): The role of adaptive mechanisms in the origin of higher levels of organization. Syst Zool 14: 272–287.

Coen ES, Meyerowitz EM (1991): The war of the whorls: genetic interactions controlling flower development. Nature 353: 31–37.

Collazo A (2000): Developmental variation, homology, and the phraryngula stage. Syst Biol 49: 3–18.

Cruden RW (1997): Implications of evolutionary theory to applied pollination ecology. In: Richards KW (ed), *Proceedings of the 7th International Symposium on Pollination.* Leiden: International Society for Horticultural Sciences, p. 27–51.

de Pinna MCC (1991): Concepts and tests of homology in the cladistic paradigm. Cladistics 7: 367–394.

de Queiroz K (1985): The ontogenetic method for determining character state polarity and its relevance to phylogenetic systematics. Syst Zool 34: 280–299.

Dickson A (1866): On the morphological constitution of the androecium of *Mentzelia*, and its analogy with that of certain Rosaceae. Trans Bot Soc Edinburgh 8: 288–298.

Diggle P (1992): Development and evolution of plant reproductive characters. In: Wyatt R (ed), *Ecology and Evolution of Plant Reproduction: New Approaches.* New York: Chapman and Hall, p. 326–355.

Doebley J, Lukens L (1998): Transcriptional regulators and the evolution of plant form. Plant Cell 10: 1075–1082.

Donoghue MJ (1992): Homology. In: Keller EF, Lloyd EA (eds), *Keywords in Evolutionary Biology.* Cambridge, MA: Harvard University Press, p. 170–179.

Endress PK (1994): *Diversity and Evolutionary Biology of Tropical Flowers.* Cambridge, MA: Cambridge University Press.

Gould SJ (1977): *Ontogeny and Phylogeny.* Cambridge: Belknap.

Green PB (1988): A theory for inflorescence development and flower formation based on morphological and biophysical analysis in *Echeveria*. Planta 175: 153–169.

Green PB (1992): Pattern formation in shoots: A likely role for minimal energy configurations of the tunica. Int J Plant Sci (Suppl) 153: S59–S75.

Green PB (1996): Transductions to generate plant form and pattern: an essay on cause and effect. Ann Bot 78: 269–281.

Harder LD, Thomson JD (1989): Evolutionary options for maximizing pollen dispersal of animal pollinated plants. Am Nat 133: 323–344.

Hempel A, Reeves PA, Olmstead RG, Jansen RK (1995): Implications of *rbc*L sequence data for higher order relationships of the Loasaceae and the anomalous aquatic plant *Hydrostachys* (Hydrostachyaceae). Plant Syst Evol 194: 25–37.

Hernández LF, Green PB (1993): Transductions for the expression of structural pattern: analysis in sunflower. Plant Cell 5: 1725–1738.

Hirmer M (1918): Beiträge zur Morphologie der polyandrischen Blüten. Flora 110: 140–192, +11 plates.

Hufford L (1988): Roles of early ontogenetic modifications in the evolution of floral morphology of *Eucnide* (Loasaceae). Bot Jahrb Syst 109: 289–333.

Hufford L (1990): Androecial development and the problem of monophyly of Loasaceae. Can J Bot 68: 402–419.

Hufford L (1996a): Ontogenetic evolution, clade diversification, and homoplasy. In: Sanderson MJ, Hufford L (eds), *Homoplasy: The Recurrence of Similarity in Evolution.* San Diego, CA: Academic Press, p. 271–301.

Hufford L (1996b): The morphology and evolution of male reproductive structures of Gnetales. Int J Plant Sci Suppl 157: S95–S112.

Hufford L (1997a): A phylogenetic analysis of Hydrangeaceae based on morphological data. Int J Plant Sci 158: 652–672.

Hufford L (1997b): The roles of ontogenetic evolution in the origin of floral homoplasies. Int J Plant Sci Suppl 158: S65–S80.

Hufford L (1998): Early development of androecia in polystemonous Hydrangeaceae. Am J Bot 85: 1057–1067.

Hufford L, Moody M, Soltis DE (in press): A phylogenetic analysis of Hydrangeaceae based on the chloroplast gene *mat*K. Int J Plant Sci.

Irish VF, Yamamoto YT (1995): Conservation of floral homeotic gene functions between *Arabidopsis* and *Antirrhinum*. Plant Cell 7: 1635–1644.

Kluge AG (1988): The characteristics of ontogeny. In: Humphries CJ (ed), *Ontogeny and Systematics.* New York: Columbia University Press, p. 57–82.

Kramer EM, VF Irish (2000): Evolution of the petal and stamen developmental programs: evidence from comparative studies of the lower eudicots and basal angiosperms. Int J Plant Sci Suppl 161: S29–S40.

Langille RM, Hall BK (1989): Developmental processes, developmental sequences and early vertebrate phylogeny. Biol Rev 64: 73–91.

Leins P, Winhard W (1973): Entwicklungsgeschichtliche Studien an Loasaceen Blüten. Oesterr Bot Z 122: 145–165.

Mabee PM (1993): Phylogenetic interpretation of ontogenetic change: sorting out the actual from the artefactual in an empirical case study of centrarchid fishes. Zool J Linn Soc 107: 175–291.

Mayr E (1960): The emergence of evolutionary novelties. In: Tax S (ed), *The Evolution of Life.* Chicago, IL: University of Chicago Press, p. 349–380.

Meier R (1997): A test and review of the empirical performance of the ontogenetic criterion. Syst Biol 46: 699–721.

Moody ML, Hufford L, Soltis DE, Soltis PS (2001): A phylogenetic analysis of Loasaceae subfamily Gronovioideae. Am J Bot 88: 326–336.

Müller GB, Wagner GP (1991): Novelty in evolution: restructuring the concept. Ann Rev Ecol Syst 22: 229–256.

Nelson G (1978): Ontogeny, phylogeny, paleontology and the biogenetic law. Syst Zool 27: 324–345.

O'Grady R (1985): Ontogenetic sequences and the phylogenetics of parasitic flatworm life cycles. Cladistics 1: 159–170.

Olmstead RG, Kim K-J, Jansen RK, Wagstaff SJ (2000): The phylogeny of Asteridae sensu lato based on chloroplast *ndhF* gene sequences. Mol phylogenet Evol 16: 96–112.

Patterson C (1982): Morphological characters and homology. In: Joysey KA, Friday AE (eds), *Problems of Phylogenetic Reconstruction.* New York: Academic Press, p. 21–74.

Payer J-B (1857): *Traité d'organogénie comparée de la fleur.* New York: Stechert-Hafner [1966 reprint].

Queller DC (1984): Pollen-ovule ratios and hermaphroditic sexual allocation strategies. Evolution 38: 1148–1151.

Remane A (1952): *Die Grundlagen des naturlichen Systems der vergleichenden Anatomie und der Phylogenetik.* Leipzig: Geest and Portig.

Ronse Decraene LP, Smets EF (1991): Morphological studies of Zygophyllaceae. I. The floral development and vascular anatomy of *Nitraria retusa.* Am J Bot 78: 1438–1448.

Ronse Decraene LP, Smets EF (1992): Complex polyandry in the Magnoliatae: definition, distribution and systematic value. Nord J Bot 12: 621–649.

Ronse Decraene LP, Smets EF (1993): The distribution and systematic relevance of the androecial character polymery. Bot J Linn Soc 113: 285–350.

Roth G, Wake DB (1985): Trends in the functional morphology and sensorimotor control of feeding behavior in salamanders: An example of the role of internal dynamics in evolution. Acta Biotheor 34: 175–192.

Sanderson MJ, MJ Donoghue (1989): Patterns of variation in levels of homoplasy. Evolution 47: 236–252.

Schlichting CD, Pigliucci M (1998): *Phenotypic Evolution: A Reaction Norm Perspective.* Sunderland, MA: Sinauer Associates.

Selker JML, Steucek GL, Green PB (1992): Biophysical mechanisms for morphogenetic progressions at the shoot apex. Dev Biol 153: 29–43.

Soltis DE, Xiang Q-Y, Hufford L (1995): Relationships and evolution of Hydrangeaceae based on *rbc*L sequence data. Am J Bot 82: 504–514.

Swofford DL (1998): *Phylogenetic Analysis Using Parsimony, version 4.0b2.* Sunderland, MA: Sinauer Associates.

Velhagen WA (1997): Analyzing developmental sequences using sequence units. Syst Biol 46: 203–210.

Weston PH (1988): Indirect and direct methods in systematics. In: Humphries CJ (ed). *Ontogeny and Systematics.* New York: Columbia University Press, p. 27–56.

Willson MF (1991): Sexual selection, sexual dimorphism and plant phylogeny. Evol Ecol 5: 69–87.

Xiang Q-Y, DE Soltis, PS Soltis (1998): Phylogenetic relationships of Cornaceae and close relative inferred from *matK* and *rbcL* sequences. Am J Bot 85: 285–297.

Xiang Q-Y (1999): Systematic affinities of Grubbiaceae and Hydrostachyaceae within Cornales—insights from *rbcL* sequences. Harv Pap Bot 4: 527–542.

3

A DETAILED SCENARIO AND POSSIBLE TESTS OF THE MOSTLY MALE THEORY OF FLOWER EVOLUTIONARY ORIGINS

Michael W. Frohlich
University of Michigan, Ann Arbor, Michigan

INTRODUCTION

The origin of flowering plants (angiosperms) has long been among the most vexatious problems in evolutionary studies. Charles Darwin famously called the sudden appearance of flowering plants, without obvious antecedents, an "abominable mystery" (Darwin and Seward, 1903). The many theories proposed to explain angiosperm origins are striking in the great range of antecedents that they suggest, in the differing features proposed for the earliest flowers, and in the extraordinarily diverse evolutionary processes posited to give rise to the features of the flower (Stebbins, 1974; Doyle, 1978; Friis et al., 1987; Friis and Endress, 1990; Endress and Friis, 1994; Hughes, 1994; Doyle, 1994, 1996; Taylor and Hickey, 1996; Krassilov, 1997). These theories have obeyed the rule of homotopy enunciated by Haeckel, which forbids structures from moving from one place to another on an organism. By contrast, heterotopy—changing the position in which a structure develops—is a central element of our new theory, the "Mostly Male" theory (Frohlich and Parker,

Beyond Heterochrony: The Evolution of Development, Edited by Miriam Leah Zelditch
ISBN 0-471-37973-5

2000). The single instance of heterotopy required by the Mostly Male theory creates the bisexual reproductive unit that becomes the flower and generates a reasonable precursor for the carpel, which is the most complex component of the flower.

Many previous theories of angiosperm origins were ultimately based on phylogenetic reasoning that identified putative relatives of angiosperms among living and/or fossil gymnosperms and on analyses of angiosperm relationships that suggested particular features as characteristic of the earliest flowers. The most recent theories based on phylogeny—the Anthophyte theory and the Neo-Pseudanthial theory—made heavy use of a supposed close relationship between flowering plants and the living gymnosperm group Gnetales. Placement of the Gnetales as extant sister to angiosperms (or even possibly as the direct ancestors of angiosperms) allowed the detailed knowledge of these living gymnosperms to be used in character state reconstruction of flowering plant ancestors and relatives (Crane, 1985; Doyle and Donoghue, 1986, 1992, 1993; Loconte and Stevenson, 1990; Doyle, 1994, 1996; Nixon et al., 1994; Hickey and Taylor, 1996). Unfortunately for these theories, monophyly of extant gymnosperms now receives extremely strong support from molecular analyses of multiple genes from the chloroplast, nucleus, and mitochondrion; therefore, Gnetales are more closely related to other living gymnosperms than to flowering plants (Samigullin et al., 1999; Bowe et al., 2000; Chaw et al., 2000; Frohlich and Parker, 2000; Donoghue and Doyle, 2000). Additional studies that do not consider all seed plant groups nevertheless support a closer relationship between Gnetales and conifers than between Gnetales and flowering plants (Qiu et al., 1999; Winter et al., 1999; Hansen et al., 1999). Living gymnosperms are so variable in their reproductive structures that, as a monophyletic group, they provide little insight for character state reconstruction on the lineage leading to angiosperms. This leaves earlier theories that place angiosperms close to the extinct pteridosperm gymnosperms ("seed ferns") as the only contenders. The morphological gap between flowering plants and pteridosperms is very large. Such a large morphological gap could be bridged by any one of a diverse set of evolutionary scenarios, with little basis to select among them. The diversity of potential scenarios renders the theories vague.

The Mostly Male theory arose from studies of modern plants. It is unique in being based not only on the morphology and the phylogeny of organisms, but also on gene function and gene loss and on gene overexpression effects. A modern pollination system supports the theory by indicating feasibility for a proposed intermediate step in flower evolution. These data suggest a particular evolutionary process in the origin of the flower. Although the theory was based on living plants, we subsequently learned that the reinterpretation of a fossil pteridosperm (in Corystospermales) gives that group all the features required for potential angiosperm ancestors. When the Mostly Male theory's evolutionary process is applied to corystosperms, the remaining morphological gap becomes greatly reduced, which allows a detailed, specific scenario to be erected, as described below. A detailed scenario allows implications of the theory to be articulated and tests to be devised, whereas a diffuse fog of ideas can be neither supported nor opposed. It is the purpose of this chapter to create an explicit scenario for the Mostly Male theory to facilitate stringent tests of the theory.

The proposed scenario is separable into distinct components that could, individually, be correct or incorrect. In this regard, it resembles Ostrom's scenario for the evolution of birds from dinosaurs (Ostrom, 1976a, 1976b, 1979). He said that his theory was intended to provide a discussion point. His proposal did, indeed, stimulate discussion. His theory's main point is now generally thought to be correct—that birds acquired flight from the ground up, rather than from the trees down (Chiappe, 1995; Padian and Chiappe, 1998; Burgers and Chiappe, 1999; but also see Garner et al., 1999; Feduccia, 1999; and Jones et al., 2000, for various other views). Other aspects are generally discounted, such as the idea that prebirds used their feathered hands to slap insects out of the air for food. I hope that the scenario presented here will stimulate advances in the understanding of the origin of flowers, as Ostrom's theory helped advance the study of the origin of birds. Unlike Ostrom's scenario, which can only be tested by studies of fossils, the Mostly Male theory can also be tested by studies of genes in modern organisms. Tests using gene expression and gene phylogeny are indirect; therefore, an explicit scenario is needed to connect the theory to potentially observable developmental-genetic phenomena.

SUMMARY OF THE MOSTLY MALE THEORY

The Mostly Male theory (Frohlich and Parker, 2000) suggests that the overall organization of the angiosperm flower derives largely from the male reproductive structures of the gymnosperm ancestor, with the minimal female structures, the ovules,[1] originally being ectopic[2] in the first flower (Fig. 3.1). The male unit that became the flower would originally have consisted of a series of microsporophylls (leaf homologues bearing pollen sacks) grouped together along a stem. The ovule antecedent would originally have been borne separately in a purely female structure. This ancestral plant would have been wind pollinated (or might have had a relatively inefficient insect pollination system based on generalized herbivory). The male unit would have become bisexual when the ovule antecedent came to be borne ectopically on the upper (distal) microsporophylls, originally to attract insects with pollination droplets[3]; this would allow efficient pollination by insects that drank the pollination droplet. Later,

[1] Ovules are the structures that develop into seeds in all seed plants, both gymnosperms and angiosperms. Terms relating to plant structures are described in more detail in the "Terminology" section of the text, and in Fig. 3.4.

[2] Ectopic structures are structures borne in the wrong place on an animal or plant. They are typically caused by mutations and have been repeatedly observed in both animals and plants.

[3] The ovules of most gymnosperms produce a droplet of liquid that protrudes from the micropyle. The droplet catches pollen that is subsequently brought into the ovule, achieving pollination.

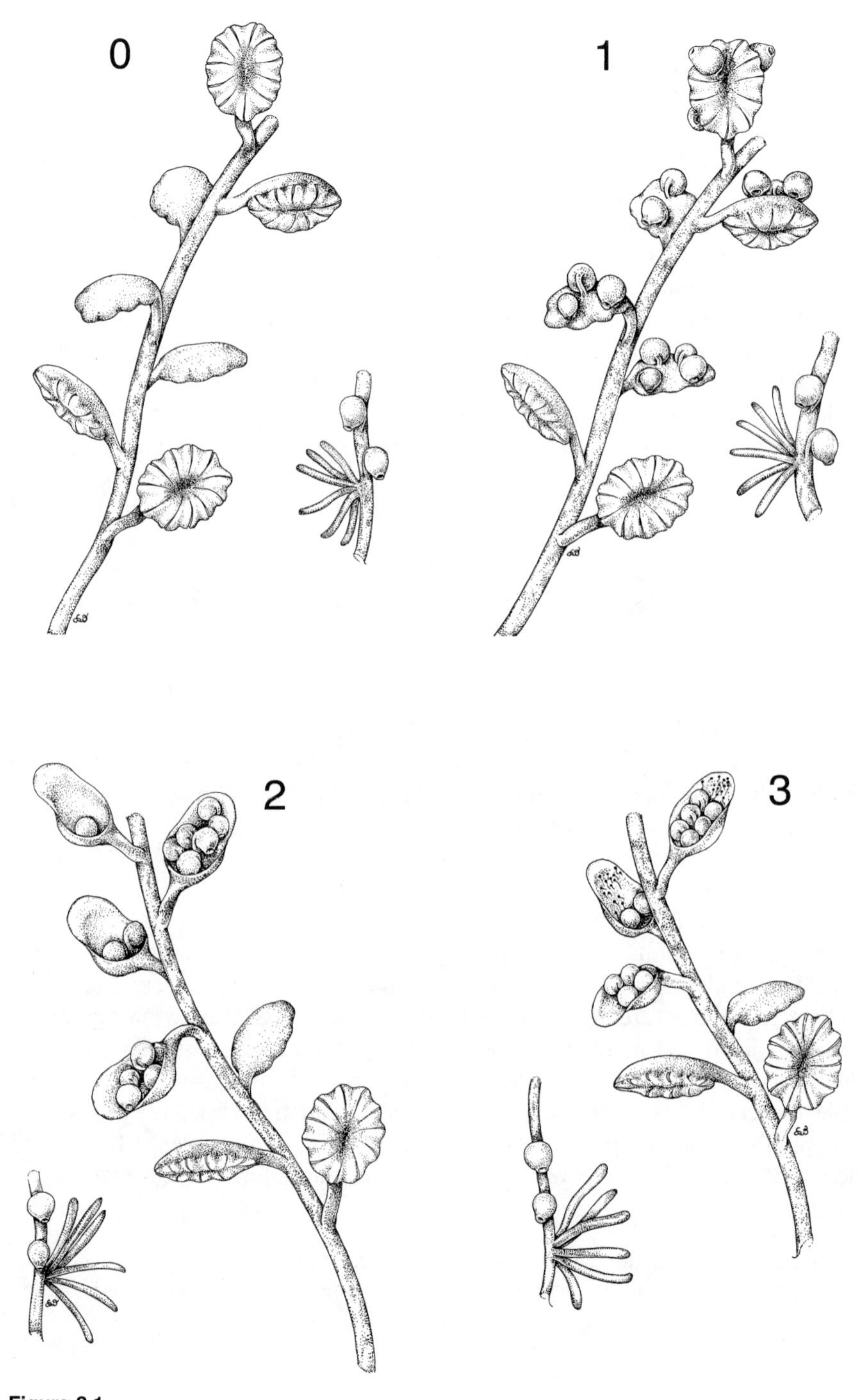

Figure 3.1.

Figure 3.1. Illustration of the evolutionary stages of the Mostly Male theory. The number of each panel corresponds to the individual evolutionary stages, as described in the text. The ancestral plant is illustrated in panel 0, showing *Pteroma* male structures and *Ktalenia* cupules. In panel 1, the male structures acquire ectopic cupules on some microsporophylls. In panel 2, the efficiency of insect pollination has increased due to various modifications. In panel 3, the ectopic ovules have become important as seed producers. (The small black marks represent pollen grains and pollen tubes and are not to scale.) In panel 4, the separately borne, pure-female cupules have been lost. In panel 5, the plant has evolved a flower and may be called an angiosperm.

the (micro)sporophyll would enclose the ovules, creating the angiosperm carpel, whereas the lower microsporophylls would be modified into angiosperm stamens, thus generating the flower. The carpel wall, though not the ovules, is ultimately derived from the *micro*sporophyll.

Supporting Evidence for the Mostly Male Theory

The supporting evidence was discussed in detail in Frohlich and Parker (2000) and is only summarized here (paragraphs 1–5).

(1) The most important evidence derives from the gene *Floricaula/LEAFY* (*LFY*). This gene is single copy in flowering plants, but all four living gymnosperm groups have two copies, due to an ancient duplication. Parsimony analysis with bootstrapping, maximum likelihood, and Wilkinson support analysis (Frohlich and Estabrook, 2000; Frohlich and Parker, 2000) all show that this gene duplication occurred before the flowering plant lineage diverged from the extant gymnosperm lineage; ergo, the flowering plant lineage originally had both copies of this gene, but one copy was subsequently lost. *LFY* in flowering plants helps to specify the apical meristem (the growing tip of the shoot) as reproductive, instead of vegetative, making it grow into a flower rather than a leafy shoot. Evidence from pine suggests that the two copies of *LFY* in gymnosperms specify the male and the female reproductive structures, respectively. Flowering plants have lost the gene copy that apparently specifies

female structures but retain the copy that apparently specifies gymnosperm male structures. *LFY* is a transcription factor that controls the activity of important cascades of downstream genes. Retention of the male-specific *LFY* homologue in flowering plants suggests retention, in flowering plants, of gene cascades active in the gymnosperm male structures; loss of the female-specific *LFY* homologue suggests that fewer of the female-specific gene cascades would have been retained in flowering plants. In classical terms, one could say that the flower is homologous to the male reproductive unit of the gymnosperm ancestor, although some mechanism outside of classical homology would have been required to explain the presence of female structures in the flower.

The essential female structure in seed plants is the ovule, which develops into the seed. In flowering plants, ovules are enclosed in carpels. In all gymnosperms (except the extinct Bennettitales and *Irania*), functional ovules are borne in structures separate from the male reproductive units. Mutations that result in ectopic structures are very well known in both plants and animals. The Mostly Male theory suggests that the ovule was originally ectopic in the flower, in particular, that ovules appeared on the adaxial surface of some microsporophylls in what had been an all-male structure. These microsporophylls eventually lost their microsporangia and became the carpel wall that encloses the ovule(s). Several additional lines of evidence are consistent with this view (see paragraphs 2–5 below).

(2) Ovules in modern flowers do not exhibit constant number or position with respect to the carpels. The number of ovules per carpel ranges from 1 (or fewer) to over 100,000 (in orchids). The position in which the ovules are borne (called "placentation") varies so much that this is a useful character for plant identification: placentation may be parietal, axile, free central, apical, or basal. These terms hide great variation in the details of ovule positioning, especially in plants with large numbers of ovules per carpel. Whatever the original number and position of the ovules in the carpel, at least some of these modified systems must represent an ectopic increase in the number of ovules and/or ectopic changes in the positions where they develop; such variations are consistent with ovules originally having been ectopic on the carpel and lacking rigidly defined positioning. In contrast, the flowering plant male structure, the anther, is much more uniform in its placement. The typical anther consists of two "thecae," each bearing a pair of microsporangia embedded within its tissues. The paired microsporangia are usually fused along a line of tissue (the stomium) that rips when the microsporangia open. A stamen bears one anther. Variations in anther structure are comparatively minor and involve loss of one theca or loss of one microsporangium in each theca, but stamens do not show an increase in the number of anthers or their component parts (with the sole possible exception of Malvales). This is consistent with ovules but not stamens originally being ectopic. (Variation in numbers of flower parts, for example, sepals, petals, stamens, and carpels, is commonly observed, but these stamens each have one anther. Many mutations in *Arabidopsis* result in an increase in the number of stamens.)

(3) Ectopic ovules are easily generated in *Petunia* by overexpression of a single *Petunia* gene, *FBP11*. These ovules are complete and perfectly formed and borne on the sepals and on petals, which have *not* been converted into carpels. This led Colombo et al. (1995) to suggest that the placenta is a flower organ on par with the sepals, petals, stamens, and carpels; this is consistent with the Mostly Male theory. Ectopic anthers, in contrast, are not observed in flowering plants. This differs from the interconversion of major flower organs (sepals, petals, stamens, and carpels) that can be achieved by manipulation of the *ABC* genes; in those cases, anthers and ovules are borne on stamens and carpels; hence, neither the anther nor the ovule is ectopic.

(3a) Recently I learned that ectopic ovules are known from the modern gymnosperm *Ginkgo biloba*, from trees growing in both China and Japan (Fujii, 1896; Sakisaka, 1929; Ma and Li, 1991). These ovules arise on occasional foliage leaves of female plants and sometimes even make normal-appearing seeds. This occurs only on some very old trees; Sakisaka (1929) suggests it is a "senile" feature of the tree. *Ginkgo* does not have cupules. *Fujii* (1896) and *Sakisaka* (1929) also report that microsporangiate structures can appear ectopically on some leaves of male plants.

(4) Strong *LFY* mutations in *Arabidopsis* completely abolish the production of stamens, but the remnant flowers still produce carpels, which are partially fertile, despite the numerous aberrations of organ number and position caused by the mutation. Furthermore, the inflorescence axis terminates in a cluster of carpels, which are not part of any flower. This is consistent with the *LFY* gene being more directly required for stamen production, whereas alternative systems can produce carpels. The recent cloning of MADS genes such as *SPATULA* and *FRUITFUL* suggests the complexity of genes that help promote carpel identity (Alvarez and Smyth, 1999; Ferrandiz et al., 2000).

(5) In modern gymnosperms, there are examples of (nonfertile) ovules borne in the male reproductive unit that presumably would originally have been ectopic on the male structures. This shows that ovules can appear on functionally male reproductive units and demonstrates a possible immediate function for ectopic ovules. Some species of all three extant genera of Gnetales show nonfertile ovules in the male structures (Hufford, 1996). These nonfertile ovules make pollination droplets that attract insects that drink the liquid. These insects also drink the pollination droplets from the fertile ovules borne on other plants and in the process transfer pollen; hence, these Gnetales are insect pollinated (Endress, 1996; Wetschnig and Depisch, 1999). Lloyd and Wells (1992) also suggested that pollination droplets were the initial reward that allowed insect pollination of flowering plant ancestors, but they based this on a supposed homology of pollination systems between Gnetales and angiosperms, in accord with the now defunct anthophyte theory (Crane, 1985; Doyle and Donoghue, 1986, 1992, 1993; Doyle, 1994, 1996). This system in Gnetales cannot be homologous to that suggested for angiosperms ancestors, as we now know that Gnetales are distantly related to angiosperms and more closely related to other extant gymnosperms.

In Frohlich and Parker (2000), we elaborated two points of this theory to explain the extreme shortness of the stem connecting the flower organs and the loss of the growing tip of that stem (i.e., morphological determinacy). Unlike flowers, gymnosperms have an elongated stem separating the components of their reproductive units (cones), and the tip of this stem has an apex that, in principle, could continue growth. At the evolutionary grade when the ecotopic ovules function only or primarily to attract pollinators, the stem between the microsporophylls would have become shortened; therefore, the ovule-bearing structures would be physically close to the pollen-bearing organs, and insects visiting to collect pollination droplets would encounter the pollen. The stem could also have lost its apical meristem to limit the number of ovule-bearing sporophylls. This way, pollination droplets would not be produced after the pollen of the associated microsporophylls is shed; to do so would attract insects away from other reproductive units that were still releasing pollen, a much more serious problem than simply wasting pollination droplets (Frohlich and Parker, 2000).

CHARACTERISTICS OF GYMNOSPERMS AND ANGIOSPERMS

Possible Fossil Gymnosperm Antecedents

The Mostly Male theory imposes several requirements for any potential ancestor of flowering plants. There must be a reasonable antecedent for the angiosperm ovule, which could be moved ectopically into the male structure, and the male structure must consist of simple microsporophylls grouped together along a stem. The corystosperms satisfy these two requirements especially well, indeed, better than the Caytoniales, which have been more often cited as potential angiosperm ancestors, and much better than other proposed groups, such as Glossopterids and Bennettitales. If corystosperms are accepted (at least provisionally) as angiosperm ancestors, then the firm starting point that they provide allows a detailed scenario to be created.

It is a truism that analyses of relationships should be based on all characters and that one should consider all of the known fossils in a search for relatives of flowering plants. Such broad and detailed studies have not yet been done in conjunction with the Mostly Male theory, and such is not the purpose of this chapter. Two Cretaceous fossils, of uncertain affinities but which have simple microsporophylls borne together on a stem, were noted in Frohlich and Parker (2000). There are few permineralized deposits of Mesozoic pteridosperms, so they are less well understood than earlier, especially Carboniferous, plants. Even some famous claimed angiosperm ancestors, such as *Sanmiguellia*, are rather poorly understood and remain of uncertain affinities among the gymnosperms. Furthermore, a large number of Cretaceous and Jurassic pteridosperm fossils are known only from impressions (J. A. Doyle, personal communication). Jurassic and early Cretaceous gymnosperm diversity may well have been substantially greater than the current state of knowledge would suggest (Hughes, 1994) and could have included poorly known plants that were close relatives of angiosperms. Additional paleobotanical studies that clarify the diversity of Jurassic and lower Cretaceous pteridosperms would be most helpful to improve understanding of the complex of organisms within which flowering plants apparently arose.

An analysis of the sort presented here must be anchored by putative relatives that are sufficiently well known to limit the range of theorizing. Because the corystosperms closely fit the Mostly Male theory, they allow a detailed, explicit scenario to be erected, which in turn facilitates analyses that could lead to tests of the theory. I will treat corystosperms as if they were flowering plant ancestors because they are the best candidates I know. I will derive flowers from the best-understood corystosperm male and female organ genera and from corystosperms that are the latest in time.

If the Mostly Male theory should prove largely correct, but with a gymnosperm other than Corystospermales as the angiosperm antecedent, I expect that this real ancestor would also exhibit the features that make corystosperms such a close fit to the theory. In particular, any real ancestor must have an ovule antecedent that could be ectopically placed on the microsporophyll and microsporophylls that were simple and grouped together along a stem specialized to bear them.

Corystospermales

Corystospermales are best known as abundant, widespread fossils from the Triassic of Gondwanaland and have left abundant fossils of nearly all parts of the plant, including permineralized wood, leaves, and male reproductive units. The female structures (cupules) have long been suggested to be possible antecedents for the angiosperm ovule (Gaussen, 1946; Doyle, 1978; Crane, 1985). The microsporangiate structure *Pteruchus* is usually found as several oval, flattened sporangia-bearing units attached to a narrow axis. It was most often interpreted as large, compound microsporophyll (Townrow, 1962; Crane, 1985; Retallack and Dilcher, 1988).

One site in Antarctica yielded permineralized materials (*Pteruchus fremouwensis*) that clearly demonstrated that the axis was a stem and that the individual oval structures were simple microsporophylls, correcting the earlier view (Yao et al., 1995). The microsporophylls were ovate and had an entire margin (Fig. 3.2). They contained a midvein with several pairs of pinnate secondary veins that extended to the margin. Microsporangia were borne in large numbers on the undersides of the sporophylls. Microsporangia were stalked and arranged in "pairs" on opposite sides of each secondary vein. Each microsporangium opened by a single longitudinal slit, with the opening slits of the paired microsporangia facing each other.

Individual microsporophylls are not found as isolated organs but rather are in groups connected by the stem. Foliage leaves are not found in organic at-

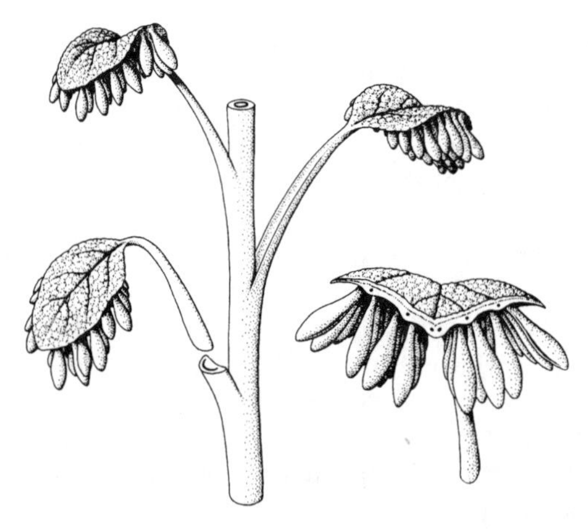

Figure 3.2. Reconstruction of *Pteruchus* male structures. [From Yao et al., 1995.]

tachment to the axes bearing the microsporophylls. This suggests that the stem with microsporophylls was shed as a unit after the pollen was dispersed. This further implies that the stem did not revert to vegetative growth after making a series of microsporophylls; if it had, then the microsporophylls would have been shed individually and/or vegetative leaves would be found distally to the microsporophylls. The absence of vegetative leaves basal to the microsporophylls implies a defined location for separation of this reproductive unit from the rest of the plant.

I accept the organ genera *Pteroma* and *Ktalenia* as corystosperms, even though the corresponding female structure is not known for the former and the male structure is not known for the latter. These two rare fossils, from the Jurassic of England and the Cretaceous Baqueró formation of Argentina, respectively, extend the temporal range of corystosperms to overlap with that of the early flowering plants.

Pteroma is known only from compression fossils, from the Jurassic Yorkshire Flora (Harris, 1964). The microsporangia-bearing unit of *Pteroma* was flattened, round to oval, and borne on a short lateral stalk. The vast majority of these units are found attached to an elongate stalk, but the elongate stalk never bears foliage leaves. Internal anatomy is not known, but one specimen clearly shows that the microsporangia-bearing units were spirally arranged (Frohlich, unpublished observations). This suggests that *Pteroma*, like *Pteruchus*, had spirally arranged simple microsporophylls borne on a specialized stem that did not revert to vegetative growth but rather was shed as a unit after pollen release (Frohlich, unpublished observations.) Unlike *Pteruchus*, *Pteroma* microsporangia were entirely embedded within the microsporophyll, with each locule extending from near the center of the sporophyll out toward the margin of the sporophyll (Fig. 3.3). Each opened by a longitudinal slit, presumably on the underside of the sporophyll.

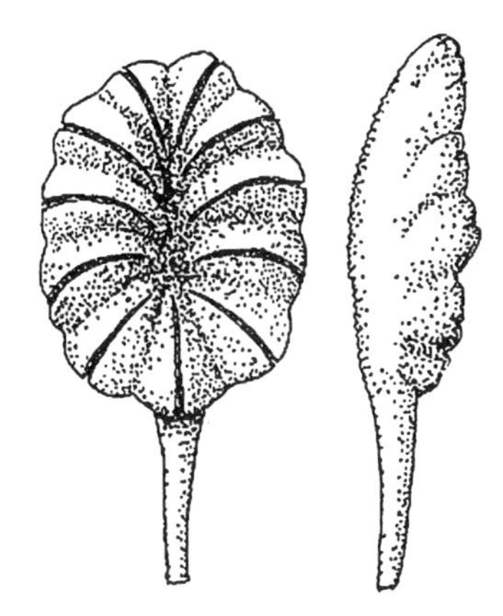

Figure 3.3. Reconstruction of *Pteroma* male structures. Left: Front view. Right: Side view. [From Harris, 1964.]

In gymnosperms, the essential female structure is the ovule (Fig. 3.4a), which consists of the stalk attached to a covering layer, the integument, that nearly surrounds the nucellus (megasporangium). Inside the nucellus develops the megaspore, which grows into the female gametophyte. The micropyle is an opening through the integument that allows entry of the pollen or pollen tube to the nucellus and ultimately to the egg. The stalk is at the opposite end of the ovule from the micropyle; such ovules are termed orthotropous.

Many pteridosperms had a specialized structure (the cupule), on (or in) which the ovules were borne. In some pteridosperms, the cupule more or less surrounded the ovule(s). Corystosperm cupules typically contained one (or two) ovules, with the cupule wall extending all the way around the ovule(s). In some, the ovule protruded beyond this opening in the cupule, as in some Triassic *Umkomasia*. In others, the cupule almost completely enclosed the ovule, except for a small opening, apparently over the micropyle, at the opposite end of the cupule from the attachment of the stalk (e.g., in *Umkomasia uniramia*; Axsmith et al., 2000). The cupule and its stalk were recurved, so the opening in the cupule (or the micropyle, if that is exposed) was positioned close to the cupule's stalk.

In most species of the cupule genus *Umkomasia* (including *Pilophorosperma*), the cupules were borne on slender branching stalks (Fig. 3.5). Small flattened

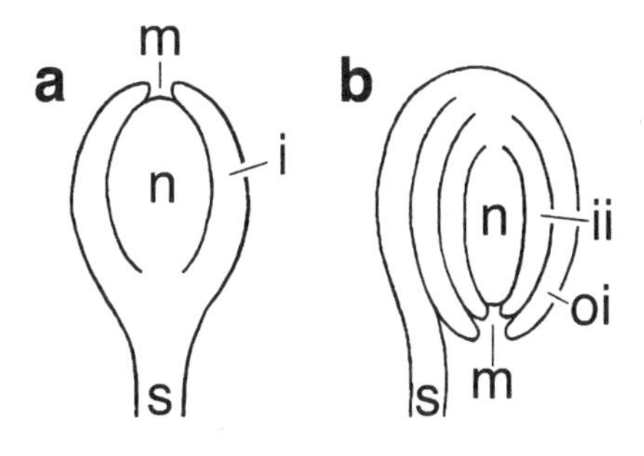

Figure 3.4. Diagrams of gymnosperm (a) and angiosperm (b) ovules. s, Stalk; n, nucellus; i, integument; m, micropyle; oi, outer integument; ii, inner integument. [From Frohlich and Parker, 2000.]

Figure 3.5. Reconstruction of *Umkomasia* (= *Pilophorosperma*) cupule bearing structure. [From Taylor and Taylor, 1993.]

structures were also present on the stalks. A long-running dispute regarding whether the cupule-bearing stalks were rachises of a compound megasporophyll or were branching stems was resolved by Axsmith et al. (2000) for *U. uniramia.* They demonstrated that the cupules were borne in whorls, implying that they were borne on a stem; therefore, each cupule represented a whole leaf that bore an ovule. This leaf was peltate because its petiole (the stalk of the cupule) was attached to the center of the lamina, which extended outward to cover virtually the whole ovule.

Leaves (and leaf homologues) are dorsiventral; that is, the two flat surfaces of a leaf typically differ morphologically. The side facing the tip of the stem that bore the leaf is the *adaxial* side (the upper side of the leaf, in most ordinary plants); the side away from the stem tip is the *abaxial* side. Because the cupule is a leaf homologue, one may ask whether the ovule was borne on cupule's adaxial side or on its abaxial side.

Axsmith et al. (2000) suggest that the *Umkomasia uniramia* ovule was borne on the *abaxial* side of the cupule. If confirmed, this would argue strongly against the corystosperm cupule as antecedent of the angiosperm ovule, in which the cupule wall is considered homologous to the angiosperm outer integument. If the angiosperm ovule was derived from the cupule, then the outer surface of the cupule must have been its abaxial side and the ovule inside the cupule must have been borne on the adaxial side.

In my view, the conclusion of Axsmith et al. (2000) is open to question. They write (p. 765) "it is likely that any enclosed ovules were borne on the abaxial sporophyll surface as the cupule itself appears to be formed by a downward

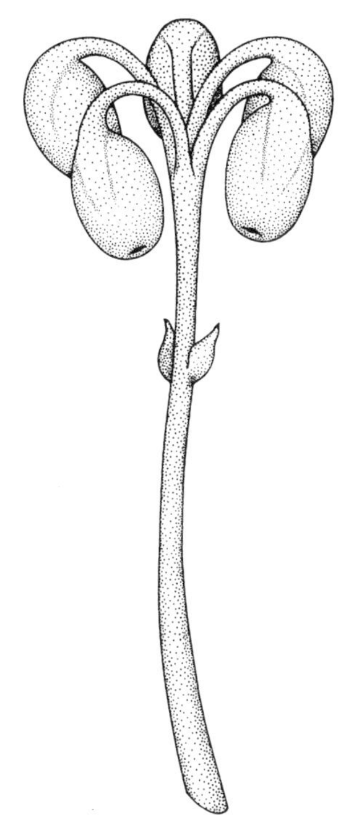

Figure 3.6. Reconstruction of *Umkomasia uniramia*, showing five cupules borne in a whorl at the tip of a stem. [From Axsmith et al., 2000.]

(abaxial) infolding of the lamina." The lower side of a leaf, in its typical orientation, would be the abaxial side. However, the petiole of a leaf can twist, reorienting the adaxial surface to face downward. Furthermore, their reconstruction (shown here in Fig. 3.6) shows the cupule wall as peltate and completely surrounding the ovule. Modern leaves (and leaf homologues) that are peltate have their petioles attached to the abaxial surface. If the *U. uniramia* cupule followed this pattern, then the ovule would have been borne on the adaxial surface. This issue can only be settled for certain by discovery of permineralized remains.

Ktalenia was the latest-occurring corystosperm, from the (early Aptian) Cretaceous Banqueró formation of Argentina. Taylor and Archangelsky (1985) described an axis bearing a number of *Ktalenia* cupules attached to a leaf, proving that it is a pteridosperm (Fig. 3.7). The cupules contained one or two ovules, which were closely covered by the cupule wall, except for a small opening at the opposite end of the cupule from the stalk. Cupules were reflexed on their stalks so the opening was positioned near the stalk. Cupules were 3–4 mm in diameter when seeds were mature, but their size at pollination is not known. There were also clusters of small, flattened appendages on the axis that bore the cupules. The leaf was twice compound and had open venation. A number of characteristics strongly suggest that this plant is a corystosperm, in

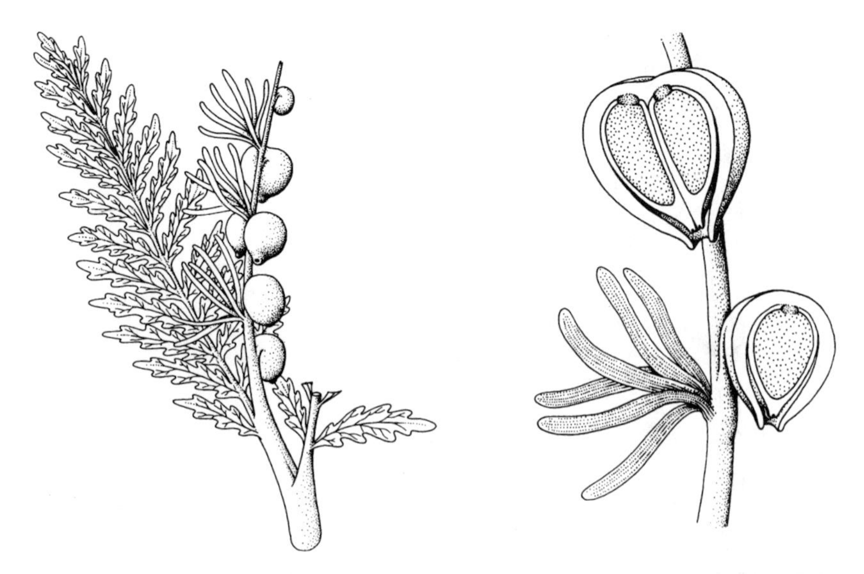

Figure 3.7. Reconstruction of *Katalenia* (the organ genus of the cupule) attached to *Ruflorina* (the organ genus of the leaf). [From Taylor and Archangelsky, 1985.]

particular (1) the presence of only one or two ovules in the cupule, (2) the cupule extending all the way around the ovules, such that the opening is bordered by cupule tissue, (3) the recurved cupule with opening rather close to the stalk, (4) the presence of small appendages on the stalks among the cupules, (5) the twice pinnately compound leaf with open venation, and (6) the veins branching twice in each leaflet. An alternative placement for *Ktalenia* might be within another pteridosperm group, the Caytoniales, but Caytoniales differed in having numerous ovules per cupule, the cupule wall not extending all the way around the opening, so part of the opening bordered the stalk rather than cupule tissue, absence of small appendages on the stalk among the cupules; leaves with two pairs of leaflets placed close together to appear palmately four-lobed, and leaves with complex net venation (Taylor and Taylor, 1993; Axsmith et al., 2000; T. N. Taylor personal communication).

Other Gymnosperms

The Caytoniales has been the pteridosperm group most widely discussed as an angiosperm precursor or relative, since Gaussen's 1946 proposal (discussed below), which derives the angiosperm ovule from the *Caytonia* cupule. In Caytoniales, the ovule was borne on the adaxial side of the cupule (Nixon et al., 1994). Caytoniales lacks features of the male reproductive unit that make corystosperms such a good fit for the Mostly Male theory. In particular, Caytoniales microsporophylls were pectinate lobed or compound, rather than ovate and entire-margined, and there is no evidence that Caytoniales microsporophylls were borne together in a group on a stem (Taylor and Taylor, 1993).

Glossopteridales were abundant in the Permian of Gondwanaland but (except for a possible occurrence in Mexico; Delavoryas and Person, 1975) were not found in the Mesozoic. They have been suggested as angiosperm ancestors (Retallack and Dilcher, 1981) and did share a number of features with angiosperms, such as net-veined leaves. Ovules were borne on the adaxial side of megasporophyll, which was itself borne, singly or in groups, on a subtending leaf. These megasporophylls varied greatly within the glossopterids, but in one (*Denkania*) the stalked megasporophyll closely surrounded a single ovule. Although not reflexed, it could be a reasonable antecedent for the angiosperm ovule. The microsporangia were borne in stalked clusters, which, like the female structures, were borne on the upper surface of a subtending leaf. The leaves bearing these male and female structures are found detached; it is not clear whether they might have been borne in groups (Retallack and Dilcher, 1981; Pigg and Trivett, 1994). Although *Denkania*, in particular, would be a good ovule antecedent, the male structures of glossopterids are a poor fit for the Mostly Male theory, and there is a very long temporal gap between the last certain glossopterids and the first flowering plants.

Several extinct gymnosperms did have bisexual reproductive structures. The most famous of these is the Mesozoic group Bennettitales, which includes such plants as *Cycadeoidea* and *Williamsoniella*. These plants had reproductive structures that were flower-like in general arrangement, with sterile structures on the outside surrounding several microsporophylls and with numerous stalked ovules borne on the stem in the center of the "flower." In between the ovules were numerous sterile scales, which apparently protected the ovules, whose micropylar tips protruded between the scales (Taylor and Taylor, 1993). Both male and female units are borne on the same axis. Meyen (1988) suggested that this bisexual reproductive unit was reorganized, so that "both sexes became monomorphic ... and modeled according to laminar male sporophylls." In other words, the ovules came to be borne on microsporophyll-like structures. This would place ovules on microsporophylls, as in the Mostly Male theory. *Cycadioidea* had highly complex microsporangiate organs, which appear very different from angiosperm stamens and from proposed carpel precursors, but other Bennettitales had flat, simple microsporophylls (Crane, 1986). It is uncertain whether the ovules were surrounded by a cupule (or some other second layer) or not (Taylor and Taylor, 1993; Crane and Kenrick, 1997). There remains a significant gap between Meyen's gametoheterotopy-modified bennettitalean and angiosperms. The adaptive advantage, suggested in the Mostly Male theory for the ectopic cupules—that the pollination droplet would attract insects for pollination—would not seem applicable to Meyen's theory because bisexual Bennettitales would already have had ovules in the "flower" that could have attracted insects with pollination droplets (although some Bennettitales did have derived, unisexual reproductive units). How the Bennettitales acquired bisexual reproductive units is not known.

Another bisexual fossil is represented by the Jurassic plant *Irania hermaphroditica*. It had several male cone-like structures attached near the base of an elongate axis that bore seeds; if the entire assemblage is considered the reproductive unit, then it was bisexual. Each of the male units had its own central axis that bore crowded sporangiophores extending out in all directions. The sporangiophores were peltate and bore two or more sporangia each. The female portion had an elongate axis that bore numerous stalked structures, which also extended in all directions from the axis. Each of the stalked female units had a pair of thin, flat structures, which have been called "cupules" (Krassilov, 1997), although their true nature is uncertain. The arrangement of parts in the male suggests that the axis is a stem and the sporangiophores are microsporophylls. *Irania* is not the intermediate predicted by the Mostly Male theory because the individual male and female units are not borne on the same axis. Also, for this reason, *Irania* is not the ideal flower precursor for other Euanthial theories (discussed below).

Attributes of Basal Angiosperm Flowers

Several recent papers all support the same arrangement of basal clades for the extant angiosperms. This is a dramatic improvement from the state of knowledge of only a few years ago, when studies conflicted profoundly in the arrangement near the base of flowering plants. Current studies place *Amborella* of the monotypic Amborellaceae as sole member of the basal clade, with Nymphaeales comprising the second clade and with a third clade containing Illiciales, Trimeniaceae, and *Austrobaileya* (Mathews and Donoghue, 1999; Qiu et al., 1999; Soltis et al., 1999; Soltis et al., 2000; Savolainen et al., 2000). Detailed analyses to reconstruct ancestral flowering plant features from these trees have not yet been published, but the taxa near the base of the angiosperms generally share a number of features (or show clear derivation from such features), which may therefore be taken as attributes of the earliest crown-group flowering plants.[4]

All have a short floral axis bearing sterile structures (tepals or sepals and petals) plus stamens (or staminodes as in *Amborella*) and separate carpels. Hence, all are bisexual or clearly derived from bisexual ancestors. The numbers of parts are often variable but are not especially large in the most basal lineages (e.g., within Nymphaeales). Parts are generally not in clear whorls and are not fused to each other. The stamens are typical of angiosperms in having two thecae, with each theca consisting of two microsporangia fused lengthwise in pairs and embedded in the stamen tissue. The stomium is a specialized tissue that bridges from one microsporangium to the other along a line extending the

[4] A "crown group" consists of all the living members of the group, all lineages down to deepest node in the cladogram straddled by living members of the group and everything (living or extinct) descended from that deepest node. The alternative is "stem group," which refers to lineages that attach to the cladogram *below* the deepest node straddled by living members.

full length of the sporangia. The microsporangia open when the stomium tissue at the line of fusion rips, creating what appears to be a single longitudinal slit-shaped opening through which the two sporangia release their pollen.

The carpels of many putative basal angiosperms (including those of the lowest clades) have been studied in detail by Igersheim and Endress (1997, 1998) and Endress and Igersheim (1997), who showed that several features traditionally ascribed to carpels of basal angiosperms (Stebbins, 1974) are in fact not present. Carpels are generally *not* folded lengthwise but are commonly ascidiate (bucket shaped). The upper edges of the carpel roll toward each other but are commonly *not* fused, at least not over the whole length along which the edges become close together. Rather, the carpel is sealed by a gelatinous secretion that fills the remaining space. Early in development, many basal carpels have a midvein and two lateral veins. Much later, as the carpel develops into a fruit, it may acquire additional veins.

The ovules of angiosperms (Fig. 3.4b) have a stalk and a nucellus, as in gymnosperms, but most angiosperms have two integuments rather than one. Also, most angiosperm ovules appear bent over (anatropous), with the opening through the integuments (micropyle) being close to the stalk. The inner integument of angiosperm ovules has typically been considered homologous to the single integument of gymnosperms. The outer integuments of *Arabidopsis* and tomato ovules express the *INO* gene in their outer surfaces. *INO* is a member of the YABBY gene family, whose expression marks the *abaxial* side of lateral organs (Baker et al., 1997; Villanneva et al., 1999); ergo, the inner integument and nucellus are borne on the *adaxial* side of the outer integument.

The new understanding of basal flowering plant clades suggests that gametophytic self-incompatibility might have been present in the earliest angiosperms (Mathews and Donoghue, 1999), contrary to earlier cladistic studies that argued against the presence of this system at the base of the angiosperms (Weller et al., 1995). Gametophytic self-incompatibility had been suggested as a possible major advantage that favored diversification and dominance of angiosperms or at least the closure of the carpel (Whitehouse, 1950; Zavada and Taylor, 1986; Dilcher, 1995).

The Morphological Gap Between Gymnosperm and Angiosperm Reproductive Units

The unique features of the flower are so different from gymnosperm structures that many homology assessments are controversial. Even the fundamental nature of the flower is disputed. Euanthial theories view the major flower organs (sepals, petals, stamens, and carpels) as leaf homologues, so the flower could be derived from a single stem bearing highly specialized leaves (Arber and Parkin, 1907). Pseudanthial theories derive the flower from a compound structure, consisting of a stem bearing leaves with axillary branches bearing other leaves and reproductive organs (Wettstein, 1935).

The carpel is the most complex component of the flower; therefore, its derivation has been the focus of many theories on angiosperm origins. To say that the carpel is a leaf homologue that bears ovules contributes little to erecting a specific scenario for the origin of the flower because ovules were borne on leaves in most gymnosperms, including the Carboniferous gymnosperm group Medullosaceae. However, no gymnosperm had ovules arranged in longitudinal rows on the adaxial side of a simple microsporophyll, as in the immediate precursor of the carpel according to euanthial theories. Carpel structure is quite complex, with ovules in some plants supplied by separate vascular traces that become distinct near the base of the carpel. This has been interpreted as a remnant of a compound origin of the carpel from the fusion of one or more leaves with an ovule-bearing structure on its upper side (Stebbins, 1974). The lack of a convincing carpel antecedent among fossil gymnosperms has been the greatest single difficulty in explaining the origin of flowers.

In pseudianthial theories, there are so many components in the ancestral compound structure that any of a variety of items could have become fused together to make the second covering of the ovule. A general hallmark of pseudanthial theories is that most components of the ancestral compound structure must be lost for it to give rise to the flower. The large number of parts in the compound structure would allow pseudanthial theories to explain almost any structure, but the many items that must be lost to yield the flower leave an especially large morphological gap.

Any scenario that explains the origin of the carpel must also explain the origin of the novel structures of the angiosperm ovule. It has long been believed that the nucellus of the angiosperm ovule is homologous to the nucellus of the gymnosperm ovule. The inner integument of the angiosperm ovule is generally thought to be homologous to the single integument of the gymnosperm ovule. The problem has been explaining the derivation of the second integument of the angiosperm ovule and the bent-over orientation of the angiosperm ovule, which brings the micropyle close to the ovule's stalk.

Gaussen (1946) proposed a specific source for the second integument, deriving it from the cupule. In the Mesozoic seed-fern group Caytoniales, the cupule almost completely enclosed the ovules, leaving a small opening between the cupule and the stalk of the cupule (Fig. 3.8). Within the cupule, numerous orthotropous ovules were oriented with micropyles facing this opening. Gaussen (1946) noted that, if the cupule contained only a single ovule, then the whole structure would have two covering layers and an opening adjacent to its stalk, resembling a flowering plant ovule. This was accepted by some prominent workers (Stebbins, 1974; Doyle, 1978) and incorporated into the Anthophyte theory (Crane, 1985; Doyle, 1994), but other features of flowers are not so directly derivable from Caytoniales. In Caytoniales, the cupules were borne on opposite sides of a narrow elongate stalk. Gaussen (1946) (and Doyle, 1978, 1994; Crane, 1985) proposed that this stalk became wide and flat and eventually enclosed the cupules to form the angiosperm carpel (or ovary). Gaussen noted that the corystosperm cupules contained only one or two ovules, and he suggested that corystosperms might have given rise to some flowering plants,

Figure 3.8. Reconstruction of cupule of Caytonia. [From Frohlich and Parker, 2000; modified from Stewart and Rothwell, 1993, according to Nixon et al., 1994.]

which he thought were polyphyletic. The branched stalks on which corystosperm ovules were borne make a poor precursor for the carpel, so more recent discussions or figures show recognizably caytonialean structures as giving rise to carpels (Doyle 1978, 1994; Crane 1985).

Nixon et al. (1994) criticized the derivation of the second integument from the *Caytonia* cupule because the stalk of the cupule forms part of the boundary for the opening that admits pollen into the *Caytonia* cupule. The homologue of this opening in flowering plants would be the outer part of the micropyle that leads past the outer integument. In many flowering plants, the outer integument completely surrounds the ovule and the micropyle, so no part of the micropyle is bounded by the ovule's stalk. This more closely resembles the structure of corystosperms, where the cupule completely surrounds the ovule. However, in other basal angiosperms, the outer integument does *not* completely surround the micropyle (Endress and Igersheim, 1997; Igersheim and Endress, 1997, 1998); indeed, in some plants, the outer integument may be so asymmetric that it does not extend between the inner integument and the stalk. As yet, there have been no studies to determine which condition is basal within flowering plants. This feature seems to change in multiple flowering plant lineages, as many groups contain representatives exhibiting each state (Doyle, 1996). *Amborella*, of the most basal angiosperm clade, has outer integuments that completely surround the ovule and form the entire outer boundary of the micropyle (Endress and Igersheim, 1997), whereas *Brasenia* of the second angiosperm clade (Nymphaeales) has the stalk appressed against the inner integument, with the stalk forming part of the boundary of the micropyle (Richardson, 1969). Basal angiosperms with orthotropous ovules (including *Barclaya* of Nymphaeales) have outer integuments that surround the ovule (Igersheim and Endress, 1998). Among other Nymphaeales, with anatropous ovules, some have outer integuments surrounding the ovule (annular outer integument, e.g.,

Victoria and *Euryale*) and others do not (semiannular outer integument, e.g., *Cabomba*) (Igersheim and Endress, 1998).

Axsmith et al. (2000) state that the *U. uniramia* cupule is an "unlikely homolog" of the angiosperm outer integument because of its relatively basal position in their phylogenetic tree (i.e., far away from angiosperms). However, their analysis was done at the time when the Anthophyte clade (grouping together Gnetales, angiosperms, Bennettitales, and Pentoxylales) was still generally accepted; their use of the Anthophyte clade as a terminal taxon was appropriate then; however, "Anthophytes" is now known to be paraphyletic or polyphyletic because it violates the monophyly of extant gymnosperms. Hence, this analysis cannot be accepted, but it could be redone with the constraint that extant gymnosperms are monophyletic and angiosperms lie outside extant gymnosperms.

The Mostly Male theory could be consistent with any origin and structure of the ovule outer integument, as long as an ovule precursor existed that might be transferred ectopically to the microsporophyll. Because corystosperms have male structures that make a good fit to the theory, I provisionally accept the corystosperm cupule as the sort of object that was ectopically transferred and became the flowering plant ovule. In the Mostly Male theory, the carpel wall is derived from the microsporophyll, which is a leaf homologue; thus, this theory is euanthial.

Insect Pollination: Source of the Crucial Selective Advantage

An evolutionary scenario must contain a selective advantage, operative at each step of the evolutionary path. This provides a rationale that allows selection to favor each new morphological feature. This is especially important here to account for the retention and subsequent elaboration of the ectopic ovule. When no selective advantage is apparent, one is reduced to claims that an organism might possibly evolve from one adaptive plateau to another, presumably very quickly and/or in a small population, with genetic drift (i.e., chance) playing the decisive role. Such a rapid change in small populations is unlikely to be confirmed by fossil discovery. The possibility of testing the Mostly Male theory by finding fossil intermediates depends on identifying selective advantages for the intermediate forms; therefore, they might reasonably have persisted for a long time.

The benefit of insect pollination was the main source of selective advantage for critical steps of the Mostly Male theory (Frohlich and Parker, 2000) and the source of the immediate advantage conferred by the ectopic ovule, causing it to be retained. Increased efficiency of insect pollination drives many subsequent changes as well. To evaluate whether this is reasonable, one must consider the history of insect pollination and the history of the insect groups that could have pollinated plants.

Plant-insect interactions are documented from very early times. Spore and/or pollen eating by insects is demonstrated as early as the Devonian, and there is evidence that insects pierced plant tissues with their mouth parts and sucked juices in the Devonian and from the Pennsylvanian on (Labandeira and Phillips, 1996; Labandeira, 1998a). There is also evidence indicating possible insect pollination of some Paleozoic pteridosperms. *Medullosa* had extremely large spores, which may have been too large for transportation by wind; hence, these plants may have been insect pollinated (Dilcher, 1979; Taylor and Millay, 1979; Retallack and Dilcher, 1988). Because the male and female structures were borne separately, the pollen itself could not be the reward that encouraged insects to visit the ovules; rather, generalized foraging on tissues of these organs was suggested as the reward. Resinous internal glands on both male and female structures could have produced rewards for insects or could have functioned to discourage herbivory (Retallack and Dilcher, 1988). However, Niklas (1992) points out that pollen size by itself is not conclusive; if the pollen had a low density, it could have been wind dispersed, resembling some wind-dispersed pollen clusters of modern plants that are as large as individual *Medullosa* pollen grains.

Taylor and Archangelsky (1985) suggest that the clustering of cupules in *Ktalenia* and of female structures in early angiosperms implies that they were pollinated through indiscriminate foraging by insects. Perhaps damage from such foraging would be detectable in fossils.

Clustering of insect-pollinated reproductive units may be favored if each individual unit provides a relatively small amount of reward for the insect. Such a cluster, in toto, might offer a substantial reward. The cost to the plant per seed might thus be reduced. Many modern flowering plants do follow such a strategy (Proctor and Yeo, 1972).

Some modern plants also have clustered male and/or clustered female structures and yet are wind pollinated. Most conifers have male cones and female cones with numerous microsporangia or ovules. Many wind-pollinated dicotyledon trees have catkins or other sorts of clusters containing numerous male or female flowers (e.g., *Betula*; Proctor and Yeo, 1972); hence, clustering need not be indicative of insect pollination. Mesozoic pteridosperms tend to have relatively small seeds (Taylor and Archangelsky, 1985), unlike many early pteridosperms which had large seeds (Taylor and Taylor, 1993); perhaps clustering was only a method to increase the number of seeds produced, made possible by the reduced unit cost of smaller seeds.

Many gymnosperms, including some modern conifers and also *Pteroma* and *Pteruchus*, had pollen grains with expanded air-filled sacks, called sacci, formed by separation of the inner and outer pollen grain walls. Each saccus may be nearly as large as the living part of the pollen grain. The sacci help the pollen grain stay airborne longer and may help keep the grain afloat on the surface film of a pollination droplet (Niklas, 1992). Presence of sacci may be evidence for wind pollination of these plants; certainly, the modern gymnosperms with sacci are wind pollinated.

The grouping of *Ktalenia* cupules is weak evidence for insect pollination. I think the presence of sacci and the absence of any obvious reward argue against insect pollination. Even so, an inefficient insect-pollination system could be consistent with the ancestral plant of the Mostly Male scenario.

A frequent conundrum for evolutionary scenarios is explaining the origin of a complex, interrelated set of features that provides obvious utility when fully elaborated, but which comprise individual elements that would each seem useless until all the *other* components of the interrelated set have appeared. It would seem that none of the elements could have arisen before the others; however, it seems impossible for all to arise simultaneously—this is the chicken-egg problem as it appears in evolutionary studies. A famous example of this problem was the putative origin of flight feathers on bird ancestors before the animal could fly. Insect pollination syndromes provide other examples because they involve elaborate specializations by both the insect and the plant. The insect must acquire behavior patterns and structures allowing it to find, collect, and process the reward from the flower. The plant must offer a reward at the proper time, must advertise the reward, must have structures that place pollen on the insect, and must collect the pollen from the insect onto the receptive female structure.

The chicken-egg conundrum is most likely answered through hypotheses of exaptation (Gould and Vrba, 1982) or preadaptation (Bock, 1959; Mayr, 1960), in which some components of the system originally had an alternative function, so that these elements can appear before other components of the system. Thus, a complex system can be separated into distinct components that could have evolved in succession. The original, preadapted function may not be obvious after the full system appears and after all components of the system have become further modified to maximize full-system functionality.

Apparently, insects made the first steps toward pollination systems by adopting plant tissues as food. The very early insects that ate spores or pollen were presumably not pollinators, but the prior existence of such insects could allow an effective pollination system to arise if a pollen-eating insect happened to carry pollen to the female structure (Labandeira, 1998b). Likewise, the earliest insects that sucked surface fluids (from the Triassic; Labandeira, 1998a) were presumably not pollinators. There may have been a variety of sources of nutritious liquids on various surfaces. It is also possible that they could have drunk pollination droplets without performing activities of benefit to the plant. Note that, among present-day wind-pollinated gymnosperms, exposed pollination droplets are found in only a few plants (e.g., *Ginkgo* and *Taxus*) and only for short periods of time; therefore, they do not constitute a useful resource for insects. In vegetation with abundant, diverse pteridosperms and other gymnosperms, exposed pollination droplets may have been easily available for a considerable period of time from a succession of taxa. Such an available resource may well have been exploited by insects. The sequestering of pollination droplets inside the cones of conifers and cycads may have arisen to deny insects access to the pollination droplets. Before this, insects might have used pollination droplets as a convenient water source if other liquid water was unavailable, due to a temporary dry spell, or simply to avoid danger from predators near other water sources.

Although pollination droplets may not have been expensive for the plant to produce, drinking by insects may have interfered with effective pollination, creating an advantage to hiding pollination droplets in cones. Other plants may have evolved mechanisms to compensate for insects' drinking, for example, by regulating the amount of liquid present in the pollination droplet and secreting more liquid to replace what is removed by insects. This ability could have been derived from mechanisms that replenished pollination-droplet liquid lost through evaporation. The pollen may have acquired surface features to make it stay on the droplet, rather than become transferred to an insect's mouthparts as the insect drank from a droplet. The plant may have acquired a signaling system involving the pollen, so the droplet could be absorbed, bringing the pollen into the micropyle as soon as pollen was present in the droplet. These mechanisms would be preadaptive features for insect pollination systems that use pollination droplet as the reward.

Although redating of the Yixian formation (Swisher et al., 1999; Barrett, 2000) moves the first occurrence of *Brachycera* flies (Ren, 1998) into the Cretaceous, there may have been other potential liquid-drinking pollinators as early as the Jurassic (Labandeira, 1997, 1998b). If they were pollinators, they could have been pollinating early Gnetaleans or Bennettitales (Labandeira, 1998b). Whether they were pollinators or not, liquid-drinking (and pollen-eating) insects would have been contemporaneous with the angiosperm ancestors (Labandeira and Sepkowski, 1993; Labandeira 1998b). The Mostly Male theory requires that insects reach this grade of evolution first, before stage one of the scenario.

ELABORATION OF THE SCENARIO OF THE MOSTLY MALE THEORY

Terminology for the Mostly Male Theory

The terminology related to plant reproductive structures can be confusing, as structures bearing one name in gymnosperms appear to have transformed into structures bearing a different name in flowering plants. The list below explains usage for the discussion of the scenario.

"Ovule" is the structure that develops into the seed in gymnosperms and angiosperms (Fig. 3.4). Gymnosperm ovules consist of the nucellus (homologous to a megasporangium), which makes the megaspore and megagametophyte. The nucellus is surrounded by one covering layer, the integument. An opening (the micropyle) in the integument allows access to the nucellus for the pollen grain. At the furthest point from the micropyle, a stalk attaches to the integument and to the nucellus. This relative position of stalk and micropyle is termed orthotropous. Gymnosperm ovules are exposed to the air at pollination.

The ovule secretes a drop of liquid that catches pollen grains; when the liquid is absorbed, the pollen grains enter the micropyle and are brought to the nucellus.

Ovules of basal angiosperms consist of the nucellus (within which form the megaspore and the megagametophyte) covered by two layers, the inner and outer integuments. The micropyle is an opening that extends through both integuments, allowing access to the nucellus for the pollen tube. The stalk is usually attached to the integuments quite close to the micropyle; depending on details of these structures, this arrangement is called anatropous (or amphitropous). Pollen grains land on the stigma (a part of the carpel) and form pollen tubes that grow into the micropyles of the ovules hidden inside the carpel.

The nucellus, stalk, and integument of the gymnosperm ovule are usually considered homologous to the nucellus, stalk, and inner integument of the angiosperm ovule. The outer integument is not considered homologous to any structure in the gymnosperm ovule; hence, the term "ovule" is used in subtly different ways for the two groups. In describing the Mostly Male scenario, I refer to the structure that was moved ectopically as a "cupule" until quite late in the scenario, by which time the term "ovule" will not be ambiguous. Where necessary, I will refer to a "gymnosperm-type ovule" or an "angiosperm-type ovule." Although cumbersome, these usages should be clear.

"Pollination droplet" refers to any liquid secreted from the micropyle by a gymnosperm-type ovule (which is inside the cupule), even if the liquid is no longer used to catch the pollen grains and bring them into the micropyle.

"Cupule" is used in the same way as traditionally used for pteridosperms, even though cupules might not be homologous across all pteridosperms. I will call the structure a cupule as long as the gymnosperm-type ovule inside makes a pollination droplet. I will call it an (angiosperm-type) ovule when the pollination droplet is no longer made, whether or not the angiosperm-type ovule is fully encased in a carpel. For cupules borne in their original position (i.e., not ectopically), I will refer to these as "pure-female cupules."

"Microsporophyll" can be confusing, as I propose that some microsporophylls acquired ectopic ovules and lost their microsporangia, converting the appendage from male to female. I will refer to these as "microsporophylls" as long as they make microsporangia, whether or not they also bear ectopic ovules. I will call the structures bearing ectopic cupules (or ovules) **"precarpels"** after the microsporangia are lost and use "pre-flower" for all of the microsporophylls and precarpels borne in close association on the same stem.

Stages in the Mostly Male Scenario

It is convenient for purposes of discussion to divide the Mostly Male scenario into a series of stages. These stages do *not* imply periodic evolutionary innovation with stasis in between. Some changes may have occurred in a different order. Figure 3.1 illustrates these stages; however, in a drawing, one cannot avoid explicit illustration of structural details that go well beyond requirements of the theory. The drawing is intended to be suggestive, not literal. Individual panels of Fig. 3.1 correspond to the stages, with the ancestral plants as panel 0. In the discussion below, some paragraphs are given numbers, for ease of reference.

Ancestral Plant (Stage 0). The ancestral plant is based on the most recent male and female corystosperm fossils, i.e., on the *Ktalenia* cupulate structure and the *Pteroma* male unit.

Stage 1. This is the initial occurrence of ectopic cupules on (at least some) microsporophylls.

Stage 2. This includes a series of smaller steps in which the male unit and the ectopic cupules become modified to improve their effectiveness.

Stage 3. At this stage, the ectopic cupules become important as seed producers, regardless of whether they had been fertile and capable of making seeds earlier.

Stage 4. At this stage, the separately borne, pure-female cupules are lost.

Stage 5. At this stage, the bisexual reproductive unit acquires additional flower features, such as the carpels becoming largely closed, if these features had not appeared before. The structure is now properly termed a flower.

Discussion of the Stages

Ancestral Plant (Stage 0). The *Pteroma* male structure and *Ktalenia* cupules are the starting points for this discussion of the Mostly Male theory (Fig. 3.1-0).

The Corystosperm structures have been described above. Corystosperms were (most likely) wind pollinated although an inefficient system of insect pollination, based on indiscriminant foraging, could (also) have been present.

Stage 1. Ectopic cupules appeared on the adaxial surface of some microsporophylls of a *Pteroma*-like reproductive unit (Fig. 3.1-1).

That ovule precursors might arise ectopically is supported by overexpression of the gene *FPB11* in *Petunia*, which causes ectopic ovules to appear on sepals and petals. The occasional presence of ectopic ovules on leaves of modern *Ginkgo* provides even stronger support. As in *Petunia*, the appearance of ectopic cupules might have been due to a single gene change. For ectopic structures to become evolutionarily significant, they must confer an *instant* selective advantage. The Gnetales demonstrate that (sterile) ovules can appear on male structures and can be involved in insect pollination. We proposed insect pollination as the selective advantage (Frohlich and Parker, 2000); however, could this have operated instantly, without any further evolution by the plant or the insect?

For an instant selective advantage, no other changes on the plant can be required. Cupules borne on the adaxial surface of the microsporophyll should not interfere with pollen release from the openings on the abaxial side. Cupules would have their usual shape, placing the micropyle close to the microsporophyll surface. Other cupules would continue to be borne in purely female units, separate from the male units.

By this time, insects that routinely drank from pollination droplets, including droplets on the pure-female units, existed. These insects should have been attracted to the pollination droplets of the ecotopic cupules on the males.

Pollen could have been deposited on these insects from the air, for example, if shaken out of the microsporangia by insects crawling on the male structures and/or picked up from surfaces on which the insect walked. Pollen loading could have been indirect, with grains first accumulating in pollination droplets on the male and the floating onto the insect's mouthparts as it drank. In *Welwitschia*, a modern insect-pollinated Gnetalean that uses a pollination droplet reward, the pollination droplet on the male does become filled with pollen (Wetschnig and Depisch, 1999).

Subsequently, when insects visited the female, pollen could have fallen into the pollination droplet or floated off of the mouthparts, if originally acquired from a pollination droplet on the male.

Even a slight increase in pollination efficiency could have provided enough selective advantage to retain genes that caused the ectopic ovules. Increased fitness may have applied both to the pollen and to the pure-female cupules of the same plant. Ectopic ovules would have increased the success of pollen in reaching pure-female cupules on any nearby plant. The chance of self-pollination of pure-female cupules of the same plant may also have increased, as its own cupules would likely have been the closest available for insects to visit. This might have been especially significant in small or dispersed populations, in which there was not enough pollen rain from the air to fertilize a large fraction of the cupules.

Seed production by the ectopic cupules would not have been necessary for them to confer a selective advantage. Even if they were fertile, in the sense of possessing all the structures of the pure-female cupules, they might have produced few seeds for various reasons. Pollination might have been inefficient due to the close proximity of the pollination droplet and the microsporophyll. The cupule might have had insufficient vascular connection to allow growth of the seed after pollination. The male structure may not have been strong enough mechanically to support ectopic cupules with developing seed. There could have been problems in dispersing the cupule after seed was produced.

An instant selective advantage could have preserved the first plants with ectopic ovules.

Stage 2. Plants with ectopic cupules underwent a series of changes, due to selection, that increased the efficiency of insect pollination (Fig. 3.1-2). These are detailed in paragraphs 1–7 below.

Selective pressure likely caused the stage 1 plant to undergo some rapid changes; many of its features would likely have been far from optimal for insect pollination, and some of these features may have been capable of easy improvement by selection. Some of the new features might be observable in fossils and/or might contribute to further changes along the path to the flower.

(1) Pollination droplets may have increased in size, to attract insects more effectively (as illustrated in the modern *Welwitschia*, which has very large pollination droplets in the male, borne on an expanded tip of the sterile ovule). With pure-female corystosperm cupules, this might be difficult because droplets might fall off the tip of the cupule. On the male, the recurved shape of the stalk and cupule should have positioned the droplet close to the microsporophyll. An enlarged droplet might be deposited on the surface of the microsporophyll, constituting "secondary pollination drop presentation," named by analogy to secondary pollen presentation. Pollen might have collected in this liquid on the microsporophyll, to be picked up on the mouthparts of insects that drank from the microsporophyll surface.

(2) The microsporophyll shape might have changed to hold a larger amount of liquid. Instead of being nearly flat, the microsporophyll may have become concave on its upper surface and subsequently might have become ascidiate (bucket shaped) to hold liquid. Perhaps the ascidiate shape of basal flowering plant carpels is a relic of this stage, originally serving to hold pollination droplet liquid. A landing platform for the insects may also have appeared.

(3) Cup-shaped microsporophylls might have interfered with dehiscence of the embedded microsporangia, which would have become curved. Microsporophylls might have become specialized, to bear either ectopic ovules or microsporangia. At this point, the structures bearing ectopic cupules may be termed "precarpels" because they no longer produce microspores.

(4) A balance must be maintained between pollination droplets available at the male and the female units. If far more were available at the male, then insects would not be tempted to visit the female. Substantially increased clustering of the pure-female cupules could have increased the volume of pollination droplets available, without the need to increase the size of droplets. This could have been a hallmark of the advent of insect pollination.

(5) Increased pollination efficiency might have been achieved by signaling, by means of color or odor, that the reward was available. The morphology of the male and/or pure female structures may have become modified to facilitate visits by insects, and such structural modifications might be observable in fossils. For example, many modern flowers have cells on the petals bearing closely spaced cuticular ridges that act as structural nonreflective surfaces, allowing light to enter the cells and be absorbed by pigments in the cell, such as flavonoids that absorb ultraviolet light. Such cuticular ridges might be preserved in fossils.

(6) Even if ectopic cupules were potentially fertile at first, they might have become sterile, as in Gnetales, yielding lineages that could not evolve flowers

but nevertheless could have persisted for a long time and thus might be found as fossils.

(7) Shortening of the flower axis and acquisition of morphological determinacy may have occurred at this stage (Frohlich and Parker, 2000).

Some of the features described for stage 2 would be recognizable in the fossil record, and the plants that bore them might have survived long enough to have left fossils.

Stage 3. At this stage, the ectopic cupules became important as seed producers, regardless of whether they had previously made significant numbers of seeds (Fig. 3.1-3). The features of stage 3 are detailed in points 1–4 below.

(1) Reproductive success is the driving force of selection; therefore, increased seed production from fertile ectopic ovules has obvious selective advantage. Because of the proximity of the fertile microsporophylls, most ectopic ovules would presumably have been self-pollinated, baring some special mechanism to prevent this. This would guarantee large seed production, even if pollinators were absent. This would have been of especial value for r-selected plants. In the southern hemisphere, at least, there seems to have been abundant sauropod dinosaurs in the fauna well into the Cretaceous, which browsed on low vegetation (Stevens and Parrish, 1999; Upchurch, 2000). Like elephants in modern Africa, they might have done substantial damage to the vegetation, perhaps creating habitat for small r-selected trees or herbs (Wing and Tiffney, 1987).

There is evidence that early flowering plants were r-selected. Paleoecological data, based on analyses of sedimentary environments, indicate that Cretaceous angiosperms lived in "unstable habitats;" furthermore, morphology of their leaves, seeds, and wood is consistent with angiosperms being early successional plants (Wing and Boucher, 1998). Modern weedy plants are noted for having low genome sizes. Analyses of the basal-most portions of the angiosperm cladogram indicate that plants on the major branches near the base of the cladogram also had small genome sizes, which is consistent with these plants being r-selected and weedy (Leitch et al., 1998).

How might have the ectopic cupules become major seed producers without losing the ability to attract insects by producing large amounts of pollination droplet? Pollen grains could not be brought into the micropyle while large amounts of pollination droplet were being secreted onto the precarpel, but the cupules might have been fertilized either before or after the bulk of the pollen was released from the adjacent microsporangia. At the start of pollen release, the cupules might have produced small pollination droplets that could be resorbed, achieving pollination. There may have been a mechanism resembling the pollen "scavenging" of some conifers, in which the pollination droplet captures grains that have already landed on adjacent surfaces (Tomlinson et al., 1991; Tomlinson, 1994; Runions and Owens, 1996; Tomlinson et al., 1997). This could approach a system of protogyny, as the female structures would effectively be receptive before the pollen is released from the associated microsporangia. Protogyny is very common among basal angiosperms (Bernhardt and Thien, 1987; Bertin and Newman, 1993); its origins might extend this far back, before the flower was fully elaborated.

The modern plant *Welwitschia* seems to continue producing droplets even after ovules are pollinated; Wetschnig and Depisch (1999) note that *Welwitschia* pollination droplets on the female are resorbed each evening and new droplets are produced each day for many days. Presumably, some of the ovules are pollinated by one of the first droplets produced, yet Wetschnig and Depisch (1999) imply that all ovules continue to make droplets for many days.

Alternatively, the cupule may have been pollinated after most pollen is released from adjacent microsporophylls, in a system resembling protandry.

(2) The dual requirements for pollination and for voluminous pollination droplet secretion could have selected in favor of pollen germination on the precarpel, with pollen tubes then growing into the cupule. Such an ancient system may be echoed in the modern angiosperm carpel; the stigma is a specialized epidermis, and pollen tubes grow through the transmitting tract of the carpel, which is also specialized epidermis. Pollen germination on the precarpel may have necessitated a chemical attractant to guide the pollen tube to the cupule; such attractants may have arisen at this time and be retained in modern plants as the secretions of the synergids, which attract pollen tubes to the angiosperm ovule. This resembles an evolutionary stage suggested by Lloyd and Wells (1992), who also proposed that pollen grains would have germinated on the adaxial side of an open carpel, before the evolution of carpel closure, in the lineage leading to angiosperms.

Pollen from modern angiosperms of many species will germinate in vitro on sugar water, and, in some modern plants, pollen tubes regularly grow through tissue other than the transmitting tract to reach the ovules (Anderson, 1980; Williams et al., 1993). In some modern gymnosperms, pollen germinates outside the ovule, on cone scale surfaces, and the pollen tubes grow into the micropyle and fertilizes the egg (Tomlinson, 1994). If the report by Zenkteler (2000) is correct, that pollen grains of various conifers will germinate when placed on the placentae of various angiosperms and that the gymnosperm pollen tubes will sometimes even grow into the angiosperm ovule and even fertilize the egg, then the evolutionary innovations discussed above may not have been difficult to achieve.

(3) If abundant seed is produced by the ectopic cupules, would the separately borne pure-female cupules be retained for very long? The ectopic cupules would have been mostly or entirely self-pollinated. Even a moderate amount of cross-pollination and genetic recombination could have been very beneficial. The seed produced from pure-female cupules should have had substantially higher rates of outcrossing and therefore be of great value to the plant. Today, there are many angiosperms that have two kinds of flowers—cleistogamous and chasmogamous—specialized for self-pollination and cross-pollination, respectively. If such double systems are evolutionarily stable today, then plants at stage 3 might have produced seed from both ectopic cupules and pure-female cupules for a long time.

Angiosperms show so many diverse mechanisms to ensure cross-pollination that this must be of great value for many plants. At least some of plants at stage 3 would likely have been under selection for increased outcrossing (cross-pollination) of the ectopic cupules. Any effective outcrossing mechanism would suffice to allow the events at the next stage in the scenario; hence, the theory does not depend on the acquisition of any particular outcrossing mechanism.

The scenario for the Mostly Male theory suggests a possible intermediate step toward one such system, that of gametophytic self-incompatibility. If this proves reasonable, it would offer some support for the Mostly Male theory, although refuting it would not oppose the theory. Gametophytic self-incompatibility has been suggested as the advantage that allowed early angiosperms to dominate the world vegetation (Whitehouse, 1950; Zavada and Taylor, 1986; Dilcher, 1995). Although the presence of gametophytic self-incompatibility at the base of the extant angiosperms was rejected by Weller et al. (1995), more recent views of angiosperm relationships suggest that it might have been present there (Mathews and Donoghue, 1999).

In gametophytic self-incompatibility systems of modern angiosperms, pollen germinates on the stigma and pollen tubes grow into the transmitting tract, but self-pollen tubes are killed (or stop growing) before they approach the ovary. The response depends on interaction between genes expressed in the pollen tubes and genes active in the carpel. This often involves a single locus, called the S locus in all self-incompatible plants, which encodes a number of proteins (McCubbin and Kao, 1999). Gametophytic self-incompatibility has been best studied in the Solanaceae, where a special RNase enzyme, expressed in the carpel, is transported into the pollen tubes and kills the developing pollen tubes of the self-pollen grains (Hiscock and Kües, 1999). Pollen tubes that contain an allele of the S locus different from those in the pistil are spared, allowing them to fertilize the ovules. The mechanism involved in gametophytic self-incompatibility has also been studied in poppies (*Papaver rhoeas*), a near-basal group within flowering plants. In poppy, the protein that stops pollen tube growth is *not* an RNase but acts through a signal transduction mechanism to induce programmed cell death in the self-pollen tube (Jordan et al., 2000). Some information is available on grasses, which have a system in which two unlinked genes must both be shared between the pollen and the stigma for rejection to occur. The molecular mechanism in grasses is not known (Baumann et al., 2000). *Illicium*, of the third-most basal angiosperm clade, also shows gametophytic self-incompatibility, but the mechanism is not known (Thien et al., 1983). Self-incompatibility systems are often lost in evolution; therefore, absence in some basal lineages would not disprove the possible presence of such a system at the base of flowering plants.

The population genetic factors required for selection to favor the S locus have received careful study (Steinbachs and Holsinger, 1999; Uyenoyama, 2000), but these analyses assume that an effective S locus has already evolved. The origin of such a system involves a chicken-egg conundrum because it is unclear how the individual genes of the S locus could provide any selective advantage until they had all been assembled into a fully functional S locus that could achieve self-incompatibility.

(4) If the angiosperm ancestor at stage 3 had a large quantity of pollen germinating on the precarpel, then a tangle of pollen tubes might result, interfering with drinking by insects. This might be prevented by a mechanism, turned on well after the protogynous cupules are pollinated, that would stop pollen tube growth or even kill pollen tubes. Most pollen on the precarpel would presumably originate from the same plant; therefore, this mechanism need not involve a generalized toxin but could involve specific interactions between the pollen tube and the pollination droplet. Because the mechanism would not become active until *after* cupule pollination, there is no loss of fitness to the pollen tubes, as they would have had *no* chance to pollinate ovules. Rather, by responding to the growth inhibitor, they facilitate pollinator access to liquid from the pollination droplet, which increases the fitness of other pollen grains from the same plant, grains that might be carried to other preflowers, where they might effect pollination. Thus, there is no need to invoke any benefit of outcrossing to realize a selective advantage if pollen responds to a growth inhibitor released by its parent plant. There are numerous mechanisms of cell-cell communication, any of which might be co-opted to stop pollen tube growth and prevent the tangle of pollen tubes.

Once a modicum of cross-pollination has been achieved, these systems become subject to improvement by selection. The system mentioned above, to prevent tangles of pollen tubes in the pollination droplet, might be a preadaptive stage that could be modified toward a gametophytic self-incompatibility system.

The mechanism proposed above, to prevent pollen tube tangles in the pollination droplet, offers a function for genes resembling the S locus that is much simpler than the prevention of self-pollination. This system would confer an advantage even when partially functional and even if a minority of individual plants have it. Relatively little pollen would arrive at a preflower from other plants; thus, prevention of tangles would not require inhibition of pollen tubes from other plants. Thus, a series of S alleles (that only affect pollen containing the same S allele) could arise even before their primary function becomes enforcement of outcrossing. Further modifications to enforce outcrossing might require nothing more than expression of the pollen tube growth inhibitor as soon as cupules are receptive. If significant amounts of pollen arrive from other preflowers on the same plant, then such early expression of the growth inhibitor might have the benefit of preventing tangles. This might then constitute a fully functional gametophytic self-incompatibility system.

Stage 4. At this stage, the separately borne cupule and the entire pure-female structure are lost (Fig. 3.1-4). Details of stage 4 are presented in points 1 and 2 below.

The last major function of the ancient, separately borne pure-female cupules was the production of outcrossed seed, but this role may have been fulfilled even more effectively by the preflower alone, as selection continued to improve its efficiency. Some of these changes may have had secondary effects that made the pure-female cupules *less* able to produce seed. For example, modifications that make the preflower more attractive to insects, whether by increased reward or simply improved advertising of the reward, would divert insect visitors away from pure-female cupules, unless the pure-female structures are also altered to become more attractive. If a relatively small proportion of seed is produced by the pure-female cupules, then selection would not act as strongly to improve their attractiveness to insects. Furthermore, the simpler structure of the pure-female cupules, compared with the whole preflower, may have made such changes difficult or impossible for evolution to achieve.

(1) Pollen and pollen tubes may have become increasingly well adapted to interact with the precarpel; these adaptations may have made them less effective in pollinating/fertilizing pure-female cupules. For example, the wings (sacci) on the pollen grains may have been lost by this time; without wings, the pollen may have sunk into liquid rather than floated on the surface (Tomlinson, 1994; Tomlinson et al., 1997).

(2) Pollen could even have become an effective reward to attract pollinators, after the preflower became able to make all the seeds needed by the plant. If pollen-eating (or spore-eating) insects began to use the preflowers as a food source and if they happened to transport enough pollen to lead to a significant amount of pollination, then selection may have begun to modify the preflower to help attract pollen-eating insects (Labandeira, 1998b). Eventually, this could lead to a complete switch in the reward offered to insects, and insects might specialize to eat this pollen. If the pure-female structures had previously acquired advertising features to help attract insects, these features would then become a detriment to the plant, as they would constitute false advertising to insects that seek pollen. This could result in selective force to abolish the pure-female structure.

Any one or a combination of such mechanisms could lead to a loss of the pure-female cupules.

Stage 5. The preflower acquired additional flower features and may now be called a true flower (Fig. 3.1-5).

The preflower would need only relatively few modifications to merit being called a flower. Some of these changes could have already occurred before or after this stage, such as modification (and sterilization) of the outer microsporophylls to provide advertising to attract insects; this could be the origin of petals or tepals. The advertising function of tepals may have preceded or accompanied their protective function (Albert et al., 1998).

The demonstration by Igersheim and Endress (1997, 1998) and Endress and Igersheim (1997) that most lower dicots do not have fully closed carpels removes that feature from the necessary attributes of the flower. All of the basal angiosperms have ovules that are largely enclosed, generally in ascidiate-shaped carpels, with any gap at the carpel top filled by a gelatinous secretion; thus, functionally, the carpel is closed. I will call this change the "sequestering" of the angiosperm-type ovules (or cupules), whether accomplished by secretion and/or by morphological carpel closure.

A prerequisite for sequestration of cupules is loss of the cupule's pollination droplet as the reward for pollinators. This could have been due to a switch to pollen as the reward, as discussed above. Alternatively, the production of liquid to reward pollinators could have moved from the cupules on the precarpels to structures at some other location. These other structures could have been cupules specialized to produce nectar (as the liquid may now be called). Alternatively, genes responsible for secretion might have been expressed in a morphologically unrelated structure, creating a nectary and/or a stigma with abundant secretion (Lloyd and Wells, 1992). These could represent additional case(s) of heterotopy, although they are not a necessary part of the Mostly Male theory. Nectaries of angiosperms are morphologically very diverse and are found in differing positions on flowers and on vegetative portions of various plants. Clearly, the morphological elements of nectaries are not homologous across the angiosperms. It is not known whether the mechanisms that actually carry out the secretion are homologous.

The stigma itself produces liquid secretions in some near basal angiosperms; Lloyd and Wells (1992) suggest that stigma exudates may have been the pollinator reward when carpels were becoming (functionally) closed. They propose that the secretion mechanism that produced the pollination droplet moved from the ovule to the surface of the precarpel and then to the region of the carpel that became the stigma.

Most basal angiosperms use pollen as a reward for pollinators, but it is uncertain whether this represents the basal condition of the flowering plant crown group or only the remnants of taxa that survived competition from eudicots and monocots. Liquid secretions are produced on flowers by members of the second- and the third-most basal clades of flowering plants, that is, by *Cabomba* in the Nymphaeales (Schneider and Jeter, 1983; Vogel, 1998) and by *Illicium* of the Illiciaceae (Thien et al., 1983). In neither case does the secretion come from a modified cupule, and structures of the two are morphologically very different from each other and from the nectaries of higher angiosperms.

Various selective advantages have been proposed that could have favored cupule/angiosperm-type ovule sequestration; most could be consistent with the Mostly Male theory. These include protection of the ovules from insect herbivores (especially beetles) (Grant, 1950), protection from drought (Axelrod, 1970), and requirements for efficient gametophytic self-incompatibility, which might be more effective if pollen tubes were forced to grow over a long stretch of carpel epidermis (Whitehouse, 1950; Zavada and Taylor, 1986; Dilcher, 1995). Additional possibilities include specific adaptations for fruit dispersal or even size and shape changes of the carpel to facilitate interactions with insects. Any of these would be consistent with the Mostly Male theory because the unique features of the theory pertain to earlier stages of the evolutionary scenario.

The gelatinous material that seals carpels of many lower angiosperms has not yet been studied in any detail to determine its functions or properties. If its sole function is to seal the carpel, then it might have been favored by any of the factors used to explain carpel closure.

At this point in the scenario, flowers have evolved. The first plant with "flowers" would have been on the stem lineage[5] well below the crown group that includes all living angiosperms. This first flowering plant may have been quite distant in time and in morphology from the base of the angiosperm crown group. There is no basis to judge the relative time at which flowers and the other features peculiar to angiosperms arose. Angiosperms appear highly distinctive precisely because we do not have fossils of the intermediates leading back to their gymnosperm ancestors.

TESTS OF THE MOSTLY MALE THEORY

The Mostly Male theory can be tested by two general classes of evidence: fossils and genes. A separate though related question is the relationship of corystosperms to angiosperms; that question can be tested only by fossils.

Fossils

The most direct evidence bearing on evolution comes from fossils. The scenario described above identifies intermediate stages that could be favored by selection and so might persist long enough to leave fossils. Discovery of fossils that show these intermediate features would strongly support the theory. Plants of stage 1 would likely exist only briefly; therefore, they should not be expected in the fossil record. Plants of stages 2–4 could have persisted for long periods and may have left fossils. Plants of stage 5 would be recognized as flowering plants but may not show much evidence bearing on the origin of flowers. Some of these fossil intermediates might be recognizable as compressions, but recognizing others would require exquisite permineralized specimens.

Attributes of the predicted intermediates are listed below:

- Cupules borne on microsporophylls (stage 2, point 1)
- Cupules borne on flat, cup-shaped shaped structures perhaps with an adjacent landing platform (stage 2, point 2)
- A great increase in clustering of the pure-female cupules (stage 2, point 4)
- Structures for attracting insects on cupules or on microsporophylls, such as closely spaced cuticular ridges of epidermal cells (stage 2, point 5)

[5] Below the deepest node in the cladogram straddled by the living angiosperms.

- Microsporophylls bearing reduced cupules, which might be interpreted as glands (stage 2, point 6)
- Flat or cup-shaped structures bearing seeds (stage 3, point 1)
- Flat or cup-shaped structures bearing ovules and covered with pollen and pollen tubes (stage 3, point 2)
- Plants that produce both flower-like structures and separately borne pure-female cupules (stage 3, point 3)
- Flat structures bearing ovules with numerous pollen grains producing only short, aborted pollen tubes (stage 3, point 4)
- Loss of sacci on pollen grains (stage 4, point 1)
- Specialized insects that eat pollen of only one identifiable plant, to the exclusion of other spores/pollen, even though others are available (stage 4, point 2)

Genes and Gene Function

Tests based on gene function and phylogeny do not depend on the vagaries of fossil discovery; therefore, they may provide the most immediately applicable tests. These tests are inherently indirect because gene interactions and gene functions can change in unexpected ways. Some genes may well have experienced unexpected shifts in function. These may appear inconsistent with the majority of genes studied. Inferences based on multiple genes are the most likely to result in historically correct interpretations, although the difficulty of such studies forces an individual researcher to study of one gene or gene family at a time.

Tests based on gene evidence derive from the fundamental assertions of the Mostly Male theory, that the overall organization of the flower derives mostly from the male reproductive unit of the gymnosperm ancestor and, in particular, that the carpel wall derives from a microsporophyll. Depending on the organizational level at which individual genes function, the first, or the second, or both of these assertions could be tested.

In gymnosperms, male and female reproductive structures were borne separately from the origin of seed plants in the Late Devonian, and megasporangia and microsporangia were borne separately even earlier in the ancestral progymnosperms. It can be assumed that some genes became specialized to function only in the male units or only in the female units before the divergence of the lineages leading to flowering plants and extant gymnosperms. In some cases, the male and female genes may be related as paralogs. The two paralogs of *LFY*, with one highly expressed only in male structures and the other only in female structures, are the first such pair discovered. In other cases, the genes active in one of the reproductive units may have no similarity to genes active in the other. Yet other genes would have been active in both types of reproductive organs.

Gene-based evidence may be divided into four somewhat overlapping categories, based on the gene's level in the genetic control hierarchy with respect to

LFY, because this major transcription factor is already known to have separate male and female paralogues. The categories are genes acting upstream of *LFY*, genes acting in parallel to *LFY*, genes acting below the level of *LFY* that show effects on multiple whorls of the flower, and genes active primarily in the carpel wall. In each case, the Mostly Male theory predicts that, if the related gymnosperm genes show specialization for male versus female reproductive units, then the angiosperm genes should have closer homology (and orthology, if it can be discerned) to genes active in the gymnosperm *male* reproductive units rather than to genes of the female. I will call this the "predicted pattern." An alternative to the predicted pattern is the "opposite pattern," in which flowering plant genes are closer to those active in the gymnosperm *female* structure. Another possibility is the "uninformative pattern," in which either the genes are phylogenetically equally far apart or the gymnosperm genes are most closely related to each other, rather than to a flowering plant gene. Individual gymnosperm genes that are active in *both* male and female structures are also uninformative, but they are excluded from the analysis because pertinent information cannot be obtained from them. To judge support for the Mostly Male theory, one would compare the number of genes showing the predicted pattern with the number showing the opposite pattern. If the former are more numerous and/or more important in the flower (not counting the ovule), then the theory is supported. If not, then the theory is opposed.

Genes upstream of *LFY* include many flowering-time genes that help regulate the level of *LFY* expression. Much active research is currently focused on these complex pathways. In most conifers, male and female structures are found in different (although sometimes overlapping) regions of the plant, and young conifers often begin making male cones before they make female cones. This implies differential regulation of the male and female reproductive units. Furthermore, cycads, *Ginkgo*, and Gnetales are dioecious. If the majority of these upstream genes show the predicted pattern, this will imply that most of the *LFY* pathway used by flowers, and not just the *LFY* pararalog, is derived from the male structure of the gymnosperm ancestor. This would constitute substantial support for the theory.

A number of genes act in parallel with *LFY* in *Arabidopsis* to help specify the flower (Hempel et al., 1997; Theissen and Saedler, 1999; Blázquez and Weigel, 2000). Many of these do interact with *LFY* as part of the circuit that becomes self-regulating when a threshold is passed, resulting in specification of the apex as floral. These genes also receive input from upstream sources, independent of *LFY* and, even in strong *lfy* mutants, can stimulate downstream cascades, generating deformed flowers. These genes would be especially interesting to examine; those that are the least tightly coupled to *LFY* would be the most significant. If these show the predicted pattern, this would be strong evidence, more or less independent of *LFY*, that flowers derive from the male structures of the gymnosperm ancestors and that female-specific upstream pathways of the gymnosperm ancestor did not become incorporated into the control systems of flowers.

Genes that act downstream of *LFY* and affect multiple whorls of the flower would also be expected to show the predicted pattern. Those genes that primarily affect stamens and petals would carry the least weight because one might claim that they are absolutely required to generate microsporangia; thus, the plant would have to evolve a method to express them no matter how the flower evolved. Still, such findings would add some support to the theory. An example is the work by Sundstrom et al. (1999), showing that the genes *DAL11*, *DAL12*, and *DAL13*, all related to the B-class gene *PI* of *Arabidopsis*, are expressed in the male but not in the female structures of spruce. If genes should be discovered that are active in the stamen and that follow the opposite pattern, rather than the predicted pattern, this would be very strong evidence against the theory because it would have been counterintuitive.

Genes affecting the carpel wall (but not directly the ovule) have the potential to provide the most convincing support for the theory. If the carpel is derived from the microsporophyll, then genes inherited from the common ancestor of extant seed plants should follow the predicted pattern rather than the opposite pattern. If the carpel (excluding the ovules) evolved from a structure of the ancestral female reproductive unit, as other theories assume, then the opposite pattern should be followed. The counterintuitiveness of the predicted pattern for the carpel wall makes this the best gene-based test of the Mostly Male theory.

The ideal gene should be clearly tied to the carpel wall and clearly distinct from genes active in the ovule (and of lesser or no importance in the stamen). To maximize the chance that gymnosperms will be found to use different genes for the male and the female reproductive structures and that the relationships of the genes can be determined, the ideal gene should be part of a medium to small size gene family. An example of an excellent candidate is the *CRABS CLAW* gene of *Arabidopsis* (Bowman and Smyth, 1999; Siegfried et al., 1999; Sessions and Yanofsky, 1999). It is part of the small "YABBY" gene family, whose members act to specify abaxial cell fate in all plant dorsiventral structures (i.e., features on the side of lateral organs *away* from the stem tip). *CRABS CLAW* acts exclusively (as far as known) in the carpel wall and in the nectary. Mutants suffer deformed carpels due to the lack of this activity, and they also lack nectaries. Although it is expressed at low levels in the stamen, *CRABS CLAW* does not appear to have an important role there, as there are no obvious defects in mutants, although its activity there could be redundant with other YABBY genes. Whereas the filament of higher plant stamens appears greatly simplified, so YABBY genes may not be important in its structure, this would not have been the case in the ancestral microsporophyll postulated in the Mostly Male theory. Hence, there would presumably have been a YABBY gene active in the microsporophyll that could have been conserved in the carpel wall. A different member of the YABBY family, the *INO* gene, is active in the outer layers of the outer integument of the ovule (Baker et al., 1997; Villanneva et al., 1999). The Mostly Male theory would predict that *CRABS CLAW* orthologs would be active in microsporophylls of gymnosperms, whereas *INO* gene orthologs would be active in the female structures. Nectaries in *Arabidopsis* are associated with the stamen. *CRABS CLAW*'s activity in the nectary could reflect a former role for *CRABS CLAW* in the stamen or could be a remnant of pollination droplet production from tissue on the precarpel (Lloyd and Wells, 1992).

The phylogenetic age of *CRABS CLAW* and the other YABBY gene paralogs is not yet known, and its function has not been elucidated in other angiosperms. If *CRABS CLAW* and *INO* do not have distinct homologues (and hopefully clear orthologs) in gymnosperms, then they might not be useful for testing the theory.

Other dorsiventrality genes could become equally promising, as could genes involving the transmission tract or other components of the carpel wall.

Organisms sometimes have duplicated genes that do not seem different from other paralogs in the capabilities of their proteins but that are expressed in limited regions of the organism. These may be examples of duplicated genes that are retained by the mechanism proposed by Force et al. (1999) and Lynch and Force (2000), in which balanced loss of expression domain forces retention of multiple paralogs; except for expression domain, they may not have functional specialization. These can include simple "housekeeping" genes that every cell must express. If genes with restricted expression domains are found that have distinct paralogs expressed in gymnosperm male and female organs, then they would also be excellent candidates to test the Mostly Male theory. These genes serve as markers for the gene cascades that turn them on.

Of the genes that might be used to test the Mostly Male theory, it is not yet known which might have distinct paralogs active only in gymnosperm male vs. female reproductive structures, as is necessary to draw inferences from them. The large number of genes that are potentially useful for testing the theory suggests that some should have appropriate paralogs to make the predicted pattern clearly distinguishable from the opposite pattern.

A complicating factor for the interpretation of gene evidence is the phenomenon of transference of function, which has occurred very commonly in plant evolution (Corner, 1958). This occurs when one structure acquires a function previously performed by another structure. Often, some of the attributes of the old structure become expressed in the structure that acquires the new function. An especially frequent example is the role of attracting pollinators by structures that are brightly colored, like petals. These structures can be sepals, bracts, or even leaves or stems. In some cases, this may be accomplished by the expression of genes in the sepal, bract, leaf, or stem that previously were expressed only in the petals. It is a semantic argument whether this makes these structures "partly homologous" to petals (Sattler and Rutishauser 1997), but such changes in gene expression could confound attempts to use gene function to reveal evolutionary history.

One may invoke transference of function in the evolution of flowers. At stage 2, both male and female structures may begin advertising their presence to help attract insects. This could have involved color and odor, as in modern flowers, but on the pure-female cupules the only large structure that could have expressed these features was the outer surface of the cupule. In modern plants, the acquisition of these features is postulated to result from expression of the B class (and A class) flower homeotic genes, which help specify petals (Albert et al., 1998). Perhaps ancient cupules also expressed such genes for the same purpose, and the expression of AP3 (a B class gene) in the *Arabidopsis* ovule outer integument may be a vestige of this. There is no known function for AP3 in the outer integument because the ovule shows no defects in *ap3* mutants.

Cupules of some pteridosperms, such as *Caytonia*, are thought to have been fleshy, perhaps allowing animal dispersal of seeds. Other cupules, such as *Ktalenia*, had hard sclerenchyma coverings. Flowering plant carpels are often fleshy for animal dispersal or hard and dry, often from sclerenchyma. Those attributes of the carpel seem to change frequently in evolution, as fleshy fruits and dry fruits occur scattered among very many families. Perhaps it is simple to evolve these features, or perhaps one or the other of these features, originally expressed in a cupule, was transferred to the carpel wall.

Because genes can change expression pattern and can change function, the most effective tests must be based on studies of many genes so that confounding evidence from those that changed function can be recognized as anomalous. By far the best approach would be a study of genes expressed in developing gymnosperm male and female structures and in the developing flowers of basal angiosperms, using expressed sequence tags (ESTs). A thorough EST study would find all the highly and moderately expressed genes and should identify a large number of genes that distinguish gymnosperm male and female structures. Expression studies based on gene microarrays could verify apparent differences between gymnosperm male and female structures. Study of ESTs from several basal angiosperms, coupled with explicit searches for expected genes that may have been missed in the EST survey, would allow rigorous testing of the genetic predictions of the Mostly Male theory.

Ultimately, such studies should allow detailed reconstruction of the networks of genetic control that specify all the reproductive organs and their cell types. It should then become possible to identify which segments of the flower's control network derive from systems active in the male or the female of the gymnosperm ancestor and to determine where the control has shifted to meld the various elements together. This would allow a far more profound understanding of the evolution of development and could generate theories of flower evolutionary origins far more powerful than the Mostly Male theory.

CONCLUSIONS

The single instance of heterotopy in the Mostly Male theory, coupled with the corystosperms as potential angiosperm ancestors, greatly reduces the remaining morphological gap that must be bridged in the evolution of flowers. This allows a detailed scenario to be proposed for the transition from gymnosperm reproductive structures to those of flowering plants. The features of intermediate stages could have conferred adaptive value; therefore, these intermediates might have existed for considerable periods of time and may have left fossils that could be discovered. Discovery of fossil intermediates leading to flowers would be the most definitive test of the theory. The Mostly Male theory makes predictions regarding the orthology relationships of genes active in gymnosperm male and female structures compared with those active in flower development. Study of these genes would provide a powerful test of the theory and is the test that is the most directly applicable. The study of genes that control development may allow other instances of heterotopy to be recognized. Heterotopy may be the underlying explanation for seemingly inexplicable evolutionary events.

ACKNOWLEDGMENTS

I thank Bruce Tiffney, Peter Wilf, and Brian Axsmith for comments on the manuscript; Y.-L. Qiu for calling to my attention the reports of ectopic *Ginkgo* ovules; Jer-Ming Hu for translating the Ma and Li (1991) paper; Karin Douthit for drawing Fig. 3.1; and Bill Anderson. This work is supported by National Science Foundation Grant DEB-9974374.

REFERENCES

Albert VA, Gustafsson MHG, Di Laurenzio L (1998): Ontogenetic systematics, molecular developmental genetics, and the angiosperm petal. In: Soltis DE, Soltis PS, Doyle JJ (eds), Molecular Systematics of Plants II DNA Sequencing. Boston, MA: Kluwer Academic Publishers, p. 349–374.

Alvarez J, Smyth DR (1999): CRABS CLAW and SPATULA, two *Arabidopsis* genes that control carpel development in parallel with AGAMOUS. Development 126: 2377–2386.

Anderson WR (1980): Cryptic self-fertilization in the Malphigiaceae. Science 207: 892–893.

Arber EAN, Parkin J (1907): On the origin of angiosperms. Bot J Linn Soc (London) 38: 29–80.

Axelrod DI (1970): Mesozoic paleography and early angiosperm history. Bot Rev 36: 277–319.

Axsmith BJ, Taylor EL, Taylor TN, Cuneo NR (2000): New perspectives on the Mesozoic seed fern order Corystospermales based on attached organs from the Triassic of Antarctica. Am J Bot 87: 757–768.

Baker SC, Robinson-Beers K, Villanueva JM, Gaiser JC, Gasser CS (1997): Interactions among genes regulating ovule development in *Arabidopsis thaliana*. Genetics 145: 1109–1124.

Barrett PM (2000): Evolutionary consequences of dating the Yixian Formation. Trends Ecol Evol 15: 99–103.

Baumann U, Juttner J, Bian X, Langridge P (2000): Self-incompatibility in the Grasses. Ann Bot 85, Suppl A: 203–209.

Bernhardt P, Thien LB (1987): Self-isolation and insect pollination in the primitive angiosperms—new evaluations of older hypotheses. Plant Syst Evol 156: 159–176.

Bertin RI, Newman CM (1993): Dichogamy in angiosperms. Bot Rev 59: 112–152.

Blázquez MA, Weigel D (2000): Integration of floral inductive signals in *Arabidopsis*. Nature 404: 889–892.

Bock W (1959): Preadaptation and multiple evolutionary pathways. Evolution 13: 194–211.

Bowe LM, Coat G, dePamphilis CW (2000): Phylogeny of seed plants based on all three genomic compartments: Extant gymnosperms are monophyletic and Gnetales' closest relatives are conifers. Proc Natl Acad Sci USA 97: 4092–4097.

Bowman JL, Smyth DR (1999): CRABS CLAW, a gene that regulates carpel and nectary development in *Arabidopsis*, encodes a novel protein with zinc finger and helix-loop-helix. Development 126: 2387–2396.

Burgers P, Chiappe LM (1999): The wing of *Archaeopteryx* as a primary thrust generator. Nature 399: 60–62.

Chaw SM, Parkinson CL, Cheng YC, Vincent TM, Palmer JD (2000): Seed plant phylogeny inferred from all three plant genomes: monophyly of extant gymnosperms and origin of Gnetales from conifers. Proc Natl Acad Sci USA 97: 4086–4091.

Chiappe LM (1995): The first 85-million years of avian evolution. Nature 378: 349–355.

Colombo L, Franken J, Van Went J, Angenent HJM, Van Tunen AJ (1995): The *Petunia* MADS box gene FBP11 determines ovule identity. Plant Cell 7: 1859–1868.

Corner EJH (1958): Transference of function. Bot J Linn Soc 56: 33–40.

Crane PR (1985): Phylogenetic analysis of seed plants and the origin of angiosperms. Ann MO Bot Gard 72: 716–793.

Crane PR (1986): The morphology and relationships of the Bennettitales. In: Spicer RA, Thomas BA (eds), *Systematic and Taxonomic Approaches in Palaeobotany*. New York: Oxford University Press and the Systematics Association, p. 163–175.

Crane PR, Kenrick P (1997): Diverted development of reproductive organs: a source of morphological innovation in land plants. Plant Syst Evol 206: 161–174.

Darwin F, Seward AC (eds) (1903): *More Letters of Charles Darwin: a Record of his Work in a Series of Hitherto Unpublished Letters*. London: John Murray, vol. 2, p. 20.

Delavoryas T, Person CP (1975): *Mexiglossa varia* gen. et sp. nov., a new genus of glossopterid leaves from the Jurassic of Oaxaca, Mexico. Palaeontograph B154: 114–120.

Dilcher DL (1979): Early angiosperm reproduction: an introductory report. Rev Palaeobot Palynol 27: 291–328.

Dilcher DL (1995): Plant reproductive strategies: using the fossil record to unravel current issues in plant reproduction. In: Hoch PC, Stephenson AG (eds), *Experimental and Molecular Approaches to Plant Biosystematics*. Monographs Syst. Bot. Missouri Bot. Gard. St. Louis, MO: Missouri Botanical Garden, vol. 53, p. 187–198.

Donoghue MJ, Doyle JA (2000). Seed plant phylogeny: demise of the anthophyte hypothesis? Curr Biol 10: R106–R109.

Doyle JA (1978): Origin of angiosperms. Annu Rev Ecol Syst 9: 365–392.

Doyle JA (1994): Origin of the angiosperm flower: a phylogenetic perspective. Plant Syst Evol Suppl 8: 7–29.

Doyle JA (1996): Seed plant phylogeny and the relationships of Gnetales. Int J Plant Sci 157: S3–S39.

Doyle JA, Donoghue MJ (1986): Seed plant phylogeny and the origin of angiosperms: an experimental cladistic approach. Bot Rev 52: 321–431.

Doyle JA, Donoghue MJ (1992): Fossils and seed plant phylogeny reanalyzed. Brittonia 44: 89–106.

Doyle JA, Donoghue MJ (1993): Phylogenies and angiosperm diversification. Paleobiology 19: 141–167.

Endress PK (1996): Structure and function of female and bisexual organ complexes in Gnetales. Int J Plant Sci 157, Suppl 6: S113–S125.

Endress PK, Friis EM (eds) (1994): *Early Evolution of Flowers.* Plant Syst Evol Suppl 8: 1–229.

Endress PK, Igersheim A (1997): Gynoecium diversity and systematics of the Laurales. Bot J Linn Soc 125: 93–168.

Feduccia A (1999): 1,2,3 = 2,3,4: Accommodating the cladogram. Proc Natl Acad Sci USA 96: 4740–4742.

Ferrandiz C, Gu Q, Martienssen R, Yanofsky MF (2000): Redundant regulation of meristem identity and plant architecture by FRUITFULL, APETALA1 and CAULIFLOWER. Development 127: 725–734.

Force A, Lynch M, Pickett FB, Amores A, Yan YL, Postlethwait J (1999): Preservation of duplicate genes by complementary, degenerative mutations. Genetics 151: 1531–1545.

Friis EM, Chaloner WG, Crane PR (1987): *The Origin of Angiosperms and Their Biological Consequences.* New York: Cambridge University Press.

Friis EM, Endress PK (1990): Origin and evolution of angiosperm flowers. Adv Plant Pathol 17: 99–162.

Frohlich MW, Estabrook GF (2000): Wilkinson support calculated with exact probabilities: an example using Floricaula/LEAFY amino acid sequences that compares three hypotheses involving gene gain/loss in seed plants. Mol Biol Evol 17: 1914–1925.

Frohlich MW, Parker DS (2000): The Mostly Male theory of flower evolutionary origins. Syst Bot 25: 155–170.

Fujii K (1896): On the different views hitherto proposed regarding the morphology of the flowers of *Ginkgo biloba* L. Bot Mag Tokyo 10: 7–8, 13–15, 104–110.

Garner JP, Taylor GK, Thomas ALR (1999): On the origins of birds: the sequence of character acquisition in the evolution of avian flight. Proc R Soc Lond B 266: 1259–1266.

Gaussen H (1946): Les Gymnospermes actuelles et fossiles. Travaux du laboratoire forestier de Toulouse Tome 2; fascicule 3.

Gould SJ, Vrba ES (1982): Exaptation—a missing term in the science of form. Paleobiology 8: 4–15.

Grant V (1950): The protection of ovules in flowering plants. Evolution 4: 179–201.

Hansen AS, Hansmann S, Samigullin T, Antonov A, Martin W (1999): *Gnetum* and the angiosperms: molecular evidence that their shared morphological characters are convergent, rather than homologous. Mol Biol Evol 16: 1006–1009.

Harris TM (1964): *The Yorkshire Jurassic Flora II Caytoniales, Cycadales and Pteridosperms*. London: British Museum (Natural History).

Hempel FD, Weigel D, Mandel MA, Ditta G, Zambryski PC, Feldman LJ, Yanofsky M (1997): Floral determination and expression of floral regulatory genes in *Arabidopsis*. Development 124: 3845–3853.

Hickey LJ, Taylor DW (1996): Origin of the angiosperm flower. In: Taylor DW, Hickey LJ (eds), *Flowering Plant Origin, Evolution and Phylogeny*. New York: Chapman Hall, p. 176–231.

Hiscock SJ, Kües U (1999): Cellular and molecular mechanisms of sexual incompatibility in plants and fungi. Int Rev Cytol 193: 165–295.

Hufford L (1996): The morphology and evolution of male reproductive structures of Gnetales. Int J Plant Sci 157, Suppl 6: S113–S125.

Hughes NF (1994): *The Enigma of Angiosperm Origins*. New York: Cambridge University Press.

Igersheim A, Endress PK (1997): Gynoecium diversity and systematics of the Magnoliales and winteroids. Bot J Linn Soc 124: 213–271.

Igersheim A, Endress PK (1998): Gynoecium diversity and systematics of the paleoherbs. Bot J Linn Soc 127: 289–370.

Jones TD, Ruben JA, Martin LD, Kurochkin EN, Feduccia A, Maderson PFA, Hillenius WJ, Geist NR, Alifanov V (2000): Nonavian feathers in a Late Triassic archosaur. Science 288: 2202–2205.

Jordan ND, Ride JP, Rudd JJ, Davies EM, Franklin-Tong VE, Franklin FCH (2000): Inhibition of self-incompatible pollen in *Papaver rhoeas* involves a complex series of cellular events. Ann Bot 85, Suppl A: 197–202.

Krassilov VA (1997): *Angiosperm Origins: Morphological and Ecological Aspects*. Sofia, Bulgaria: Pensoft Publishers.

Labandeira CC (1997): Insect mouthparts: ascertaining the paleobiology of insect feeding strategies. Annu Rev Ecol Syst 28: 153–193.

Labandeira CC (1998a): Early history of arthropod and vascular plant associations. Annu Rev Earth Planet Sci 26: 329–377.

Labandeira CC (1998b): How old is the flower and the fly? Science 280: 57–59.

Labandeira CC, Phillips TL (1996): Insect fluid-feeding on upper Pennsylvanian tree ferns (Palaeodictyoptera, Marattiales) and the early history of the piercing-and-sucking functional feeding group. Ann Entomol Soc Am 89: 157–183.

Labandeira CC, Sepkowski Jr JJ (1993): Insect diversity in the fossil record. Science 261: 310–315.

Leitch I, Chase MW, Bennett MD (1998): Phylogenetic analysis of DNA c-values provides evidence for a small ancestral genome size in flowering plants. Ann Bot 82, Suppl A: 85–94.

Loconte H, Stevenson DW (1990): Cladistics of the spermatophyta. Brittonia 42: 197–211.

Lloyd DG, Wells MS (1992): Reproductive-biology of a primitive angiosperm, *Pseudowintera colorata* (Winteraceae), and the evolution of pollination systems in the anthophyta. Plant Syst Evol 181: 77–95.

Lynch M, Force A (2000): The probability of duplicate gene preservation by subfunctionalization. Genetics 154: 459–473.

Ma F-S, Li J-X (1991): *Ginkgo biloba*: the ovuliferous leaf and its phylogenetic implication. Acta Phytotaxon Sin 29: 187–189.

Mathews S, Donoghue MJ (1999): The root of angiosperm phylogeny inferred from duplicate phytochrome genes. Science 286: 947–950.

Mayr E (1960): The emergence of evolutionary novelties. In: Tax S (ed), *Evolution After Darwin: the University of Chicago Centenial. The Evolution of Life*. Chicago, IL: University of Chicago Press, vol. 1, p. 349–380.

McCubbin AG, Kao T-H (1999): The emerging complexity of self-incompatibility. Sex Plant Reprod 12: 1–5.

Meyen SV (1988): Origin of the angiosperm gynoecium by gametoheterotopy. Bot J Linn Soc 97: 171–178.

Niklas KJ (1992): *Plant Biomechanics*. Chicago, IL: University of Chicago Press.

Nixon KC, Crepet WL, Stevenson D, Friis EM (1994): A reevaluation of seed plant phylogeny. Ann MO Bot Gard 81: 484–533.

Ostrom JH (1976a): Archaeopteryx and the origin of birds. Biol J Linn Soc 8: 91–182.

Ostrom JH (1976b): Some hypothetical anatomical stages in the evolution of avian flight. Smithson Contrib Paleobiol 27: 1–27.

Ostrom JH (1979): Bird flight—how did it begin? Am Sci 67: 46–56.

Padian K, Chiappe LM (1998): The origin and early evolution of birds. Biol Rev Camb Philos Soc 73: 1–42.

Pigg KB, Trivett ML (1994): Evolution of the glossopterid gymnosperms from Permian Gondwana. J Plant Res 107: 461–477.

Proctor M, Yeo P (1972): *The Pollination of Flowers*. New York: Taplinger Publishing Co.

Qiu YL, Lee JH, Bernasconi-Quadroni F, Soltis DE, Soltis PS, Zanis M, Zimmer EA, Chen ZD, Savolainen V, Chase MW (1999): The earliest angiosperms: evidence from mitochondrial, plastid and nuclear genomes. Nature 402: 404–407.

Ren D (1998): Flower-associated Brachycera flies as fossil evidence for Jurassic angiosperm origins. Science 280: 85–88.

Retallack GJ, Dilcher DL (1981): Arguments for glossopterid ancestry of angiosperms. Paleobiology 7: 54–67.

Retallack GJ, Dilcher DL (1988): Reconstructions of selected seed ferns. Ann MO Bot Gard 75: 1010–1057.

Richardson FC (1969): Morphological studies of the Nymphaeaceae. IV. Structure and development of the flower of *Brasenia schreberi* Geml. University of California Pub. Bot., vol 47.

Runions CJ, Owens JN (1996): Pollen scavenging and rain involvement in the pollination mechanism of interior spruce. Can J Bot 74: 115–124.

Sakisaka M (1929): On the seed-bearing leaves of *Ginkgo*. Jpn J Bot 4: 219–236, plates XXIII–XXV.

Samigullin TKh, Martin WF, Troitsky AV, Antonov AS (1999): Molecular data from the chloroplast rpoC1 gene suggest a deep and distinct dichotomy of contemporary spermatophytes into two monophyla: gymnosperms (including Gnetales) and angiosperms. J Mol Evol 49: 310–315.

Sattler R, Rutishauser R (1997): The fundamental relevance of morphology and morphogenesis to plant research. Ann Bot 80: 571–582.

Savolainen V, Chase MW, Hoot SB, Morton CM, Soltis DE, Bayer C, Fay MF, de Bruijn AY, Sullivan S, Qiu Y-L (2000): Phylogenetics of flowering plants based on combined analysis of plastid atpB and rbcL gene sequences. Syst Biol 49: 306–362.

Schneider EL, Jeter JM (1983): Morphological studies of the Nymphaeaceae. XII. The floral biology of *Cabomba caroliniana*. Am J Bot 69: 1410–1419.

Sessions A, Yanofsky MF (1999): Dorsoventral patterning in plants. Gen Dev 13: 1051–1054.

Siegfried KR, Eshed Y, Baum SF, Otsuga D, Drews GN, Bowman JL (1999): Members of the YABBY gene family specify abaxial cell fate in *Arabidopsis*. Development 126: 4117–4128.

Soltis DE, Soltis PS, Chase MW, Mort ME, Albach DC, Zanis M, Savolainen V, Hahn WH, Hoot SB, Fay ME, Axtell M, Swensen SM, Prince LM, Kress WJ, Nixon KC, Farris JS (2000): Angiosperm phylogeny inferred from 18s fDNA, fbcL, and atpB sequences. Bot J Linn Soc Lond: 133: 381–461.

Soltis PS, Soltis DE, Chase MW (1999): Angiosperm phylogeny inferred from multiple genes as a tool for comparative biology. Nature 402: 402–404.

Stebbins GL (1974): *Flowering Plants Evolution Above the Species Level*. Cambridge, MA: Harvard Univ. Press.

Steinbachs JE, Holsinger KE (1999): Pollen transfer dynamics and the evolution of gametophytic self-incompatibility. J Evol Biol 12: 770–778.

Stevens KA, Parrish JM (1999): Neck posture and feeding habits of two Jurassic sauropod dinosaurs. Science 284: 798–800.

Sundstrom J, Carlsbecker A, Svensson ME, Svenson M, Johanson U, Theissen G, Engstrom P (1999): MADS-box genes active in developing pollen cones of Norway spruce (*Picea abies*) are homologous to the B-class floral homeotic genes in angiosperms. Dev Genet 25: 253–266.

Swisher III CC, Wang Y-Q, Wang X-L, Xu X, Wang Y (1999): Cretaceous age for the feathered dinosaurs of Liaoning, China. Nature 400: 58–61.

Taylor DW, Hickey LJ (eds) (1996): *Flowering Plant Origin, Evolution and Phylogeny*. New York: Chapman Hall.

Taylor TN, Archangelsky S (1985): The cretaceous pteridosperms *Ruflorinia* and *Ktalenia* and implications on cupule and carpel evolution. Am J Bot 72: 1842–1953.

Taylor TN, Millay MA (1979): Pollination biology and reproduction in early seed plants. Rev Palaeobot Palynol 27: 329–355.

Taylor TN, Taylor EL (1993): *The Biology and Evolution of Fossil Plants*. Englewood Cliffs, NJ: Prentice Hall.

Theissen G, Saedler H (1999): The golden decade of molecular floral development (1990–1999): a cheerful obituary. Dev Genet 25: 181–193.

Thien LB, White DA, Yatsu LY (1983): The reproductive-biology of a relict *Illicium floridanum* Ellis. Am J Bot 70: 719–727.

Tomlinson PB (1994): Functional-morphology of saccate pollen in conifers with special reference to Podocarpaceae. Int J Plant Sci 155: 699–715.

Tomlinson PB, Braggins JE, Rattenbury JA (1991): Pollination drop in relation to cone morphology in Podocarpaceae—a novel reproductive mechanism. Am J Bot 78: 1289–1303.

Tomlinson PB, Braggins JE, Rattenbury JA (1997): Contrasted pollen capture mechanisms in Phyllocladaceae and certain Podocarpaceae (Coniferales). Am J Bot 84: 214–223.

Townrow JA (1962): On *Pteruchus*, a microsporophyll of the Corystospermaceae. Bull Br Mus (Nat Hist) Geol 6: 287–320.

Upchurch P (2000): Neck posture and feeding habits of two Jurassic sauropod dinosaurs. Science 284: 798–800.

Uyenoyama MK (2000): A prospectus for new developments in the evolutionary theory of self-incompatibility. Ann Bot 85A: 247–252.

Villanneva JM, Broadhvest J, Hauser BA, Meister RJ, Schneitz K, Gasser CS (1999): INNER NOOUTER regulates abaxial-adaxial patterning in *Arabidopsis* ovules. Genes Dev 13: 3160–3169.

Vogel S (1998): Remarkable nectaries: structure, ecology, organophyletic perspectives. II. Nectarioles. Flora 193: 1–29.

Weller SG, Donoghue MJ, Charlesworth D (1995): The evolution of self-incompatibility in flowering plants: a phylogenetic approach. In: Hoch PC, Stephenson AG (eds), *Experimental and Molecular Approaches to Plant Biosystematics*. Monogr. Syst. Bot. Missouri Bot. Gard. St. Louis, MO: Missouri Botanical Garden, vol. 53, p. 355–182.

Wetschnig W, Depisch B (1999): Pollination biology of *Welwitschia mirabilis* Hook. f. (Welwitschiaceae, Gnetopsida). Phyton-Annales Rei Bot 39: 167–183.

Wettstein RR (1935): *Handbuch der systematischen Botanik*. Vienna: Franz Deuticke.

Whitehouse HLK (1950): Multiple-allelomorph incompatibility of pollen and style in the evolution of the angiosperms. Ann Bot NS 14: 199–216.

Williams EG, Sage TL, Thien LB (1993): Functional syncarpy by intercarpellary growth of pollen tubes in a primitive apocarpous angiosperm, *Illicium floridianum* (Illiciaceae). Am J Bot 80: 137–142.

Wing SL, Boucher LD (1998): Ecological aspects of the Cretaceous flowering plant radiation. Annu Rev Earth Planet Sci 26: 379–421.

Wing SL, Tiffney BH (1987): The reciprocal interaction of angiosperm evolution and tetrapod herbivory. Rev Palaeobot Palynol 50: 179–210.

Winter KU, Becker A, Munster T, Kim JT, Saedler H, Theissen G (1999): MADS-box genes reveal that gnetophytes are more closely related to conifers than to flowering plants. Proc Natl Acad Sci USA 96: 7342–7347.

Yao X, Taylor TN, Taylor EL (1995): The Corystosperm pollen organ *Pteruchus* from the Triassic of Antarctica. Am J Bot 82: 535–546.

Zavada MS, Taylor TN (1986): The role of self-incompatibility and sexual selection in the gymnosperm-angiosperm transition—a hypothesis. Am Nat 128: 538–550.

Zenkteler M (2000): In vitro pollination of angiosperm ovules with gymnosperm pollen grains. In Vitro Cell Dev Biol Plant 36: 125–127.

4

ALLOMETRIC PATTERNING IN TRILOBITE ONTOGENY: TESTING FOR HETEROCHRONY IN *NEPHROLENELLUS*

Mark Webster

Department of Earth Sciences, University of California, Riverside, California

H. David Sheets

Department of Physics, Canisius College, Buffalo, New York

Nigel C. Hughes

Department of Earth Sciences, University of California, Riverside, California

INTRODUCTION

Realization of the large evolutionary potential of shifts in rates or timing of developmental events has made heterochrony a keyword in evolutionary theory (Gould, 1992). Heterochrony is the displacement in time of an ontogenetic event or a change in the rate of development in a descendant relative to the ancestral ontogeny (e.g., Gould, 1977, 2000; Alberch et al., 1979; McKinney and McNamara, 1991). In recent years, the concept of heterochrony has become widened to the point that it has become a "catch-all" description applied to

Beyond Heterochrony: The Evolution of Development, Edited by Miriam Leah Zelditch
ISBN 0-471-37973-5

almost any morphological differences between homologous structures among ancestor and descendant species (McKinney and McNamara, 1991; but see Gould, 2000). However, ontogeny can be modified in other ways, such as in spatial patterning (heterotopy) or the amount (heterometry) or discrete state of a feature (heterotypy) (e.g., Zelditch and Fink, 1996; Arthur, 2000). Such alternatives to heterochrony are rarely considered in evolutionary studies, largely due to lack of a rigorous and testable hypothesis that allows for rejection of heterochrony. Perhaps accordingly, heterochronic modification of developmental trajectories has been claimed to offer the greatest evolutionary potential (Klingenberg, 1998). Heterochrony has been proposed as being possibly responsible for the Cambrian radiation (McKinney and McNamara, 1991) and as a major factor in metazoan evolution ever since (Clark, 1964).

Trilobites have formed a linchpin in paleontological examples of heterochrony for almost a century (see reviews in McNamara, 1988; Chatterton and Speyer, 1990, 1997). Indeed, heterochrony has been claimed to have been a dominant control in the evolutionary history of trilobites (McNamara, 1986, 1988; Fortey and Owens, 1990). However, most (if not all) of the claims of heterochrony in trilobites are either speculative or inadequately tested (see below). Alternative explanations for the morphological evolution of trilobites are rarely considered [see Sundberg (2000) for an exception].

After a brief review of reported cases of heterochrony in trilobites, this chapter discusses a new, putative case of peramorphic heterochrony in the evolution of *Nephrolenellus geniculatus* from *N. multinodus*, two species of olenelloid trilobites from the Lower Cambrian shelfal deposits of the Great Basin, USA. This case ranks among the most strongly supported examples of heterochrony under the traditional criteria for identifying it in the fossil record. However, a more rigorous morphometric analysis designed to provide a testable hypothesis for the identification of heterochrony demonstrates that this example is not a simple case of evolution by global peramorphosis. The species follow unique ontogenetic trajectories in terms of patterns of morphological differentiation, suggesting that *N. geniculatus* evolved from *N. multinodus* through a more complex pattern of spatial repatterning and that any heterochronic shifts were localized and of minor importance. This example highlights the differences between concepts of heterochrony among workers and suggests that many purported cases of heterochrony in trilobites may require revision when more rigorously tested. Cases of evolution diagnosed as heterochrony are likely to have been produced by changes to developmental parameters other than (or in addition to) simple timing of events. The prevalence of heterochrony in trilobite evolution may be considerably exaggerated.

HETEROCHRONY IN TRILOBITE EVOLUTION

Few groups offer more potential for evolutionary insight than trilobites. The calcified exoskeleton of trilobites was easily fossilized, which, with their ecological abundance, make trilobite remains among the most familiar fossils in Paleozoic marine deposits: the record provides detailed documentation of an evolutionary history spanning some 250 million years. Periodic exoskeletal molting allowed each individual to leave a series of potential fossils recording its morphological appearance at almost all ontogenetic stages. Assessment of growth patterns in trilobites is possible by examination of morphological changes between conspecific individuals of different sizes. Interspecific ontogenetic comparison has led to claims of rife heterochronic evolution. [When assessing morphological change as a function of size rather than absolute age, "allometric heterochrony" is being measured (McKinney and McNamara, 1991; Zelditch and Fink, 1996; O'Keefe et al., 1999; but see Chapter 8).]

The regulatory mechanism(s) by which changes in developmental timing or rates arise lies within the field of developmental biology and is beyond the reasonable limits of paleontological insight. Claims of heterochronic evolution in trilobites therefore fall into a nonmechanistic concept of heterochrony, relying on observations of parallelism between ontogeny and phylogeny (heterochrony sensu Gould, 1977). Identification of heterochrony in this sense requires certain criteria to be met (see also Chapter 8). First, the evolutionary relationship between the putative heterochronic taxa must be adequately demonstrated (a sister species or anagenetic relationship in ideal circumstances). Second, the taxa must differ only in the timing of events along a conserved ontogenetic trajectory (thus producing a parallelism between ontogeny and phylogeny). Differences in ontogenies that are not explained by differences in rates or timing of common events cannot be interpreted as heterochronic in nature, and alternative explanations must be invoked (e.g., heterotopy).

Despite their amenity for the rigorous testing of evolutionary hypotheses, most (if not all) proposed cases of heterochrony in trilobites are poorly supported or are speculative. These inadequacies result from failures to meet the basic criteria required to identify heterochrony and can be broadly divided into two classes: (1) failure to satisfactorily demonstrate an ancestor-descendant or sister taxon relationship between the taxa in question and (2) inadequate demonstration of conservation of ontogenetic trajectory.

Phylogenetic Relationship Between Taxa. Heterochronic species should have a strongly supported ancestor-descendant or sister species relationship (Fink, 1982, 1988; Chatterton and Speyer, 1997). With notable exceptions (Edgecombe and Chatterton, 1987; Ramsköld, 1988), studies of heterochrony in trilobites are rarely conducted within a well-defined phylogenetic framework. Many are based on analysis at the superspecific level (e.g., Beecher, 1897; Jaekel, 1909; Stubblefield, 1936, 1959; Størmer, 1942; Robison, 1967; Robison and Campbell, 1974; Jell, 1975), taking single species as representative of larger clades. Analysis at such a level entails extremely loose requirements for conservation of ontogenetic trajectory between putatively heterochronic taxa and is rather subjective in nature.

Evolutionary relationships between fossil species have been proposed on stratigraphic, biogeographic, or morphological grounds (e.g., McNamara, 1988). The stratigraphic order of appearance of morphologically similar forms can be used to infer anagenetic or cladogenetic relationships between those taxa (McNamara, 1986; Edgecombe and Chatterton, 1987; Zhou et al., 1994; Crônier et al., 1998). Care must be taken to ensure that the rank order of appearance of the taxa is well supported: stratigraphic sections can often contradict each other in detail (especially in high-resolution studies at low taxonomic levels). Stratigraphically inferred phylogenetic relationships are "testable" in that the lowest first occurrence of a presumed descendant should not be lower than that of a presumed ancestor. Stratigraphically intermediate forms of nonintermediate morphology can falsify putative heterochronoclines (of two or more species) inferred purely on stratigraphic grounds (contra McNamara, 1988).

Biogeography can be used to infer phylogenetic relationships if there is evidence supporting endemic evolution of the taxa (McNamara, 1988). Dangers of basing phylogenetic hypotheses on biogeographic data are exemplified by the studies of McNamara (1978) and Adrain and Chatterton (1994). McNamara (1978) used the geographic isolation of the Scottish olenelloid assemblage to infer in situ evolution of a single paedomorphocline. However, subsequent cladistic analysis of the Olenelloidea (Lieberman, 1998, 1999) placed the Scottish species into evolutionary disparate clades, thus falsifying direct ancestor-descendant relationships among them. Similarly, Adrain and Chatterton (1994) used a cladistic analysis of all adequately known Silurian species of *Otarion* to falsify a putative peramorphocline of three *Otarion* species that could be inferred from a stratigraphic section in the Mackenzie Mountains.

Ultimately, all hypotheses of trilobite phylogeny are founded on morphological similarities between taxa. Whereas cladistics offers the most rigorous and replicable manner of phylogenetic inference, well-supported hypotheses can be made from traditional descriptive or phenetic methods. Strong cases for heterochrony should involve species-level analysis (in which homology of morphology and conservation of ontogenetic trajectory between taxa are more likely) and should be as comprehensive as possible in species representation. Very few purported cases of heterochrony in trilobites meet these criteria; they aim at patterns of evolution at superspecific levels (see above) or infer phylogenetic relationships without adequate support (Hupé, 1954; Fortey, 1974, 1975; McNamara, 1978, 1981a, 1981b; Ludvigsen, 1979; Fortey and Rushton, 1980; Chatterton and Perry, 1983; Müller and Walossek, 1987; Fortey and Owens, 1987; Zhou et al., 1994; Crônier et al., 1998).

Conservation of Ontogenetic Trajectory. For parallelism between ontogeny and phylogeny to exist, it is necessary that the ontogenetic trajectory of ancestor and descendant taxa is conserved, with all differences attributable to modification of timing of events along that shared trajectory. It is doubtful whether this can be assessed in the absence of preadult ontogenetic information (Whittington, 1981; Chatterton and Speyer, 1997). However, claims of heterochronic evolution in trilobites have been made based largely or solely on adult morphology, where knowledge of preadult form is lacking (e.g., McNamara,

1978; Henry et al., 1992). Such claims typically compare aspects of the "general morphology" of adults of the taxa in question and base heterochronic conclusions on some broad concept of which features are "juvenile" in nature. These are static morphological comparisons of individuals rather than dynamic comparisons of ontogenetic features (Alberch et al., 1979).

Shape changes during trilobite ontogeny (either across the entire organism or a single structure) were often complex but must be analyzed to assess the degree to which ontogenetic trajectories were conserved between taxa. Most morphometric analyses of trilobite ontogeny to date do not extend far beyond simple plots of width vs. length of various sclerites or structures (e.g., Palmer, 1957; Hunt, 1967; Chatterton et al., 1990; Chatterton and Speyer, 1997; Nedin and Jenkins, 1999). Such analyses give an impression of overall proportion but ignore complexities of trilobite morphology and lead to an extremely simplified quantification of growth.

Many purported cases of heterochrony in trilobites are founded on simple one-dimensional analysis of morphology, such as number of thoracic segments, degree of advancement of spines, or size [or, where several variables are cited, each is treated individually (e.g., Edgecombe and Chatterton, 1987; Nedin and Jenkins, 1999)]. Because such variables can only vary along that one dimension, analyses are guaranteed to assign any between-species differences to heterochrony (Zelditch and Fink, 1996; Zelditch et al., 2000). Conservation of overall ontogenetic trajectory is assumed rather than demonstrated. Statements that discrete events such as the disappearance of spines or the appearance of furrows, visual surfaces, and functioning sutures during ontogeny serve as useful time markers against which interspecific tests for heterochrony can be made (Chatterton and Speyer, 1997) are based on inherently one-dimensional variables and similarly do not rigorously demonstrate the operation of heterochrony. If heterochrony is to be anything more than a truism, comparisons must take into account the complexities of ontogenetic growth (e.g., patterns and rates of shape change across the organism).

Identification of heterochrony by morphological transformation should rely on observations of whole organisms rather than parts thereof, since developmental rate and timing are properties of an entire individual (Gould, 1977; Zelditch and Fink, 1996). Several claims of heterochrony in trilobites have been based on particular morphological structures, such as the eyes (Clarkson, 1971, 1975, 1979; Henry et al., 1992; Feist, 1995; Zhou et al., 1994), hypostome (Fortey, 1990), or genal spines (Lieberman, 1998). In other cases (including the present study), single sclerites are used because of the infrequency of preservation and the successful recovery of fully articulated specimens. Empirical documentation of dissociated heterochrony between conspecific cephala and pygidia (Edgecombe and Chatterton, 1987) warns against extrapolation of results based on single sclerites to entire organisms.

It can be seen from the above discussion that few, if any, reported cases of heterochrony in trilobites meet the basic criteria required for formulating a meaningful evolutionary argument. However, during an ongoing study into the paleobiology and evolutionary history of olenelloid trilobites by the senior author, a putative example of heterochrony was discovered that fulfills these criteria. This chapter assesses the case for heterochronic evolution in *Nephrolenellus* using both traditional evidentiary criteria (which it meets) and more stringent criteria designed to provide a rigorous testable hypothesis for the existence of heterochrony (which it does not meet).

CASE STUDY: THE EVOLUTION OF *NEPHROLENELLUS*

Phylogenetic Relationship Among *Nephrolenellus* Species

Nephrolenellus was originally diagnosed by small holaspid (adult) size (sagittal cephalic length of typically less than 12 mm), prominent constriction of the glabella at L2, subglobular L4, short ocular lobes with posterior tips widely separated from the glabella, a prothorax of 13 segments, and an opisthothorax of many segments (at least 22) (Palmer and Repina, 1993, 1997; Palmer, 1998). [Terminology for olenelloid morphology follows that of Whittington and Kelly (1997) with modifications proposed by Palmer (1998) and Webster (1999); see also Fig. 4.1.] A cladistic analysis by Lieberman (1998) listed a 45° angle of divergence of the ocular lobe from the sagittal axis as the only autapomorphy of *Nephrolenellus* (his character 20, state 2; homoplasic in outgroup taxon *Daguinaspis*), but the number of prothoracic segments was not considered in his analysis.

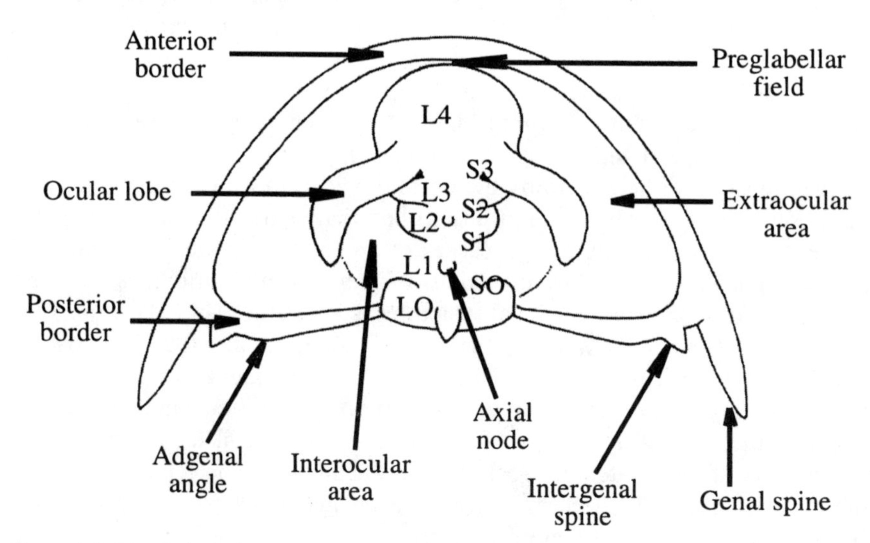

Figure 4.1. Morphological terms applied to the olenelloid cephalon (shown on an outline of the cephalon of *Nephrolenellus multinodus*). The glabella consists of lobes LO, L1, L2, L3, and L4 separated by furrows SO, S1, S2, and S3, respectively. The preglabellar field (the area between the anterior of L4 and the anterior border) is absent in many species (such as *N. geniculatus*).

As currently defined, the genus *Nephrolenellus* consists of two species (*N. multinodus* and *N. geniculatus*). A third species (to be described elsewhere) differs from other *Nephrolenellus* species because it possesses a slit-like (rather than pit-like) S3, 14 (rather than 13) prothoracic segments, and an axial spine on the first opisthosomal segment. This species is referred to here as *Nephrolenellus*? n. sp., although its exact phylogenetic relationship to *Nephrolenellus* requires further analysis. *Nephrolenellus*? n. sp. is characterized primarily by a strong anterior deflection of the posterior cephalic margin at the adgenal angle and location of the base of the genal spines opposite L1. It is stratigraphically older than the other species, known only from the Bristolia Zone Delamar Member (Pioche Formation) of east-central Nevada. The ontogeny of this species is presently unknown.

Nephrolenellus multinodus (Palmer in Palmer and Halley, 1979) is found across the Great Basin in a stratigraphic position consistently higher than *Nephrolenellus*? n. sp., in the lower portions of the Lower Cambrian Comet Shale Member (Pioche Formation), Pyramid Shale Member (Carrara Formation), and equivalents. Adults are characterized by separation of the glabella from the anterior border, only a weakly developed adgenal angle along the posterior margin of the cephalon, and only slight advancement of the genal spines (Fig. 4.2). The glabella of holaspid *N. multinodus* bears axial nodes on LO, L1, L2, and rarely L3.

Lieberman (1999) assigned a single cephalon collected 10 m above the Gog Group in Jasper Park (western Alberta) to a new species, *N. jasperensis*. This specimen was previously identified as *N. multinodus* (Palmer and Halley, 1979, pl. 4, Fig. 6). Lieberman (1998, 1999) disputed this earlier assignment on the basis of differences in the form of the posterior cephalic border (more rounded adgenal angle situated approximately two-thirds of the distance between the distal tip of the ocular lobes and the genal angle in the Canadian form but located directly behind the genal spine in *N. multinodus*), in the conformation of glabellar furrows, and in the degree of vaulting of the extraocular area. However, this specimen seems to fall within the range of morphological variance exhibited by *N. multinodus* based on comparisons to new collections of the latter species, particularly toward the top of its stratigraphic range. *Nephrolenellus jasperensis* is herein considered a junior synonym of *N. multinodus*.

Nephrolenellus geniculatus Palmer, 1998 is the stratigraphically highest species, occurring abundantly in the uppermost portions of the Lower Cambrian Comet Shale Member (Pioche Formation), Pyramid Shale Member (Carrara Formation), and equivalents. Although it occurs in the same stratigraphic sections, *N. geniculatus* appears consistently above *N. multinodus*, the ranges in juxtaposition but never overlapping. *Nephrolenellus geniculatus* differs from *N. multinodus* in that its glabella impinges the anterior border, in having a sharper adgenal angle and slightly more advanced genal spines, and in having an axial node on glabellar lobe LO (occasionally also L1, extremely rarely L2) only (Fig. 4.2).

Although it shares morphological similarity to well-known taxa such as

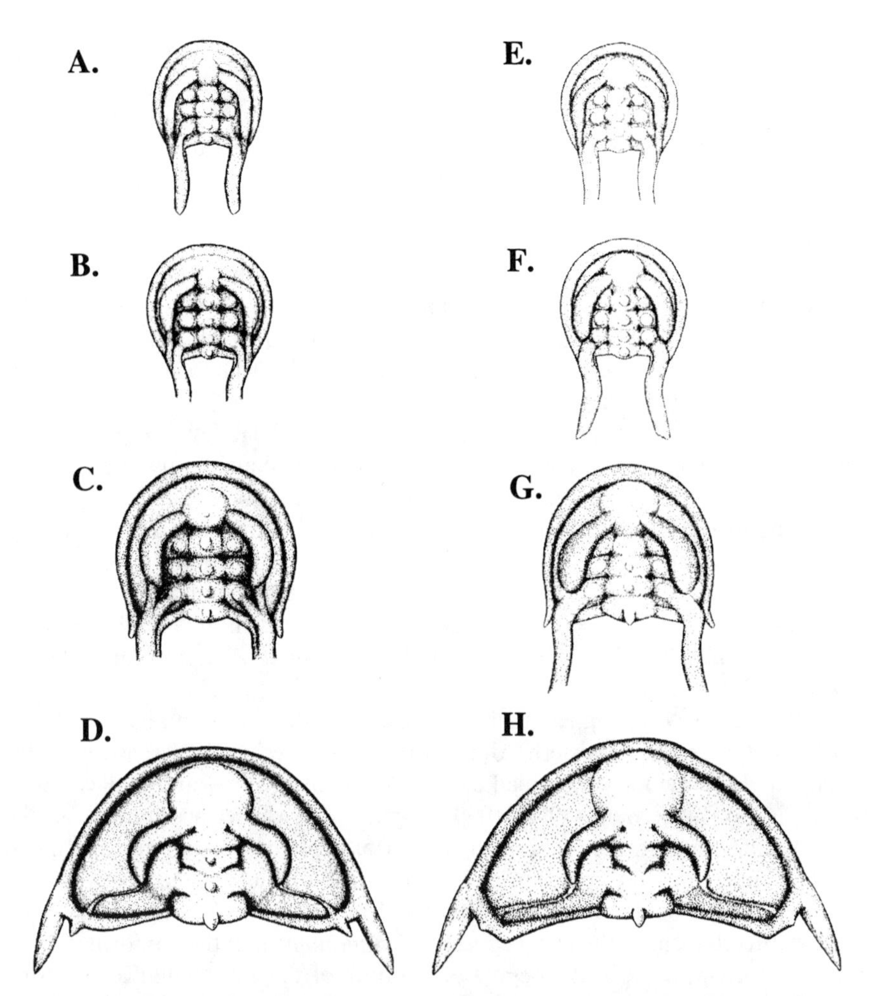

Figure 4.2. Reconstructions of *Nephrolenellus multinodus* (A–D) and *Nephrolenellus geniculatus* (E–H) at each stage of development. A: phase 1, sagittal cephalic length = 0.6 mm. B: phase 2, sagittal cephalic length = 0.75 mm. C: phase 3, sagittal cephalic length = 1.6 mm. D: phase 4, sagittal cephalic length = 7 mm. E: phase 1, sagittal cephalic length = 0.6 mm. F: phase 2, sagittal cephalic length = 0.86 mm. G: phase 3, sagittal cephalic length = 1.5 mm. H: phase 4, sagittal cephalic length = 7 mm. See text for details.

Bristolia within the Biceratopsiidae (Lieberman, 1998), the immediate phylogenetic affinities of *Nephrolenellus* probably lie with very poorly known genera, such as *Arcuolenellus*. A major revision of all biceratopsiid taxa by the senior author is in progress, based on extensive new collections. For the present paper, phylogenetic relationships among *Nephrolenellus* species were assessed using three species of *Bristolia* as outgroup taxa. *Bristolia bristolensis* (Resser, 1928), the type species of the genus, is a form with strongly advanced genal spines.

Table 4.1. Character states for each of the six taxa included in the phylogenetic analysis. Characters and states are described in Appendix. ? = state unknown (missing data); W, X, Y, and Z represent polymorphism in taxon for that character (W = states 0 or 1; X = states 1 or 2 or 3; Y = states 0 or 1 or 2; Z = states 2 or 3). See Figure 4.3 and text.

	00000 12345	00001 67890	11 12
Bristolia bristolensis	X2000	00000	01
Bristolia harringtoni	X1000	00000	01
Bristolia anteros	42111	101??	01
Nephrolenellus? n.sp.	W1111	00100	00
Nephrolenellus multinodus	00111	2Z111	10
Nephrolenellus geniculatus	00111	2Y111	00

Stratigraphically older species such as *Bristolia harringtoni* Lieberman, 1999 possess less advanced genal spines and may represent a less derived condition more appropriate for comparison with *Nephrolenellus*. A third species, *Bristolia anteros* Palmer in Palmer and Halley, 1979, is included because it shows a glabellar morphology, general holaspid size, and cephalic relief markedly similar to *Nephrolenellus* (compare pl. 1, Fig. 10 to pl. 4, Fig. 4 in Palmer and Halley, 1979), although it has always been assigned to *Bristolia* because of its extremely advanced genal spines and sharp adgenal angle.

Phylogenetic relationships among the *Nephrolenellus* species, *Bristolia anteros*, *B. bristolensis*, and *B. harringtoni* were hypothesized by cladistic analysis of 12 characters (see Appendix, Table 4.1), each coded according to the state expressed in the holaspid stage. An exhaustive search of these data [multistate characters treated as unordered, outgroup taxa not specified, conducted using PAUP version 4.0 (Swofford, 1998)] yielded a single most parsimonious tree of length 24 steps and consistency index of 0.958 (Fig. 4.3).

This analysis demonstrates monophyly of *Nephrolenellus* (diagnosed by moderately short genal spines) and a sister-taxon relationship between *N. multinodus* and *N. geniculatus* (supported by adgenal angle of less than 45°, pit-like S3, retention of axial nodes on glabellar lobes anterior to the occipital ring, prothorax of 13 segments, and loss of an axial spine on the first opisthothoracic segment). Under the criterion of parsimony, the analysis places *B. anteros* as sister taxon to the *Nephrolenellus* clade, rendering *Bristolia* (as presently defined) paraphyletic. Firmer resolution of the phylogenetic affinities of this species requires discovery of articulated specimens bearing thoracic information.

The phylogenetic hypothesis presented above proposes that *N. multinodus* and *N. geniculatus* are sister taxa. The stratigraphic occurrence of the taxa suggests that *N. geniculatus* may have evolved anagenetically from *N. multinodus*.

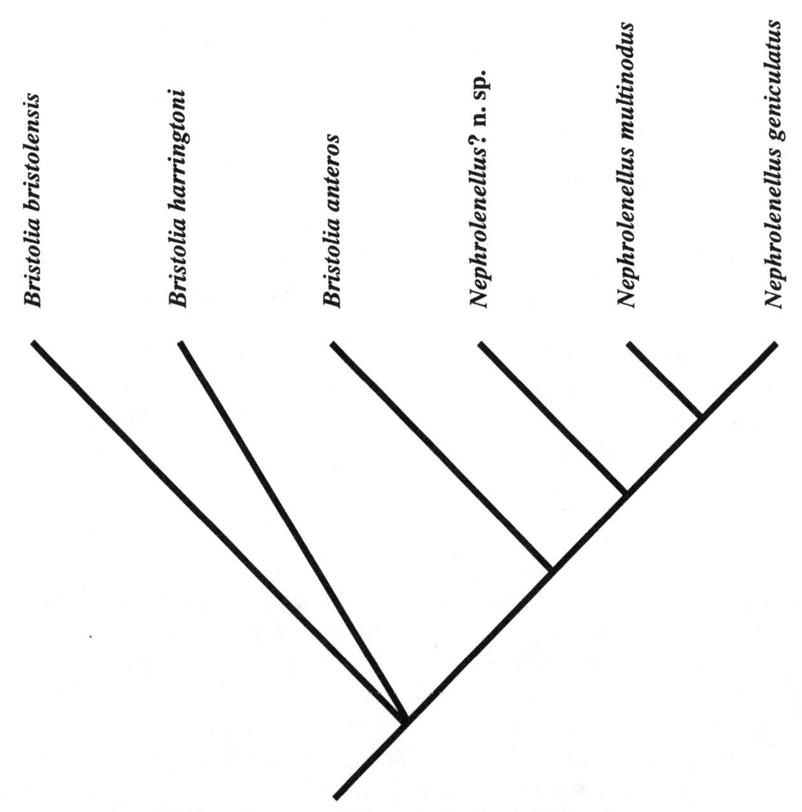

Figure 4.3. Single most parsimonious cladogram depicting hypothesized relationship between *Nephrolenellus* species and selected outgroup taxa. Based on analysis of 12 characters (Appendix, Table 4.1). Tree length = 24 steps; consistency index = 0.958. See text for details.

A Case for Peramorphic Evolution

Ontogenetic features suggest that *N. geniculatus* evolved from *N. multinodus* by extrapolation of the ancestral ontogenetic trajectory. This is evinced by the fact that generation of all interspecific differences (the distribution of nodes on the glabella axis, the shape of the adgenal angle, and the impingement of L4 into the anterior border) can all be explained as simple extensions of trends established in the ontogeny of *N. multinodus*. This is peramorphic heterochrony (Alberch et al., 1979; McKinney and McNamara, 1991).

Ontogenetic development in *N. multinodus* and *N. geniculatus* is very similar (Fig. 4.2). Both species initially possess axial nodes on glabella lobes LO, L1, L2, and L3 (the former species also initially possess a tiny node on L4). During ontogeny, these nodes are sequentially reduced in size and lost, anterior-most first (Figs. 4.2 and 4.4). *Nephrolenellus multinodus* typically loses the nodes on L4 and L3, retaining the others into maturity. *Nephrolenellus geniculatus*,

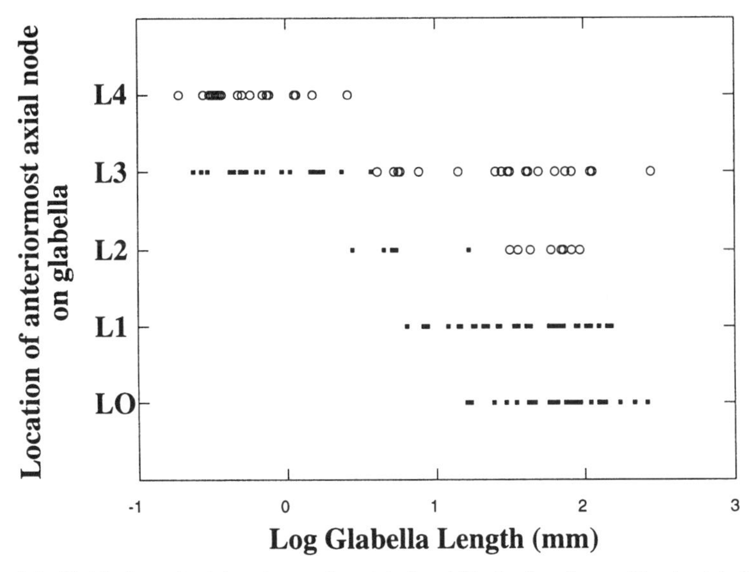

Figure 4.4. Distribution of axial nodes on the glabella of *Nephrolenellus multinodus* (circles) and *N. geniculatus* (squares) plotted against log glabella length. Data points indicate anteriormost glabella lobe on which an axial node is developed (each more posterior lobe also bears an axial node). Such one-dimensional analysis of morphology can be interpreted as evidence of heterochrony. See text for details.

however, typically loses the nodes on L3, L2, and L1 during development. The ontogeny of *N. multinodus* (and other olenelloids) demonstrates that adgenal angles are progressively developed during ontogeny (the genal spines becoming more advanced) and that the distance between the anterior lobe of the glabella and the anterior border (the preglabellar field) decreases (Webster, 1999).

The characters distinguishing adults of the species (differences in the number of glabellar axial nodes, the form of the adgenal angle, and the proximity of the glabella to the anterior border) are all ontogenetically dynamic features, which follow predictable courses through development. Each of these trends progresses further during development of *N. geniculatus* than during development of *N. multinodus*. Furthermore, specimens of *N. multinodus* toward the top of its stratigraphic range show an adult morphology trending toward that of *N. geniculatus*: the adgenal angle is more pronounced and the genal spines are slightly more advanced, but the glabella does not impinge into the anterior border and axial nodes are retained on glabellar lobes LO, L1, and L2.

Heterochrony has been claimed to have been rife among Cambrian trilobites, with cases of paedomorphosis being numerically dominant over cases of peramorphosis [McNamara, 1978, 1981b, 1983, 1986; 1988; although peramorphosis has been inferred for some aspects of olenelloid phylogeny (McNamara, 1986, p. 146–147)]. However, as discussed above, few previously published cases meet the basic requirements to demonstrate evolution by heterochrony even in a qualitative sense. Under these criteria, the argument for peramorphic evolution of *N. geniculatus* from *N. multinodus* presented above may represent the most strongly supported case of heterochrony in trilobite evolution published to date.

However, the features used as evidence for peramorphosis in *Nephrolenellus* (extent of node loss, angularity of the adgenal angle, and impingement of the glabella into the anterior border) have only been examined within a one-dimensional framework. When such variables are plotted individually against size, any interspecific differences can only be interpreted as heterochrony (see above; Fig. 4.4). In reality, these differences may relate (at least in part) to other factors such as heterotopy. These alternative explanations may reflect the operation of fundamentally different underlying mechanisms. To make heterochrony a meaningful model of evolution, hypotheses of its action must be testable.

FRAMING A TESTABLE HYPOTHESIS OF HETEROCHRONY

A key distinction between heterochrony and alternative ontogenetic evolutionary mechanisms is that evolution by pure heterochrony requires the patterns of ontogenetic shape change to be shared between ancestor and descendant: the two differ only in the rate at which these shape changes occur or in the timing of entry (or termination) of otherwise identical ontogenetic stages. If ancestral and descendant ontogenies differ in patterns of shape differentiation, then they do not share the same ontogenetic trajectory, and the hypothesis of evolution by pure heterochrony is rejected (Zelditch and Fink, 1996; Rice, 1997; O'Keefe et al., 1999). Under the hypothesis that heterochrony requires conservation of ontogenetic trajectory (*allometric conservation*), the claims of evolution by global heterochrony (McKinney and McNamara, 1991) become rigorously testable.

Allometric conservation can be tested at the organismal scale or at the level of discrete organismal parts. Any differences in patterns of shape differentiation (*allometric repatterning*) may result from localized, dissociated heterochrony (which in turn requires allometric conservation on a more local scale) and/or from alternative developmental mechanisms (such as heterotopy).

The case for peramorphic evolution in *Nephrolenellus* clearly meets the traditional evidentiary criteria for the identification of heterochrony. However, the one-dimensionality of the evolving features (above) ensured that the interspecific differences could be explained in no other way. It remains to compare the ontogenetic trajectories of the species under the null hypothesis that there is no difference in the patterns of ontogenetic shape change between the species. [Contrary to tradition, here the relationship of interest (evolution by heterochrony) is true when the null hypothesis is accepted.] Rejection of the null hypothesis is evidence that the species do not share a conserved ontogenetic trajectory and that the case for evolution by peramorphosis is an oversimplification.

MATERIALS AND METHODS

Materials

Preservation of completely articulated specimens of olenelloid trilobites is extremely rare. This probably relates to the simple articulation between thoracic segments (see Whittington, 1989), which would dissociate soon after death of the individual. The analysis below assesses allometric patterning of the cephalon alone. Therefore, any "global" patterns detected relate to integrated growth patterns in the head rather than across the entire organism. Restriction of the analysis to the cephalon is a function of preservation of the fossils (which drastically reduces sample size of articulated specimens) but can be somewhat justified on the grounds that all detected interspecific differences lie in cephalic features: the thorax of *N. geniculatus* is indistinguishable from that of *N. multinodus*. (The pygidium of both species is unknown.)

Over 400 cephala of *N. geniculatus* in late stages of ontogenetic development were examined, collected from shale units in the Ruin Wash, Oak Springs Summit, Hidden Valley (Lincoln County, Nevada; Palmer, 1998) and the Marble Mountains (San Bernardino County, California). Well-preserved specimens from Ruin Wash were selected for morphometric analysis. Approximately 100 late ontogenetic stage cephala of *N. multinodus* were obtained from fine-grained clastics and silty carbonates at Emigrant Pass, Eagle Mountain, Titanothere Canyon, Pyramid Peak, Echo Canyon, Frenchman Mountain, Grassy Spring [see Palmer and Halley (1979) and Palmer (1998) for locations], and the Marble Mountains. Unfortunately, most of these specimens were tectonically deformed.

Silicified cephala representing the early growth stages of each species were picked from insoluble residues remaining after dissolution of the encasing carbonate nodules in dilute acetic acid. Nodules from 1.6 m below the base of the Middle Cambrian at Hidden Valley yielded specimens of *N. geniculatus* for morphometric analysis. Nodules from 22 m below the base of the Middle Cambrian in the Groom Range, from Oak Springs Summit, and from lower in the Hidden Valley section yielded specimens of *N. multinodus* for morphometric analysis. Additional specimens of *N. geniculatus* from Antelope Canyon were used for descriptive purposes but were not included in the morphometric analyses. Lists of specimens examined for descriptive and morphometric analysis, and details of their stratigraphic and geographic provenance, are available from the senior author on request.

A complication with using geometric morphometric methods on trilobites is that their morphology may become modified through taphonomic and diagenetic processes. Tectonically deformed specimens were excluded from all analyses. However, the problem remains when specimens with different taphonomic histories are included in the same analysis (e.g., silicified and nonsilicified specimens). Potential biases associated with differences in preservational mode (discussed below) are not found to influence the interpretation of the results.

A Test for Allometric Conservation

Patterns of integrated ontogenetic shape change were assessed using geometric morphometric techniques. Morphometric analysis was conducted in three stages. First, specimen shape was summarized in a set of discrete points (landmarks), presumed to be homologous on all individuals included in the study. The x, y coordinates of each of these landmarks were digitized for each specimen [oriented dorsal side up with ocular lobes horizontal, following the standard of Shaw (1957)]. Second, differences in shape were isolated by superimposing landmark configurations of specimens. This involved rescaling and rotating the raw configurations to show minimal discrepancy in landmark locations. Finally, vectors of integrated components of shape change were calculated from patterns of landmark migration in the superimposed data. This provided the data from which patterns of ontogenetic shape change were quantified. Each step of the analysis is described in detail below. When these steps were followed for the ontogenies of *N. multinodus* and *N. geniculatus*, it became possible to statistically compare their patterns of morphological differentiation and therefore test the hypothesis of allometric conservation.

Because the age of particular individuals (in terms of elapsed time since hatching) is impossible to determine for trilobites (even when they can be assigned to discrete instars), some other variable must be used as a proxy for age when assessing patterns of development. Traditionally, this takes the form of some measure of size. The dangers of inferring absolute age from size are well known (Strauss, 1987; Blackstone, 1987a, 1987b; McKinney, 1988; Blackstone and Yund, 1989; Schweitzer and Lohmann, 1990; Godfrey and Sutherland, 1995a, 1995b, 1996) but typically relate to identification of specific processes of heterochrony rather than simply of heterochrony itself (Blackstone, 1987a; Zelditch et al., 2000; also see Chapter 8). Size may be measured in several ways. Univariate linear measures such as glabellar or cephalic length have traditionally been used in analyses of trilobite growth. However, univariate measures are more vulnerable to intraspecific variation and allometry than are multivariate size measures, in which such problems are likely to be averaged out by inclusion of several variables. Centroid size (the square root of the summed square distances between each landmark and the centroid of the form; Bookstein, 1991) is the most widely used multivariate measure. Centroid size is uncorrelated with shape where growth is purely isometric (Bookstein, 1991) and therefore offers the most satisfactory measure of size.

When comparing patterns of shape differentiation between species in this way, it is important that each of the ontogenies under comparison is linear (i.e., follows a uniform trajectory). A shift in pattern of shape change during the ontogeny of a species (a "kink" in its ontogenetic trajectory) will inflate the within-species variance when ontogenetic vectors are calculated from linear regressions to the data. Morphometric analysis of the olenelloid cephalon has revealed the existence of five discrete phases of development during olenelloid ontogeny, homologous in all taxa so far examined. Each phase of this common cephalic developmental framework (cCDF) corresponds to the appearance of a discrete morphological structure or relationship between structures. *Nephrolenellus* species terminate their ontogenetic development during cCDF phase 4 and do not undergo the marked lateral expansion of L2, which signifies entry into phase 5. Pronounced shifts in allometric patterning of the glabella are often associated with transitions between these developmental phases, suggesting that olenelloid ontogenetic trajectories are nonlinear. This necessitates detailed examination of within-species modes of differentiation in *N. multinodus* and *N. geniculatus* before between-species comparisons.

Landmark Selection

A total of 23 landmarks were selected to summarize cephalic shape (Table 4.2, Fig. 4.5). These landmarks are operationally homologous on each species (Smith, 1990, and references therein). Most selected landmarks are either points of intersection of three structures, points of maximal curvature, or tips of discrete morphological structures and represent type 1 or 2 landmarks (Bookstein, 1991). Landmarks along the sagittal axis (1–3) can be considered to represent type 2 or type 3 (extremal points) landmarks (Bookstein, 1991).

Table 4.2. Locations of cephalic landmarks selected for analysis. See Figure 4.5. Landmarks 4 to 13 inclusive have homologues on the left and right sides of the sagittal axis. Note that landmarks 11 and 13 are applicable only to later ontogenetic stages (cCDF Phases 3 and 4; see text). Landmark 12 was digitized as the adaxial side of the base of the intergenal spine in Phase 1 and 2 specimens, but was calculated or digitized as the midpoint of the intergenal spine or node in Phase 3 and 4 specimens (as the spine is expressed as a node in large Phase 4 individuals, only the midpoint of which can be reliably digitized). No analysis compared Phases 1 or 2 with Phases 3 or 4 using this landmark, so nonhomology of this landmark between early and late ontogenetic stages does not influence results. Landmark 13 was digitized as the adaxial side of the base of the genal spine.

Landmark	Description
1	Anteriormost point of cephalon on sagittal axis.
2	Anteriormost point of glabella on sagittal axis.
3	Midpoint of posterior edge of occipital ring on sagittal axis.
4	Juncture of axial furrow with posterior cephalic margin.
5	Juncture of axial furrow with occipital furrow.
6	Juncture of axial furrow with S1.
7	Juncture of axial furrow with S2.
8	Juncture of axial furrow with posterior margin of ocular lobe.
9	Juncture of axial furrow with anterior margin of ocular lobe.
10	Posterior tip of ocular lobe.
11	Adgenal angle.
12	Base of intergenal spine.
13	Base of genal spine.

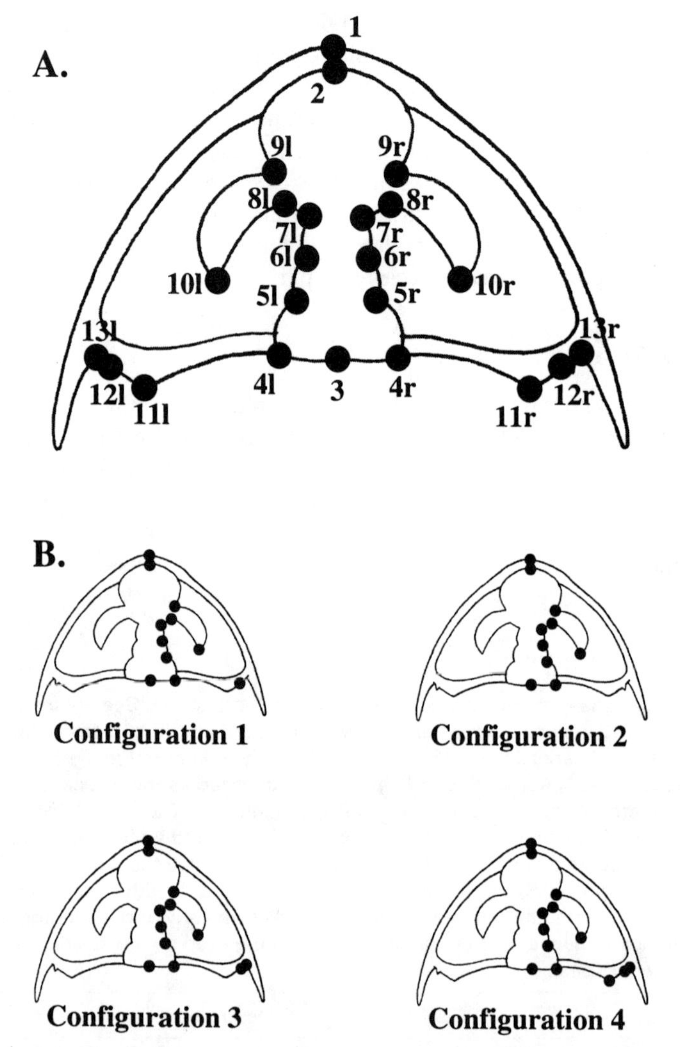

Figure 4.5. Locations of cephalic landmarks selected for analysis. See Table 4.2. A: total set of 23 landmarks. Homologous landmarks on the left and right halves are given the same number followed by l or r, respectively. The coordinates of these landmark pairs were averaged after reflection across the sagittal axis (see text). B: configurations of landmark subsets used in the analyses.

Cephala of many olenelloids are notably vaulted throughout ontogeny. The present analysis considers the location of landmarks in two-dimensional space only, and information in the *z*-plane is lost. However, given the potential biases introduced through taphonomic compaction of the fossils (see below), the third dimension is not likely to yield reliable data.

Coordinates of homologous (bilaterally symmetrical) landmarks on the left and right sides of the cephalon were averaged across the sagittal axis (Fig. 4.5).

It is assumed that there is no systematic asymmetry of the organism, which would necessitate use of landmarks on both sides of the sagittal axis [a reasonable assumption for trilobites: see Smith (1998)].

Morphometric analysis was restricted to silicified specimens and internal molds and parts of nonsilicified specimens. Specimens missing data or showing evidence of distortion were omitted from the analysis. Final sample size for the present morphometric study was 34 for *N. multinodus* ($n = 16$, 5, 4, and 9 for phases 1, 2, 3, and 4, respectively) and 66 for *N. geniculatus* ($n = 5$, 10, 11, and 40 for phases 1, 2, 3, and 4, respectively).

Not all landmarks are present on individuals of all age classes. For example, the genal spine does not appear until cCDF phase 3, and landmark 13 is therefore absent from all phase 1 and 2 individuals. For this reason, morphometric analyses were conducted on subsets of the full 23 landmarks (Fig. 4.5). Each configuration represents a balance between coverage of cephalic morphology and the number of specimens included in the study. Some configurations are applicable only to specific stages of development (configuration 1 to phase 1 and 2 only; configurations 3 and 4 to phases 3 and 4). Configuration 2 can be used for comparison of any ontogenetic stages but omits abaxial landmarks and therefore does not summarize overall cephalic shape well. In all analyses, the landmark configuration that offered maximal morphological coverage for the particular ontogenetic stage(s) under comparison was used.

Quantification of the Pattern of Shape Differentiation

Superimposition Techniques. In landmark-based morphological comparisons, shape is the residual geometric information remaining once location, scale, and rotational effects are removed from an object (Kendall, 1977). Shape differences therefore depend on the method by which landmark configurations of the forms under consideration are superimposed (i.e., how location, scale, and rotational effects are removed). Two general methods exist for the superimposition of two or more sets of landmark data. Although widely used in morphometric analyses, each is prone to produce biologically unsound superimpositions under certain conditions. Here, we discuss limitations of each method and propose a third method of superimposition that can be used when the general methods fail.

Procrustes superimposition methods seek to minimize the sum of squared errors (i.e., the offset between corresponding landmarks) between forms. A variety of criteria can be used to minimize the error, giving several methods of Procrustes superimposition (Bookstein, 1991; Dryden and Mardia, 1998, and references therein). Procrustes superimposition is used to align forms before thin-plate spline (TPS) analysis (see below) and is widely used in morphometric studies. However, Procrustes methods can be unsatisfactory under certain conditions. In forms with a clear axis of symmetry, it is usual to reflect landmark data from one side of the form onto the other side across the axis of symmetry and average the coordinates of paired homologous landmarks. The axis of

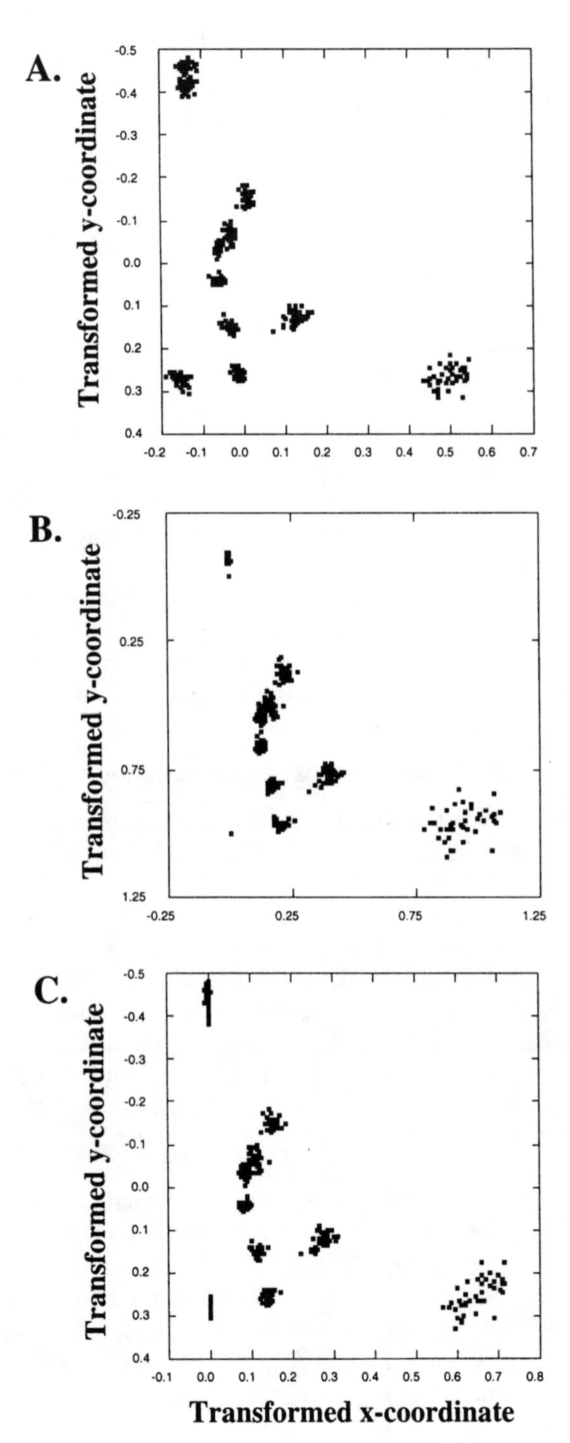
A.
B.
C.
Transformed y-coordinate
Transformed y-coordinate
Transformed y-coordinate
Transformed x-coordinate
-0.5
-0.4
-0.3
-0.2
-0.1
0.0
0.1
0.2
0.3
0.4
-0.2 -0.1 0.0 0.1 0.2 0.3 0.4 0.5 0.6 0.7
-0.25
0.25
0.75
1.25
-0.25 0.25 0.75 1.25
-0.1 0.0 0.1 0.2 0.3 0.4 0.5 0.6 0.7 0.8

symmetry then lies along an edge of the configuration. This axis is a biologically determined feature with a preferred direction that must be respected in any superimposition manipulation: it must remain parallel and coincident across forms. Procrustes methods regard all rotations and translations of landmark data as equivalent during superimposition and do not constrain common orientation of axes of symmetry across forms (Fig. 4.6A).

Superimposition by two-point registration avoids this problem (Bookstein, 1986, 1991; "Bookstein registration" of Dryden and Mardia, 1998). Superimposition is achieved by fixing the positions of two landmarks at the points $0, 0$ and $0, 1$. Choosing points along the axis of symmetry to lie at these points results in forms being aligned along that baseline (Fig. 4.6B). However, Bookstein registration (BR) does not constrain the two forms to have the same centroid size but rather the same baseline length. (The advantage of avoiding linear measures of size was discussed above.) Furthermore, in fixing the coordinates of the baseline landmarks, any variation in the positions of these landmarks is mapped onto the positions of the other landmarks as a function of their perpendicular distance from the baseline (Dryden and Mardia, 1998). This can artificially inflate estimates of shape variance between forms.

Where landmarks are to be reflected and averaged across an axis of symmetry and where it is desirable to minimize overall variance between forms, the sliding baseline registration (SBR) method (described herein) is the most appropriate superimposition technique. This method does not permit relative rotation of baseline or translation perpendicular to the baseline between forms, thus respecting priority of the axis of symmetry. Translation parallel to the baseline ("baseline sliding") and rescaling operations are permitted during superimposition. Forms are scaled to unit centroid size, a more robust proxy for size than the linear baseline measure (see above). The degree of baseline sliding between forms is determined by requiring either (1) that the component of centroid position along the baseline is fixed (i.e., if the axis of symmetry lies along the x axis, then the x component of the centroid position is constrained to a fixed value in all forms) or (2) that the sum of squared distances between corresponding landmarks be minimized. In the present study, the symmetry axis was constrained to lie along the x axis, and the x component of centroid position was set to zero by sliding the configuration of landmarks along this axis (Fig. 4.6C). Note the clearer definition of the ontogenetic vector of landmark 13 obtained using SBR compared with that obtained using BR of the same data (Fig. 4.6, C and B).

Little developmental work has been done to date on the sum of squared

◄

Figure 4.6. Methods of superimposition of landmark data. Landmark configurations of 40 specimens of *Nephrolenellus geniculatus* (in phase 4 of development, using landmarks 1–10 and 13) are superimposed using generalized least-squares Procrustes (A), Bookstein registration (B), and sliding baseline registration (C) methods. Baseline for Bookstein and sliding baseline registrations is sagittal axis, defined by landmarks 2 and 3. See text for details.

errors approach to the SBR method. Distribution models that allow inference based on the sliding baseline method have not yet been developed. In this study, we rely on bootstrap methods (Efron, 1982; Efron and Tibshirani, 1993), which require no model of the distribution. Any other use of SBR for inference would presently require similar numerical techniques. SBR should not be regarded as a general method, and its robustness is not as well established as that of the generalized least-squares Procrustes methods. SBR may be a useful adjunct to the other methods, particularly in visualizing the nature of shape changes or differences in a biologically reasonable way.

Quantifying Shape Changes: Vectors of Growth. Growth vectors summarizing the magnitude and orientation of isolated patterns of shape change during ontogeny can be calculated following superimposition of forms using BR, SBR, or Procrustes (through thin-plate spline analysis) methods.

Vectors Derived from BR or SBR. With superimposition by BR or SBR methods (using the sagittal axis as baseline, defined by landmarks 2 and 3), the landmark configuration of each specimen is scaled to unit size (baseline length or centroid size, respectively). Homologous landmarks form distinct clouds of data points in shape space. Where growth is isometric, the clouds will be essentially equidimensional, with data points scattered randomly around the centroid a landmark. Where growth is allometric, the data points will form an ellipse or line, tracing the relative migration of that landmark through ontogeny. The major axis of the ellipse or average orientation of the line gives a vector of ontogenetic movement for the landmark, with an x and y component. A growth vector is calculated by regressing the x and y components of the trajectory of each landmark on log centroid size. The regression coefficients for each component (two per landmark) form the vectors to be used in subsequent comparisons of ontogenetic growth patterns.

Vectors Derived from TPS analysis. TPS analysis quantifies shape differences as a deformation between landmarks and can be used to identify integrated shape changes between forms. Multivariate statistical methods such as principal components analysis (PCA) can also be used to identify integrated patterns of shape change (e.g., Tissot, 1988; Fink, 1988; Hughes, 1994; O'Keefe et al., 1999). However, TPS provides more empirical and easily interpretable results than do these other methods (Zelditch et al., 1992; Zelditch and Fink, 1996). The precise and replicable descriptors of shape difference that TPS produces are in strictly geometric terms and are not equatable with discrete biological processes. Rather, TPS serves to summarize net shape change resulting from these processes. Technical details of TPS and its decomposition into uniform and partial warps are given by Bookstein [1989, 1991; see Zelditch et al. (1992, 1993), Swiderski (1993), and Zelditch and Fink (1995) for less mathematical introductions]. Only a brief summary of the principles of TPS is given here.

The landmark configuration of a target form is superimposed using the generalized least-squares Procrustes superimposition method [see Bookstein (1991) and references therein] on the landmark configuration of a designated reference form. A hypothetical, infinitely thin plate (represented as an orthogonal grid) is fitted over the landmark configuration of the reference form. At each landmark, the plate is raised vertically (in the z plane) according to the degree of mismatch (in the x, y plane) of that landmark between the reference and target form. Relief of the plate between the "pinned" landmark points is determined by a mathematical function serving to minimize the amount of "bending energy" required to warp it out of its orthogonal shape at each landmark (a function used to model the bending behavior of real steel plates). The bending energy required to flex the plate between landmarks is inversely proportional to the spatial distance between those landmarks. Warping of the grid therefore reflects offset in the relative positions of homologous landmarks between the target and reference forms.

Projection of this grid back into the x, y plane results in a two-dimensional picture of the minimal deformation (in terms of bending energy) of the reference form required to obtain the configuration of the target form. Shape differences in landmark position are extrapolated to regions (tissues) between landmarks.

It is possible to mathematically decompose the total amount of deformation in the grid into a series of geometrically independent bending motions. One of these motions is an affine transformation, wherein deformation of the grid is homogenous (each square of the grid is affected equally): originally parallel grid lines remain parallel and no bending energy is required. This is termed the uniform warp. The remaining deformation of the grid (the nonaffine or nonuniform component) can be decomposed into a series of more localized independent bending motions. These are known as principal warps and are ordered according to their bending energy. Principal warps are calculated from the pairwise distances among all landmarks in the reference form (the number of warps is equal to three fewer than the total number of landmarks in the configuration). The principal warps with low bending energy correlate to large-scale patterns of shape change where deformation occurs over a larger space. Higher principal warps with greater bending energy correlate to smaller-scale, more localized patterns of shape change where deformation must be taken up between closely spaced landmarks. Principal warps have no magnitude or direction; they represent idealized modes of bending determined entirely by the landmark configuration of the reference form.

The total deformation of the grid, reflecting shape differences between reference and target forms, is partitioned into each of these principal warps (after removing the uniform component). The contribution of each principal warp to the total deformation is calculated as a vector (with an x and y component), indicating how strongly and in what orientation that mode of bending is expressed. A principal warp multiplied by its vector is termed a partial warp. The net deformation of the grid caused by the sum of all partial warps plus the uniform warp accounts precisely for the observed offset in landmarks of the target and reference forms.

Principal warps are determined by the reference form. Patterns of shape change between different specimens are made directly comparable by calculating the partial warp scores for each away from a common reference form. In all TPS analyses conducted for the present study, reference form was designated as the generalized least-squares Procrustes consensus form of all phase 3 and 4 *N. geniculatus* individuals (except for comparisons using landmark configuration 1, for which reference form was the consensus of all phase 1 and 2 *N. geniculatus* specimens). This form lies close to the mean of all specimens in the study and therefore has a negligible effect on analyses.

Differences in shape between specimens will be reflected as differences in partial warp scores for the same principal warp. Allometric growth during ontogeny can be recognized by the correlation between size and shape (e.g., Zelditch and Fink, 1995). Differences in patterns of ontogenetic shape change are reflected in differences in coefficients when the x and y components of partial warp scores and uniform warps are regressed against log centroid size. This allows identification of shifts in patterns of shape change during the ontogeny of individual species (see below) or on an evolutionary scale (Zelditch et al., 2000; see below). The regression coefficients for each component (two per warp) form the vectors to be used in subsequent comparisons of ontogenetic growth patterns.

Angle Between Ontogenies. The degree of similarity in net patterns of shape change between ontogenies (either different stages in the ontogeny of a single species or comparable stages of development in different species) may be quantified as an angle between the "ontogenetic vectors" that describe those ontogenies. An ontogenetic vector summarizes the direction of ontogenetic shape change (a function of the underlying patterns of shape change) over the size range of specimens representing that ontogeny. It is composed of the regression coefficients obtained by regressing each component of shape change (warp scores or landmark vectors) on log centroid size (described above). The angle between two ontogenetic vectors is the inverse cosine of the dot products of the normalized ontogenetic vectors. Normalization sets the magnitude of a vector to 1, and the dot product of two normalized vectors is the vector correlation (the inverse cosine of which represents the angle between them). An angle of 0° (corresponding to a vector correlation of 1) implies that the ontogenies share precisely the same underlying patterns of shape differentiation, whereas larger angles imply greater differences in the mode of shape change between the ontogenies.

The ontogenetic vectors involved in the calculation are geometric descriptors of integrated shape changes occurring across the organism during its ontogeny. In the present study, vectors were calculated using each of the three manipulations of the raw landmark data (BR, SBR, and TPS; described above). Angles between ontogenetic stages within each species and between comparable ontogenetic stages of *N. multinodus* and *N. geniculatus* were calculated using each of these methods.

One potential criticism of this method is that the angle of 0° between ontogenies required to demonstrate conservation of allometric patterning (and therefore the existence of pure heterochrony) is "too strict" and will never be met in empirical data. For this reason, the angle between two ontogenies (or ontogenetic stages) is statistically compared with the range of angles expressed within each ontogeny (or stage) to ascertain a confidence limit on the degree of similarity in patterns of shape change between groups. Conservation of allometric patterning is demonstrated when the angle between ontogenies is 0° or where it is statistically indistinguishable from 0° (i.e., if within-group variance exceeds between-group values). This considerably relaxes the requirements for identification of heterochrony, while maintaining a rigorously testable hypothesis.

Within-group variance in ontogenetic vectors was calculated by bootstrapping residuals of the regression procedure used to generate the ontogenetic vectors, using the bootstrap resampling methods of Efron and Tibshirani (1993) (see also Zelditch et al., 2000). Four hundred replications were drawn for each sample under comparison. Confidence intervals proved to be stable under repeated calculations for 400 bootstrap sets; therefore, larger numbers of bootstrap sets were not necessary. Within-group variance was computed from these bootstrap values at 95%, 90%, and 80% confidence limits (only the 95% limits are listed in the results). The angle between ontogenies is considered significant if it exceeds the bootstrapped within-group variance at 95% confidence.

RESULTS

Within-Species Ontogenetic Trajectories

Data. Comparison of the patterns of shape differentiation between developmental phases in *Nephrolenellus geniculatus* (Table 4.3; Fig. 4.7) demonstrates that the overall ontogenetic trajectory of this species is nonlinear. When the pattern of shape differentiation followed during cCDF phase 1 is compared with that followed during phase 2 using BR of landmark data, a between-phase angle of 105° is calculated (Table 4.3). This value is larger than the bootstrapped within-phase variance of either of the individual phases (95.2° and 97.9° for phases 1 and 2, respectively). This demonstrates that the mode of shape differentiation followed during phase 1 is significantly different (at 95% confidence) from that followed during phase 2. Similar results are obtained when vectors of shape change are calculated using SBR. The between-phase angle generated by TPS analysis of the landmark data is significant to 90% confidence (Table 4.3). Significant differences in patterns of shape differentiation between phases 2 and 3 are generated using all three superimposition

Table 4.3. Results of within-species comparison of patterns of shape differentiation between developmental phases of *Nephrolenellus geniculatus*. Comparisons based on vectors calculated using Bookstein registration (BR), sliding baseline registration (SBR), and thin-plate spline analysis (TPS) of landmark data. Within-group angles calculated from 400 bootstraps, given at 95% confidence limit (earliest phase first). Config. = landmark configuration used in comparison. Sig. = Statistical significance of between-phase angle. See text for details.

		BR			SBR			TPS		
Phase Transition	Config.	Between Group Angle	Within Group Angle	Sig. (%)	Between Group Angle	Within Group Angle	Sig. (%)	Between Group Angle	Within Group Angle	Sig. (%)
1 to 2	1	105	95.2/97.9	95	107.4	93.1/96.8	95	97.3	97.4/93.9	90
2 to 3	2	67.9	52.7/36.1	95	60.7	57.1/40.9	95	67.1	55.2/46.6	95
3 to 4	3	23.4	18.7/102.2	<80	24.1	18.3/99.8	<80	21.4	22.3/88.8	<80

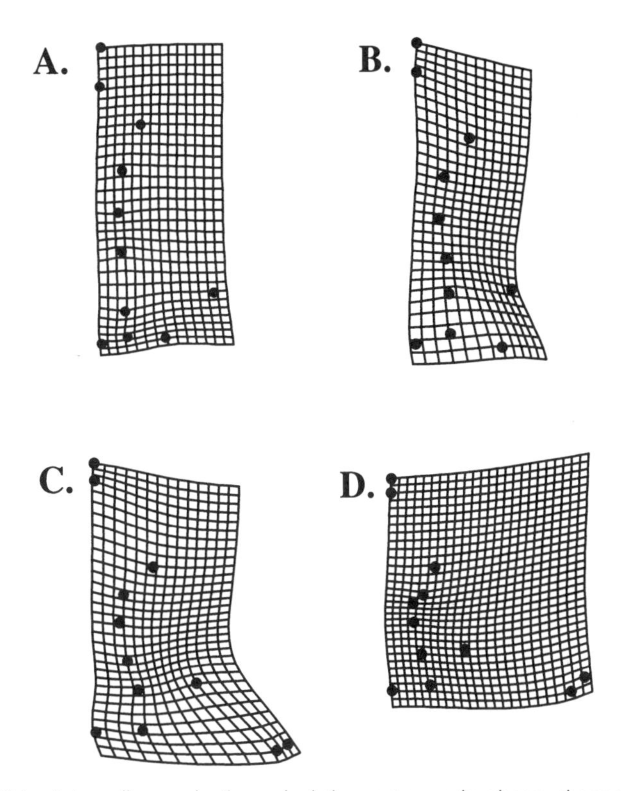

Figure 4.7. Thin-plate spline projections depicting ontogenetic shape changes in *Nephrolenellus geniculatus.* A: phase 1, landmark configuration 1. B: phase 2, landmark configuration 1. C: phase 3, landmark configuration 3. D: phase 4, landmark configuration 3. Reference and target forms are consensus of two smallest and largest specimens in each phase, respectively.

methods to 95% confidence (Table 4.3). However, the pattern of shape differentiation is not significantly different between phases 3 and 4 (Table 4.3) due to large within-phase variance in phase 4.

A similar nonlinear trajectory is followed by *N. multinodus* (Fig. 4.8; Table 4.4). Phases 1 and 2 show different patterns of shape differentiation (to 95% confidence, using any superimposition method). Large within-phase variance in phase 2 renders the patterns of shape change in phase 2 statistically indistinguishable from those in phase 3. However, statistically significant differences in allometric patterning are found between phases 3 and 4 of development in this species (using BR and SBR methods).

Interpretation and Implications. *Nephrolenellus multinodus* and *N. geniculatus* show broadly similar trends in morphological differentiation. Each undergoes a marked shift in its mode of shape differentiation on entry into phase 2 and on entry into either phase 3 (*N. geniculatus*) or phase 4 (*N. multinodus*). These

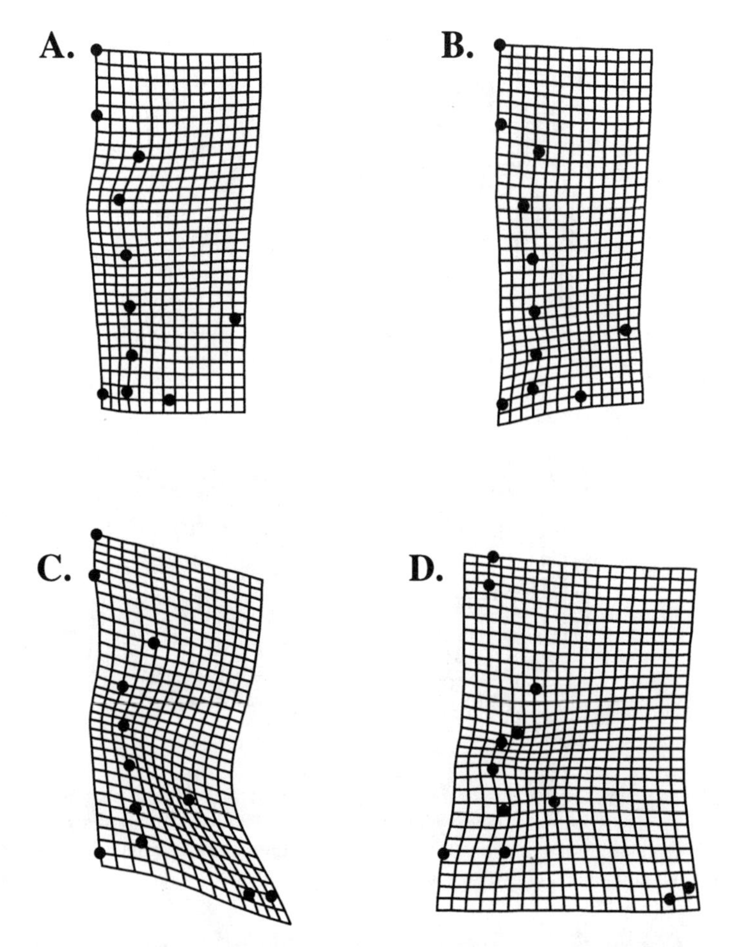

Figure 4.8. Thin-plate spline projections depicting ontogenetic shape changes in *Nephrolenellus multinodus.* A: phase 1, landmark configuration 1. B: phase 2, landmark configuration 1. C: phase 3, landmark configuration 3. D: phase 4, landmark configuration 3. Reference and target forms are consensus of two smallest and largest specimens in each phase, respectively (except for phase 3 and small phase 4 forms, which are based on single specimens).

analyses support similar findings for the ontogenetic trajectories of these species based on morphometric analysis of glabellar landmarks only (unpublished data).

Confirmation of the nonlinearity of ontogenetic trajectories for these species constrains interspecific comparison of developmental patterns of shape differentiation (to test for allometric conservation) to a phase-by-phase basis. Comparison of entire trajectories will bias results in that nonlinear vectors would be fitted with a linear regression model (the only model available), thus inflating calculated within-trajectory variance and decreasing the significance of any between-trajectory differences. Because arguments of heterochrony rely on

Table 4.4. Results of within-species comparison of patterns of shape differentiation between developmental phases of *Nephrolenellus multinodus*. Comparisons based on vectors calculated using Bookstein registration (BR), sliding baseline registration (SBR), and thin-plate spline analysis (TPS) of landmark data. Within-group angles calculated from 400 bootstraps, given at 95% confidence limit (earliest phase first). Config. = landmark configuration used in comparison. Sig. = Statistical significance of between-phase angle. See text for details.

		BR			SBR			TPS		
Phase Transition	Config.	Between Group Angle	Within Group Angle	Sig. (%)	Between Group Angle	Within Group Angle	Sig. (%)	Between Group Angle	Within Group Angle	Sig. (%)
1 to 2	1	126	78.4/83.1	95	104.7	79.5/100.3	95	103	80.4/100.1	95
2 to 3	2	73.3	112.3/30.6	<80	71.2	118.2/35.9	<80	72.8	118.2/36.2	<80
3 to 4	3	58.5	20.4/55.6	95	49.1	27.1/46.0	95	37.1	21.9/58.3	<80

conservation of a shared trajectory between ancestor and descendant, an entire-ontogeny comparison is deemed unsuitable.

Between-Species Comparison

Statistically significant (to 95% confidence) patterns of shape differentiation between *N. multinodus* and *N. geniculatus* during phase 1 of development were found when ontogenetic vectors were compared using the TPS method (Table 4.5). The BR and SBR superimposition methods found those differences significant to 90% confidence. Large within-phase variance in both species renders patterns of shape differentiation statistically indistinguishable between *N. multinodus* and *N. geniculatus* during phase 2 (Table 4.5). Patterns of shape differentiation differ significantly between the species during phase 3 using any method of vector calculation. The TPS method found significant differences between the species during phase 4 (to 95% confidence), but the BR and SBR methods found those differences to be significant to only 80% and 90% confidence, respectively (due to large within-phase variance exhibited by *N. geniculatus*).

INTERPRETATION

The case for evolution of *N. geniculatus* from its sister species *N. multinodus* by global peramorphic extension of the ancestral ontogeny is rather convincing at first glance. Three features serve to differentiate the species (fewer nodes on the glabella axis, closer proximity of the frontal lobe of the glabella to the anterior border, and more angular adgenal angle in *N. geniculatus*), and each of these is an extension of a general ontogenetic trend in the ancestor (see above). Furthermore, the two species share outwardly similar ontogenetic trajectories in that each undergoes significant changes in its mode of allometric shape differentiation on entry into developmental phase 2 and a later phase (see within-species ontogenetic trajectories). However, the key question when testing for the existence of heterochrony is whether the patterns of shape differentiation during development are shared between the species, since heterochrony requires that the ontogenetic trajectory be conserved between ancestor and descendant.

After quantification of the degree of similarity in ontogenetic trajectories between these species, the null hypothesis of allometric conservation is rejected. *Nephrolenellus geniculatus* and *N. multinodus* differ significantly in their modes of shape differentiation during phases 1, 3, and 4 of development (to 95% confidence using the TPS method of vector comparison). The BR and SBR methods support this argument equally strongly for phase 3 differences, but less so for phase 1 and 4 differences (typically to 90% confidence).

Very large angles were generated between the species at all developmental phases using any method of vector calculation (Table 4.5). Had the identification of heterochrony (i.e., allometric conservation) required an angle of 0° between the species, then allometric repatterning would have been found across

Table 4.5. Results of between-species comparison of patterns of shape differentiation between *Nephrolenellus multinodus* and *Nephrolenellus geniculatus*. Comparisons based on vectors calculated using Bookstein registration (BR), sliding baseline registration (SBR), and thin-plate spline analysis (TPS) of landmark data. Within-group angles calculated from 400 bootstraps, given at 95% confidence limit (*N. multinodus* first). Config. = landmark configuration used in comparison. Sig. = Statistical significance of between-phase angle. See text for details.

		BR			SBR			TPS		
Phase Comparison	Config.	Between Group Angle	Within Group Angle	Sig. (%)	Between Group Angle	Within Group Angle	Sig. (%)	Between Group Angle	Within Group Angle	Sig. (%)
1	1	93.8	78.4/94.8	90	90.2	79.5/93.0	90	99.4	80.4/97.5	95
2	1	54.3	82.7/97.9	<80	49.3	102.2/96.8	<80	48.5	102.6/93.9	<80
3	3	39.3	20.4/32.8	95	43.1	27.1/32.1	95	38.8	21.9/38.6	95
4	4	70	69.4/95.2	80	94.2	56.9/98.0	90	86.7	54.7/76.0	95

the board. However, the requirements for demonstration of allometric conservation were considerably relaxed by allowing within-phase variance to place confidence limits on the angles: such relaxation rendered some interspecific angles as large as 94° significant to only 90% confidence and angles of 70° to only 80% confidence. The analytical method employed herein provides a rigorously testable hypothesis of allometric conservation (and therefore heterochrony) but should not be considered "too strict."

The ontogeny of *N. geniculatus* is not an extrapolation of a shared trajectory with *N. multinodus*. Rather, a significant divergence between the ontogenetic trajectories of these species is apparent at least as early as phase 3 of development and (depending on choice of analytical method) perhaps even as early as phase 1. Evolution of *N. geniculatus* from *N. multinodus* cannot be explained by a global (in terms of the cephalon) peramorphic extension of the ancestral ontogeny. Instead, the explanation must account for localized modification of cephalic growth fields.

Dissociated Heterochrony?

Significantly different modes of allometric patterning between *N. multinodus* and *N. geniculatus* are demonstrated in the above analyses. Such analyses were founded on morphological coverage of the entire cephalon. Interspecific differences therefore cannot be attributed to heterochrony affecting the cephalon "globally." It could still be argued that they result from localized (dissociated) heterochrony, in which growth of individual structures or regions is modified by heterochrony. However, vectors of landmark movement on the cephalon of each species (Fig. 4.9) show that interspecific differences are not limited to single morphological regions or structures. Allometric repatterning is evident across the entire cephalon, producing highly complex differences in shape that are more parsimoniously interpreted as modifications in spatial patterning during development (heterotopy) rather than a large number of local heterochronic modifications.

Biological Signal or Taphonomic Noise?

An obvious cause for concern when interpreting results from morphometric analyses of fossils relates to how much the perceived difference in shape between specimens is "nonbiological" in nature. Taphonomic processes such as compaction have been demonstrated to significantly alter the appearance of fossils in both vertical and bedding-parallel planes (e.g., Webster and Hughes, 1999). Basic "biological" trends may not be obscured if the pattern and magnitude of compaction-related deformation is uniform across all specimens in a sample (the effect of such deformation might be to increase the degree of observed scatter of data points around a regression line). However, spurious results may be obtained if the pattern or magnitude of compaction-related deformation is nonuniform across the sample. Such conditions may be met if

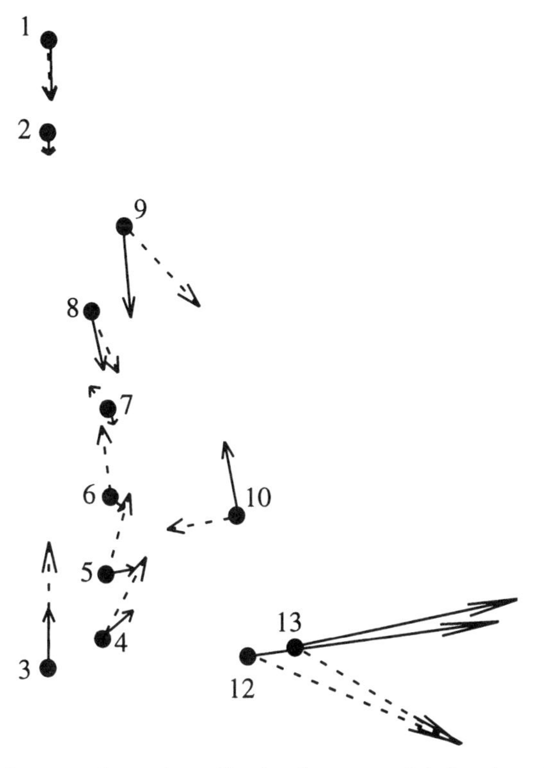

Figure 4.9. Plots of comparative vectors of landmark movement during phase 3 of development in *Nephrolenellus multinodus* (dashed arrows) and *N. geniculatus* (solid arrows). Vectors calculated using the sliding baseline registration superimposition method. Numbers identify landmarks (see Fig. 4.5). See text for details.

the sample is nonhomogeneous in the mode of preservation (i.e., some specimens preserved in an uncompacted, silicified state and others preserved in a compacted state in shales).

There appears to be a size-related bias in the preservation of cephala in a silicified state (Webster and Hughes, 1999) in that cephala in phase 4 of development are rarely found intact. Conversely, small cephala in developmental phase 3 or younger are rarely found in shales (most likely owing to difficulty of discovery). Compaction has been demonstrated to cause significant lateral movement of cephalic features in olenelloid trilobites preserved in shales when compared with uncompacted conspecific cephala of similar size preserved in a silicified state (Webster and Hughes, 1999). However, the observed differences in allometric patterning both within and between the ontogenetic trajectories of the species are unlikely to result from differences in taphonomic history. All specimens preserved in a compacted state in shales were of phase 4 morphology. No significant difference in allometric patterning can be found between phase 4 specimens of *N. geniculatus* preserved in a silicified state and those preserved in shales (unpublished data), suggesting that compaction-related deformation introduces only uniform bias at most. There are therefore no grounds for suspecting that the shifts in allometric patterning detected within the ontogenetic trajectories of *N. geniculatus* or *N. multinodus* are artifacts of preservation.

CONCLUSIONS

The hypothesis that *N. geniculatus* evolved from *N. multinodus* by pure global peramorphosis is not supported by detailed morphometric analyses. Rather than extrapolating an otherwise shared ontogenetic trajectory (in terms of patterns of growth), *N. geniculatus* follows a unique trajectory from at least phase 3 of development and arguably as early as phase 1. Differences in allometric patterning are not easily explained as a number of local heterochronic modifications. This indicates that differences in spatial patterning during development (heterotopy) exist.

The results presented here take on added significance because the case for evolution of *N. geniculatus* from *N. multinodus* by global peramorphosis is probably the best supported case for heterochronic evolution in trilobites. No other purported cases meet the basic criteria that the phylogeny and ontogenetic trajectories of the taxa in question must be known (with as much certainty as the fossil record allows). Here, the species were demonstrably sister taxa (arguably anagenetically related) and showed very similar ontogenies. The characters used to diagnose the species were all easily attributed to extension of a shared ontogenetic pathway. The fact that even this case fails to pass the strict test for evolution by pure heterochrony suggests that many of the less strongly supported cases of heterochronic evolution in trilobites reported in the literature should be questioned. Claims that heterochrony played a dominant role in trilobite phylogeny (McNamara, 1986, 1988; Fortey and Owens, 1990) may be founded on gross overestimates of the abundance of heterochronic evolution in these organisms. Certainly, previous examples must be reevaluated in light of, and new cases framed within, testable hypotheses that allow for alternative explanations.

It is hoped that this study will increase awareness that information pertinent to the field of evolutionary development can be gleaned from the fossil record and that paleontological data should not be ignored as necessarily speculative. This study is the first to rigorously test for heterochrony in trilobites using a multivariate, dynamic comparison of ancestor and descendant ontogenies within a well-constrained phylogenetic framework. Multivariate analyses of the ontogenetic development of other trilobite species offer the potential for more such comparisons, but the few taxa for which such data exist are phylogenetically too disparate for detailed testing of heterochronic hypotheses (Hughes, 1994; Hughes and Chapman, 1995; Crônier et al., 1998; Hughes et al., 1999).

Clearly, there is room for much research. Empirical studies of trilobite paleobiology offer an important resource of evolutionary data for the future.

In regard to the suite of factors involved in organismal adaptation to various environments, Whittington (1957, p. 460) wrote "The operation of these varied factors is extremely complex, and difficult to unravel from the fossil record, but the attempt will lead to a better understanding than will the tendency to overemphasize the importance of any one process." Heterochrony has been used as an elegant and persuasive explanation for many evolutionary changes in the fossil record. However, absence of rigorous tests to assess hypotheses of heterochrony has undoubtedly led to overestimation of the frequency of this mode of evolution at the expense of other potential factors, such as heterotopy. Morphometric techniques now exist that enable the complexities of ontogenetic growth to be unraveled. With use of such techniques comes the hope of more profound insights into the mode of evolution and development.

ACKNOWLEDGMENTS

The authors express their gratitude to Miriam Zelditch for the opportunity to include this work in the present volume and Carol Waddell-Sheets for encouraging H. D. Sheets to enter the field of mathematical paleontology. Character states for species in the cladistic analysis were independently coded by A. R. (Pete) Palmer and Ed Fowler to check consistency and robustness. Comments from Pete Sadler helped improve the manuscript. Research was funded in part by NSF Grant EAR-9980372.

REFERENCES

Adrain JM, Chatterton BDE (1994): The aulacopleurid trilobite *Otarion*, with new species from the Silurian of northwestern Canada. J Paleontol 68: 305–323.

Alberch P, Gould SJ, Oster GF, Wake DB (1979): Size and shape in ontogeny and phylogeny. Paleobiology 5: 296–317.

Arthur W (2000): The concept of developmental reprogramming and the quest for an inclusive theory of evolutionary mechanisms. Evol Dev 2: 49–57.

Beecher CE (1897): Outline of a natural classification of the trilobites. Am J Sci 3: 89–106, 181–207.

Blackstone NW (1987a): Allometry and relative growth: pattern and process in evolutionary studies. Syst Zool 36: 76–78.

Blackstone NW (1987b): Size and time. Syst Zool 36: 211–215.

Blackstone NW, Yund PO (1989): Morphological variation in a colonial marine hydroid: a comparison of size-based and age-based heterochrony. Paleobiology 15: 1–10.

Bookstein FL (1986): Size and shape spaces for landmark data in two dimensions (with comments and rejoinder). Stat Sci 1: 181–242.

Bookstein FL (1989): Principal warps: thin-plate splines and the decomposition of deformations. IEEE Trans Pattern Anal Machine Intelligence 11: 567–585.

Bookstein FL (1991): *Morphometric Tools for Landmark Data: Geometry and Biology.* New York: Cambridge University Press.

Chatterton BDE, Perry DG (1983): Silicified Silurian odontopleurid trilobites from the Mackenzie Mountains. Palaeontogr Can 1: 1–126.

Chatterton BDE, Siveter DJ, Edgecombe GD, Hunt AS (1990): Larvae and relationships of the Calymenina (Trilobita). J Paleontol 64: 255–277.

Chatterton BDE, Speyer SE (1990): Applications of the study of trilobite ontogeny. In: Mikulic DG (ed), *Short Courses in Paleontology, Arthropod Paleobiology.* Knoxville, TN: Paleontological Society, no 3, p 116–136.

Chatterton BDE, Speyer SE (1997): Ontogeny. In: Kaesler RL (ed), *Treatise on Invertebrate Paleontology. Part O. Arthropoda 1: Trilobita, Revised.* Boulder and Lawrence: Geological Society of America and University of Kansas, vol 1, p 173–247.

Clark RB (1964): *Dynamics of Metazoan Evolution.* Oxford: Oxford University Press.

Clarkson ENK (1971): On the early schizocroal eyes of *Ormathops* (Trilobita, Zeliszkellinae). Mem Bur Recherches Geol Miniéres 73: 51–63.

Clarkson ENK (1975): The evolution of the eye in trilobites. Fossils Strata 4: 7–31.

Clarkson ENK (1979): The visual system of trilobites. Palaeontology 22: 1–22.

Crônier C, Renaud S, Feist R, Auffray J-C (1998): Ontogeny of *Trimerocephalus lelievrei* (Trilobita, Phacopida), a representative of the Late Devonian phacopine paedomorphocline: a morphometric approach. Paleobiology 24: 359–370.

Dryden IL, Mardia KV (1998): *Statistical Shape Analysis.* New York: John Wiley and Sons.

Edgecombe GD, Chatterton BDE (1987): Heterochrony in the Silurian radiation of encrinurine trilobites. Lethaia 20: 337–351.

Efron B (1982): *The Jackknife, the Bootstrap and Other Resampling Plans.* Philadelphia, PA: Society for Industrial and Applied Mathematics.

Efron B, Tibshirani RJ (1993): *An Introduction to the Bootstrap.* New York: Chapman and Hall.

Feist R (1995): Effect of paedomorphosis in eye reduction on patterns of evolution and extinction in trilobites. In: McNamara KJ (ed), *Evolutionary Change and Heterochrony.* New York: John Wiley and Sons, p 226–244.

Fink WL (1982): The conceptual relationship between ontogeny and phylogeny. Paleobiology 8: 254–264.

Fink WL (1988): Phylogenetic analysis and the detection of ontogenetic patterns. In: McKinney ML (ed), *Heterochrony in Evolution: A Multidisciplinary Approach.* New York: Plenum Press, p 71–91.

Fortey RA (1974): The Ordovician trilobites of Spitsbergen. I. Olenidae. Norsk Polarinstitutt Skrifter 160: 1–181.

Fortey RA (1975): The Ordovician trilobites of Spitsbergen. II. Asaphidae, Nileidae, Raphiophoridae and Telephinidae of the Valhallfonna Formation. Nor Polarinst Skr 162: 1–207.

Fortey RA (1990): Ontogeny, hypostome attachment and trilobite classification. Palaeontology 33: 529–576.

Fortey RA, Owens RM (1987): The Arenig Series in South Wales. Bull Br Mus (Nat Hist) Geol 41: 69–307.

Fortey RA, Owens RM (1990): Trilobites. In: McNamara KJ (ed), *Evolutionary Trends*. London: Belhaven Press, p 121–142.

Fortey RA, Rushton AWA (1980): *Acanthopleurella* Groom 1902: origin and life-habits of a miniature trilobite. Bull Br Mus (Nat Hist) Geol 33: 79–89.

Godfrey LR, Sutherland MR (1995a): Flawed inference: why size-based tests of heterochronic processes do not work. J Theoret Biol 172: 43–61.

Godfrey LR, Sutherland MR (1995b): What's growth got to do with it? Process and product in the evolution of ontogeny. J Hum Evol 29: 405–431.

Godfrey LR, Sutherland MR (1996): Paradox of peramorphic paedomorphosis: heterochrony and human evolution. Am J Phys Anthropol 99: 17–42.

Gould SJ (1977): *Ontogeny and Phylogeny*. Cambridge: Harvard University Press.

Gould SJ (1992): Heterochrony. In: Keller EF, Lloyd EA (eds), *Keywords in Evolutionary Biology*. Cambridge: Harvard University Press, p 158–165.

Gould SJ (2000): Of coiled oysters and big brains: how to rescue the terminology of heterochrony, now gone astray. Evol Dev 2: 241–248.

Henry J-L, Vizcaïno D, Destombes J (1992): Evolution de l'oeil et hétérochronie chez les Trilobites ordoviciens *Ormathops* DELO 1935 et *Toletanaspis* RABANO 1989 (Dalmanitidae, Zeliszkellinae). Palaeontol Z 66: 277–290.

Hughes NC (1994): Ontogeny, intraspecific variation, and systematics of the Late Cambrian trilobite *Dikelocephalus*. Smithson Contrib Paleobiol 79.

Hughes NC, Chapman RE (1995): Growth and variation in the Silurian proetide trilobite *Aulacopleura konincki* and its implications for trilobite palaeobiology. Lethaia 28: 333–353.

Hughes NC, Chapman RE, Adrain JM (1999): The stability of thoracic segmentation in trilobites: a case study in developmental and ecological constraints. Evol Dev 1: 24–35.

Hunt AS (1967): Growth, variation, and instar development of an agnostid trilobite. J Paleontol 41: 203–208.

Hupé P (1954): Classification des trilobites. Ann Paleontol 39: 61–168.

Jaekel O (1909): Über die Agnostiden. Z Dtsch Geol Ges 61: 380–401.

Jell PA (1975): Australian Middle Cambrian eodiscoids with a review of the superfamily. Palaeontographica (Abt. A) 150: 1–97.

Kendall DG (1977): The diffusion of shape. Adv Appl Probab 9: 428–430.

Klingenberg CP (1998): Heterochrony and allometry: the analysis of evolutionary change in ontogeny. Biol Rev 73: 79–123.

Lieberman BS (1998): Cladistic analysis of the early Cambrian olenelloid trilobites. J Paleontol 72: 59–78.

Lieberman BS (1999): Systematic revision of the Olenelloidea (Trilobita, Cambrian). Bull Peabody Mus Nat Hist 45: 1–150.

Ludvigsen R (1979): *Fossils of Ontario. Part 1: The Trilobites*. Ontario: Life Sciences Miscellaneous Publications, Royal Ontario Museum, p 1–96.

McKinney ML (1988): Classifying heterochrony: Allometry, size, and time. In: McKinney ML (ed), *Heterochrony in Evolution: A Multidisciplinary Approach*. New York: Plenum Press, p 17–34.

McKinney ML, McNamara KJ (1991): *Heterochrony: The Evolution of Ontogeny.* New York: Plenum Press.

McNamara KJ (1978): Paedomorphosis in Scottish olenellid trilobites (Early Cambrian). Palaeontology 21: 635–655.

McNamara KJ (1981a): Paedomorphosis in Middle Cambrian xystridurine trilobites from northern Australia. Alcheringa 5: 209–224.

McNamara KJ (1981b): The role of paedomorphosis in the evolution of Cambrian trilobites. USGS Open-file Report 81–743: 126–129.

McNamara KJ (1983): Progenesis in trilobites. Spec Pap Palaeontol 30: 59–68.

McNamara KJ (1986): The role of heterochrony in the evolution of Cambrian trilobites. Biol Rev 6: 121–156.

McNamara KJ (1988): The abundance of heterochrony in the fossil record. In: McKinney ML (ed), *Heterochrony in Evolution: A Multidisciplinary Approach.* New York: Plenum Press, p 287–325.

Müller KJ, Walossek D (1987): Morphology, ontogeny and life-habit of *Agnostus pisiformis* from the Upper Cambrian of Sweden. Fossils Strata 19: 1–124.

Nedin C, Jenkins RJF (1999): Heterochrony in the Cambrian trilobite *Hsuaspis.* Alcheringa 23: 1–7.

O'Keefe FR, Rieppel O, Sander PM (1999): Shape dissociation and inferred heterochrony in a clade of pachypleurosaurs (Reptilia, Sauropterygia). Paleobiology 25: 504–517.

Palmer AR (1957): Ontogenetic development of two olenellid trilobites. J Paleontol 31: 105–128.

Palmer AR (1998): Terminal Early Cambrian extinction of the Olenellina: documentation from the Pioche Formation, Nevada. J Paleontol 72: 650–672.

Palmer AR, Halley RB (1979): Physical stratigraphy and trilobite biostratigraphy of the Carrara Formation (Lower and Middle Cambrian) in the southern Great Basin. United States Geological Survey Professional Paper 1047: 1–131.

Palmer AR, Repina LN (1993): Through a glass darkly: taxonomy, phylogeny, and biostratigraphy of the Olenellina. Univ Kans Paleontol Contrib New Series 3.

Palmer AR, Repina LN (1997): Suborder Olenellina. In: Kaesler RL (ed). *Treatise on Invertebrate Paleontology. Part O. Arthropoda 1: Trilobita, Revised.* Boulder and Lawrence: Geological Society of America and University of Kansas, vol. 1, p 405–429.

Ramsköld L (1988): Heterochrony in Silurian phacopid trilobites as suggested by the ontogeny of *Acernaspis.* Lethaia 21: 307–318.

Resser CE (1928): Cambrian fossils from the Mohave Desert. Smithson Misc Collect 81: 1–14.

Rice SH (1997): The analysis of ontogenetic trajectories: when a change in size or shape is not heterochrony. Proc Natl Acad Sci USA 94: 907–912.

Robison RA (1967): Ontogeny of *Bathyuriscus fimbriatus* and its bearing on the affinities of the corynexochid trilobites. J Paleontol 41: 213–221.

Robison RA, Campbell DP (1974): A Cambrian corynexochid trilobite with only two thoracic segments. Lethaia 7: 273–282.

Schweitzer PN, Lohmann GP (1990): Life-history and the evolution of ontogeny in the ostracode genus *Cyprideis.* Paleobiology 16: 107–125.

Shaw AB (1957): Quantitative trilobite studies II. Measurement of the dorsal shell of non-agnostidean trilobites. J Paleontol 31: 193–207.

Smith GR (1990): Homology in morphometrics and phylogenetics. In: Rohlf FJ, Bookstein FL (eds), *Proceedings of the Michigan Morphometrics Workshop. Special Publication Number 2.* Ann Arbor, MI: University of Michigan Museum of Zoology, p 325–338.

Smith LH (1998): Asymmetry of Early Paleozoic trilobites. Lethaia 31: 99–112.

Størmer L (1942): Studies on trilobite morphology. Part II: The larval development, the segmentation and the sutures, and their bearing on trilobite classification. Nor Geol Tidsskr 21: 49–164.

Strauss RE (1987): On allometry and relative growth in evolutionary studies. Syst Zool 36: 72–75.

Stubblefield CJ (1936): Cephalic sutures and their bearing on current classifications of trilobites. Biol Rev 11: 407–440.

Stubblefield CJ (1959): Evolution in trilobites. Q J Geol Soc Lond 115: 145–162.

Sundberg FA (2000): Homeotic evolution in Cambrian trilobites. Paleobiology 26: 258–270.

Swiderski DL (1993): Morphological evolution of the scapula in tree squirrels, chipmunks, and ground squirrels (Sciuridae): an analysis using thin-plate splines. Evolution 47: 1854–1873.

Swofford DL (1998): *PAUP*. Phylogenetic Analysis Using Parsimony* (*and other methods). Version 4. Sunderland: Sinauer Associates.

Tissot BN (1988): Multivariate analysis. In: McKinney ML (ed), *Heterochrony in Evolution: A Multidisciplinary Approach.* New York: Plenum Press, p 35–51.

Webster M (1999): Paleobiologic aspects of olenelloid trilobites from the uppermost Dyeran C-Shale Member of the Pioche Formation, Nevada. Unpublished MS Thesis, University of California, Riverside, CA.

Webster M, Hughes NC (1999): Compaction-related deformation in Cambrian olenelloid trilobites and its implications for fossil morphometry. J Paleontol 73: 355–371.

Whittington HB (1957): The ontogeny of trilobites. Biol Rev 32: 421–469.

Whittington HB (1981): Paedomorphosis and cryptogenesis in trilobites. Geol Mag 118: 591–602.

Whittington HB (1989): Olenelloid trilobites: type species, functional morphology and higher classification. Phil Trans R Soc Lond B 324: 111–147.

Whittington HB, Kelly SRA (1997): Morphological terms applied to Trilobita. In: Kaesler RL (ed), *Treatise on Invertebrate Paleontology. Part O. Arthropoda 1: Trilobita, Revised.* Boulder and Lawrence: Geological Society of America and University of Kansas, vol 1, p 313–329.

Zelditch ML, Bookstein FL, Lundrigan BL (1992): Ontogeny of integrated skull growth in the cotton rat *Sigmodon fulviventer*. Evolution 46: 1164–1180.

Zelditch ML, Bookstein FL, Lundrigan BL (1993): The ontogenetic complexity of developmental constraints. J Evol Biol 6: 621–641.

Zelditch ML, Fink WL (1995): Allometry and developmental integration of body growth in a piranha, *Pygocentrus nattereri* (Teleostei: Ostariophysi). J Morphol 223: 341–355.

Zelditch ML, Fink WL (1996): Heterochrony and heterotopy: stability and innovation in the evolution of form. Paleobiology 22: 241–254.

Zelditch ML, Sheets HD, Fink WL (2000): Spatiotemporal reorganization of growth rates in the evolution of ontogeny. Evolution 54: 1363–1371.

Zhou Z, McNamara KJ, Yuan W, Zhang T (1994): Cyclopygid trilobites from the Ordovician of northeastern Tarim, Xinjiang, northwest China. Rec West Aust Mus 16: 593–622.

APPENDIX

Cladistic Analysis of *Nephrolenellus*

Characters and character states used in the cladistic analysis of *Nephrolenellus*. States are ordered arbitrarily; "0" does not imply primitive state. Terminal taxa are coded according to the state expressed in the holaspid stage.

1. Transverse line between posterior margins of genal spines at points of contact with cephalic border crosses glabella at (0) occipital ring; (1) L1; (2) L2; (3) L3; (4) L4. (This character partly determines cephalic outline. Where line crosses glabella at occipital furrow, S1, S2, or S3, code for L1, L2, L3, and L4, respectively.)
2. Adgenal angle deflects distal portion of posterior cephalic margin anteriorly (relative to transverse line) by (0) less than 45°; (1) 45° to 60°; (2) approximately 90°. (This character partly determines cephalic outline but is considered independent of character 1.)
3. L4 and ocular lobes: (0) separated by shallow furrow; (1) merge smoothly together. (Must be coded from specimens showing minimal compactional deformation. Where contact is marked by a change in slope but no distinct furrow, code 1.)
4. Terminal ontogenetic stage: (0) cCDF phase 5; (1) cCDF phase 4. (This character determines features such as isolation of S2 from axial furrow, shape of L2, and location of maximal glabellar constriction.)
5. Axial furrow at abaxial margins of L1: (0) well defined; (1) absent, L1 merges into interocular area. (Must be coded from specimens showing minimal compactional deformation.)
6. Orientation of S3, measuring away from sagittal axis, directed: (0) anterolaterally; (1) posterolaterally; (2) pit-like, no clear orientation.
7. Glabellar nodes present on (0) occipital ring only; (1) occipital ring and L1; (2) occipital ring, L1, and L2; (3) occipital ring, L1, L2, and L3. (Spines and nodes are here considered homologous and are not differentiated.)
8. Ocular lobes diverge from glabellar axis: (0) at about 10–20° angle, interocular area relatively narrow (tr.); (1) at about 45° angle, interocular

area relatively broad (tr.). (The angle is measured between the sagittal axis and a chord along the arc of the ocular lobe.)

9. Number of prothoracic segments: (0) 14; (1) 13.
10. Axial spine on first opisthothoracic segment: (0) present; (1) absent.
11. L4: (0) impinges on anterior border; (1) does not impinge on anterior border.
12. Length of genal spine: (0) short, less than two-thirds sagittal cephalic length; (1) long, greater than sagittal cephalic length.

5

THE SPATIAL COMPLEXITY AND EVOLUTIONARY DYNAMICS OF GROWTH

Miriam Leah Zelditch
Museum of Paleontology, University of Michigan, Ann Arbor, Michigan

H. David Sheets
Department of Physics, Canisius College, Buffalo, New York

William L. Fink
Museum of Zoology, Department of Biology, University of Michigan, Ann Arbor, Michigan

INTRODUCTION

Evolutionary biologists often study growth as a means to an end—the end being a causal explanation for morphological novelties. But growth is not merely a means to an (adult) end, it is also an interesting and complex phenomenon in its own right, involving interactions among numerous genes, gene products, cells, and tissues, all of which are organized in both space and time. Taken separately, growth processes may be simple, but, taken together, they form a complex system. For that reason, growth is a useful model for the evolution of

Beyond Heterochrony: The Evolution of Development, Edited by Miriam Leah Zelditch
ISBN 0-471-37973-5

complex systems as well as for the evolution of development. Moreover, growth is often studied quantitatively, in terms commensurate with the formalisms of evolutionary theory. Because of this common language, studies of growth offer a bridge between developmental and evolutionary theory. However, remarkably few studies focus on the specifically developmental aspects of allometry.

One neglected developmental aspect of growth is its spatial organization. Huxley (1932), who pioneered quantitative studies of growth, concluded that growth rates typically reveal an orderly spatial pattern. For example, growth rates of appendicular structures tend to grow along a linear proximodistal gradient, gradients coordinated from appendage to appendage along the anteroposterior body axis. Such regular patterns are also found in studies of whole organisms, suggesting a global coordination of growth rates. In a study of herring (*Clupea harengus*) near the transition from larval to juveniles phases, Huxley found that growth rates are highest at the two ends of the anteroposterior body axis, decelerating toward the middle, a pattern called a U-shaped gradient (Fuiman, 1983). Processes responsible for such orderly and spatially integrated patterns can have profound evolutionary consequences, like other sources of developmental integration.

Whether such orderly patterns are common and whether such highly integrated patterns are constraints on evolution are open questions. A study of larval teleosts suggests that the U-shaped gradient is widely shared (Fuiman, 1983), a conclusion tentatively supported by several more recent studies (e.g., Koumoundouros et al, 1999; Sarpédonti et al., 2000). However, this gradient is not found in larval sculpins (Cottidae) (Strauss and Fuiman, 1985), so it is not universally present in larvae. When it occurs, it may be special to larval growth, as Fuiman found no evidence of gradients in juvenile growth. However, he did find a pattern almost as orderly: nearly isometric growth of lengths over the whole body. Other studies have found neither gradients nor isometry; although they suggest that growth rates are far from haphazard, they nevertheless indicate that growth rates have a spatially complex distribution, with localized accelerations and decelerations (Zelditch and Fink, 1995; Reis et al., 1998). If indeed complex, juvenile growth might be historically labile, being less constrained by factors responsible for spatial integration. This inference follows the line of reasoning of the decoupling hypothesis: the greater the number of independent components, the greater potential for diversification (e.g., Liem, 1973; Vermeij, 1973; Lauder, 1981). This expectation seems consistent with the pattern found in sculpins; not only are juvenile allometries quite diverse, they are also more diverse than those of larvae (based on data in Strauss and Fuiman, 1985; reanalyzed).

Paradoxically, an extensive review of the literature concludes that later phases of growth are typically conservative, in contrast to early ones (Klingenberg, 1998), so the less spatially integrated (juvenile) phase seems the more constrained in its evolutionary transformations. At least three explanations can be suggested to reconcile these apparently contradictory patterns. First, juvenile growth may actually be coherent and spatially orderly. Second, juvenile growth may be less conservative than recognized. Third, the spatial integration of growth rates may be immaterial for the evolutionary dynamics of growth. After all, spatial organization itself might evolve; on purely a priori grounds, there is no reason to suspect it is conserved (Zelditch and Fink, 1996).

Our two objectives herein are first to determine whether juvenile growth rates fit the simple gradient model found in teleost larvae or other equally simple models of globally graded rates. Using these simple models, we examine deviations from them for regularities suggesting more realistic but equally orderly models. Second, we determine whether spatial patterns of growth rates are conservative, being modified solely by extension or truncation or, instead, reveal structural changes in their spatial patterns.

PHYLOGENETIC CONTEXT

The animals used in this research are piranhas, South American fresh-water characiform fishes. Piranhas are morphologically and ecologically diverse, with bodies ranging from shallow and elongate to discoid and diets ranging from fins and scales to fruits and seeds (Goulding, 1980). There are about 60 nominal species of piranhas, but the actual number remains unclear. Those used in this study were selected to represent body shape diversity in the group as well as phylogenetic relationships at different levels in the phylogenetic hierarchy. Sample sizes, and the range of body sizes within each sample, are shown in Table 5.1.

The phylogeny we accept (Fig. 5.1) is supported by numerous morphological features (but see below). *Pygopristis denticulata* (Cuvier), the most morphologically generalized member of the group, is the sister group to the others. Our

Table 5.1. Sample sizes and range of body sizes (in mm), measured as standard length (between landmarks 1 and 7, Fig. 5.2)

Species	Sample Size	Size Range
Pygopristis denticulata	32	23.1–182.1
Serrasalmus elongatus	46	28.4–227.0
S. gouldingi	38	28.7–282.8
S. manueli	40	39.9–190.0
S. spilopleura	37	74.0–176.0
Pygocentrus piraya	34	16.0–322.0
P. nattereri	89	20.5–277.5
P. cariba	21	24.5–193.9

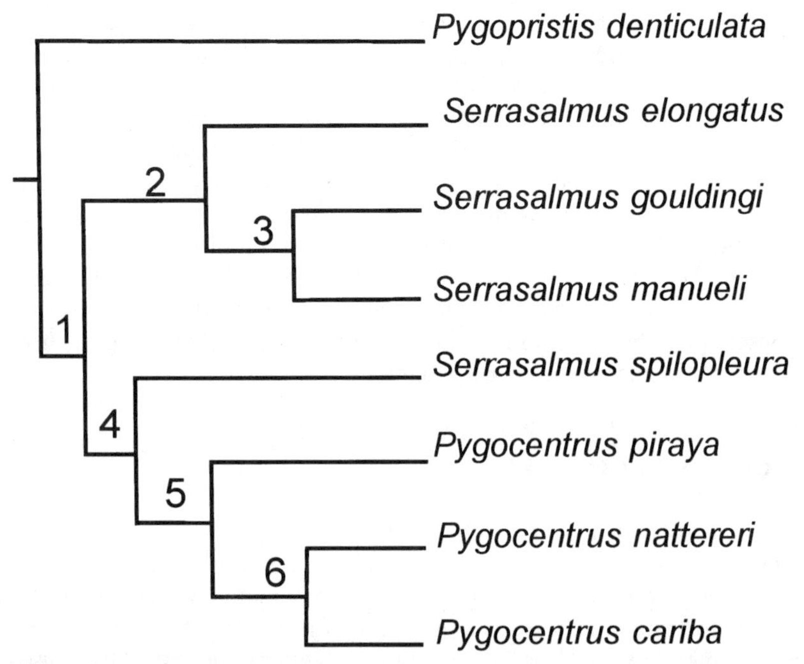

Figure 5.1. Cladogram depicting relationships among the eight species analyzed herein, as inferred from morphological data.

sampling of this species includes specimens from both the Amazon and Orinoco river drainages; there is no evidence that more than one species is represented here. *Pygocentrus* is a well-supported clade and includes three species traditionally considered very similar to each other; sampled specimens are listed in Fink (1993). *Pygocentrus nattereri* Kner is distributed widely in the Amazon and Paraguay/Parana river basins, *P. cariba* (Humboldt) is found in the Orinoco drainage and tributaries, and *P. piraya* (Cuvier) is restricted to the Rio São Francisco of Brazil. There are four species of *Serrasalmus* in the analysis. One group, clade 2 in Figure 5.1, represents a sample from about 20 species. *S. elongatus* Kner is included because it has the shallowest body form of all piranhas, and it is a specialized scale and fin eater; our specimens come from both the Amazon and Orinoco river drainages. *S. gouldingi* Fink and Machado-Allison and *S. manueli* (Fernandez-Yepez and Ramirez), two deep-bodied species thought to be sister taxa, are also members of clade 2. These two species are included in this study because they are morphologically indistinguishable as small juveniles but diverge in shape during ontogeny. Our samples of both species come from a relatively restricted area in the Anavilhanas archipelago in the lower Rio Negro of Brazil (Fink and Machado-Allison, 1992). *Serrasalmus spilopleura* Kner, representing a complex of perhaps five species, is of uncertain relationships; indeed, our recognition of these specimens as *S. spilopleura* is tentative, pending revision of the group. These specimens are almost all from the lower Rio Madeira of Brazil. Previous studies based primarily on osteology place *S. spilopleura* well within the genus (Machado-Allison, 1985); some soft morphology, such as the gas bladder, and some molecular evidence (Orti, et al., 1996) suggest that this species may be more closely related to *Pygocentrus* than to other *Serrasalmus*. For the purposes of this paper, we accept the latter hypothesis but do not make nomenclatural changes to match it as this is not the appropriate venue.

METHODS

Size and Shape Data

To infer the spatial distribution of growth rates from measures of size and shape, we use three sets of variables, all based on the locations of discrete, homologous anatomical landmarks, which are either visible or palpable on all specimens (Fig. 5.2A). From the coordinates of these landmarks, we calculate traditional measurements (distances between pairs of landmarks) following the protocol of Fink, 1993 (Fig. 5.2B). We also analyze shape coordinates from the two-point registration (Bookstein, 1986; 1991), which involves assigning two landmarks the coordinates 0,0 and 1,0, respectively, and calculating coordinates of the remaining landmarks relative to these. The two fixed points are at landmarks 1 and 7. The third set comprises parameters of the thin-plate spline (partial warp scores, including scores on the uniform component). These are geometric components of a deformation, which are useful for statistical and graphical analyses but have no particular biological interpretation (for more details, see Chapter 4; for technical accounts, see Bookstein, 1989, 1991). Forms are superimposed to minimize the Procrustes distance between them for purposes of statistical analyses; for purposes of graphical depiction, superimpositions are based on the sliding baseline registration, which avoids biologically unreasonable rotations of the anteroposterior body axis without fixing the length of the baseline (for more details, see Chapter 4).

Allometric coefficients for all three data sets are estimated by multivariate regression. For analyses of traditional measurements, the entire set of measurements is treated as the dependent variable and regressed on the logarithm of one widely used measure of body size, standard length [SL; the distance between landmarks 1 and 7 (Fig. 5.2)]. Allometric coefficients greater than 1.0 are positively allometric, meaning they lengthen at a greater rate than body length; conversely, coefficients lower than 1.0 are negatively allometric, meaning that they lengthen at a lesser rate than body length; coefficients of 1.0 are isometric, meaning that they increase at the same rate as body length. For analyses of shape coordinates, the full set of landmarks (other than the two fixed endpoints

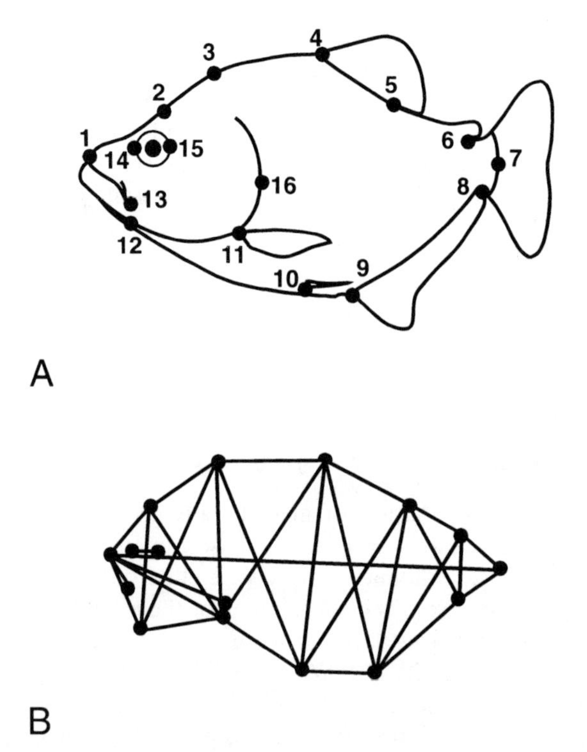

Figure 5.2. A: landmarks. 1, Snout tip, anteroventral junction of anteromedial borders of premaxillaries; 2, anterior border of epiphyseal bridge bone at the dorsal midline (an insect pin was inserted into the top of the cranium to detect the border and was left in place for digitization); 3, posterior tip of supraoccipital bone where it lies adjacent to epaxial musculature and the median dorsal septum; 4, dorsal fin origin, not including anterior modified fin rays, marking the anterior junction of the fin and the dorsal body midline; 5, posterior end of dorsal fin base at the dorsal body midline; 6, posterior end of adipose fin base, where it joins with the skin of the posterior back on the dorsal midline; 7, posterior border of hypural bones (identified as the bending axis of the caudal-fin base); 8, posterior end of anal fin base, at the ventral midline; 9, anal fin origin, marking the junction of the fin and the ventral body midline; 10, pelvic fin insertion, where the fin projects laterally from the pelvic girdle; 11, pectoral fin insertion, where the pectoral fin extends laterally from its joint with the pectoral girdle; 12, mandible/quadrate joint (usually marked by an insect pin placed in the middle of the joint), marking the junction between the lower jaw and "face"; 13, posterior border of maxillary bone, where it intersects the third infraorbital (cheek) bone; 14, anterior border of bony orbit, along the horizontal body axis; 15, posterior bony border of orbit, along the horizontal body axis; 16, posterior border of bony operculum, at most posterior point from the snout tip. B: protocol for traditional measurements.

of the baseline) comprise the dependent variable, which is regressed on a measure of geometric scale, centroid size (the square root of the summed squared distances of each landmark to the centroid of the form). Because shape changes rapidly during early growth, we log transform centroid size. For analyses of the

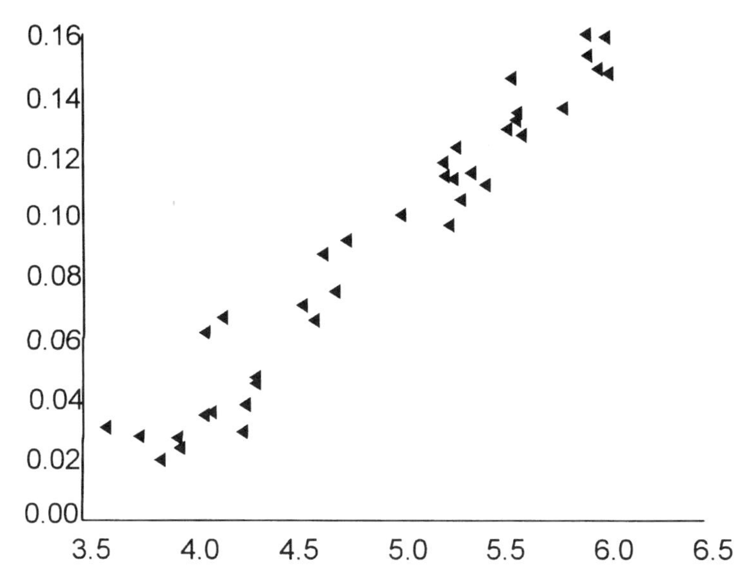

Figure 5.3. Linearity of the ontogenetic change in shape (measured as the Procrustes distance between each specimen and the average small juvenile shape) plotted on size (measured as the logarithm of centroid size).

parameters of the thin-plate spline, the complete set of partial warp scores (including uniform component scores) comprise the dependent variable, which is regressed on log-transformed centroid size. Considering that all our analyses are based on linear regression, it is important to check the linearity of the relationship between shape and log-transformed centroid size; that assumption is reasonably well met by the data, as shown for one sample in Figure 5.3.

Constructing and Interpreting Growth Profiles

Growth Profiles Based on Traditional Morphometric Measurements. To examine the spatial patterns of allometric coefficients, we construct growth profiles, plotting allometric coefficients as a function of the location of a measurement on a body axis. For example, Figure 5.4 shows the allometric coefficients first plotted on the organism and then a growth profile for dorsal allometric coefficients relative to position on the anteroposterior body axis. In this plot, we simply order the measurements from anterior to posterior end, so the positions on the body axis are ordinal numbers not coordinates of position. This approach can be problematic, especially when measurements are unequally spaced, but we use it because it does not require having coordinates for endpoints of measurements; therefore, it is applicable to published data sets lacking that information.

Two classical models are plotted in Figure 5.4, along with an unusual but perhaps equally realistic one. A linear gradient (Fig. 5.4A) is characterized by the smooth decrease (or increase) of growth rates from one end to the other. This is the pattern detected by Huxley (1932) in several analyses, especially of measurements along proximodistal axes. Also, this is the pattern found in larval growth rates of one species of sculpins (*Enophrys bison*), based on the allometric coefficients published by Strauss and Fuiman (1985). A U-shaped gradient (Fig. 5.4B) is characterized by growth rates that gradually decrease toward the middle of the body and then gradually increase toward the posterior end. This is the pattern found in larvae of numerous species of teleosts (Fuiman, 1983). The wave-like pattern (Fig. 5.4C) has two peaks of equal amplitude, separated by troughs, also of equal amplitude. A variant of this pattern is found in sculpins (Fuiman and Strauss, 1985).

One problem with growth profiles based on these data is that several measurements do not parallel any body axis, for example, the most anterior and posterior measurements in Figure 5.3. As a result, their allometric coefficients potentially confound growth rates along both axes. A second problem is that growth rates change from measurement to measurement, but we cannot see where such changes occur nor whether they change abruptly or smoothly. A third problem is that the profile is based on measurements along one transect of one dimension of the body, for example, the most dorsal measurements along the anteroposterior axis. Of course, profiles can be constructed for ventral measurements as well and also for measurements along the dorsoventral axis, but the relationship among these profiles may be difficult to perceive because it requires integrating information from several separate plots. Using geometric methods, we can address these problems because shape coordinates allow us to estimate growth rates along each body axis and the reconstruction of ontogeny as a deformation allows us to interpolate between landmarks over the whole body.

Growth Profiles Based on Shape Coordinates. Growth profiles can be constructed for shape coordinates, just as for traditional measurements, but their interpretation can seem counterintuitive. Also, the interpretation depends on whether profiles represent growth parallel or perpendicular to the baseline. Given a baseline oriented along the anteroposterior body axis, a change in the location of a point on that axis is described by its x coefficient. When the landmark is displaced anteriorly (toward the 0,0 point), a vector representing that change points to the left; its x coefficient has a negative sign (Fig. 5.5A). Put in morphological terms, the region between the landmark and the 0,0 point is shortened relative to the region posterior to the landmark. The growth profile depicts these coefficients as a function of the landmark location, as this information is available. In this case, the slope of the profile over the first segment is negative, whereas it is positive over the second. Conversely, if a landmark is

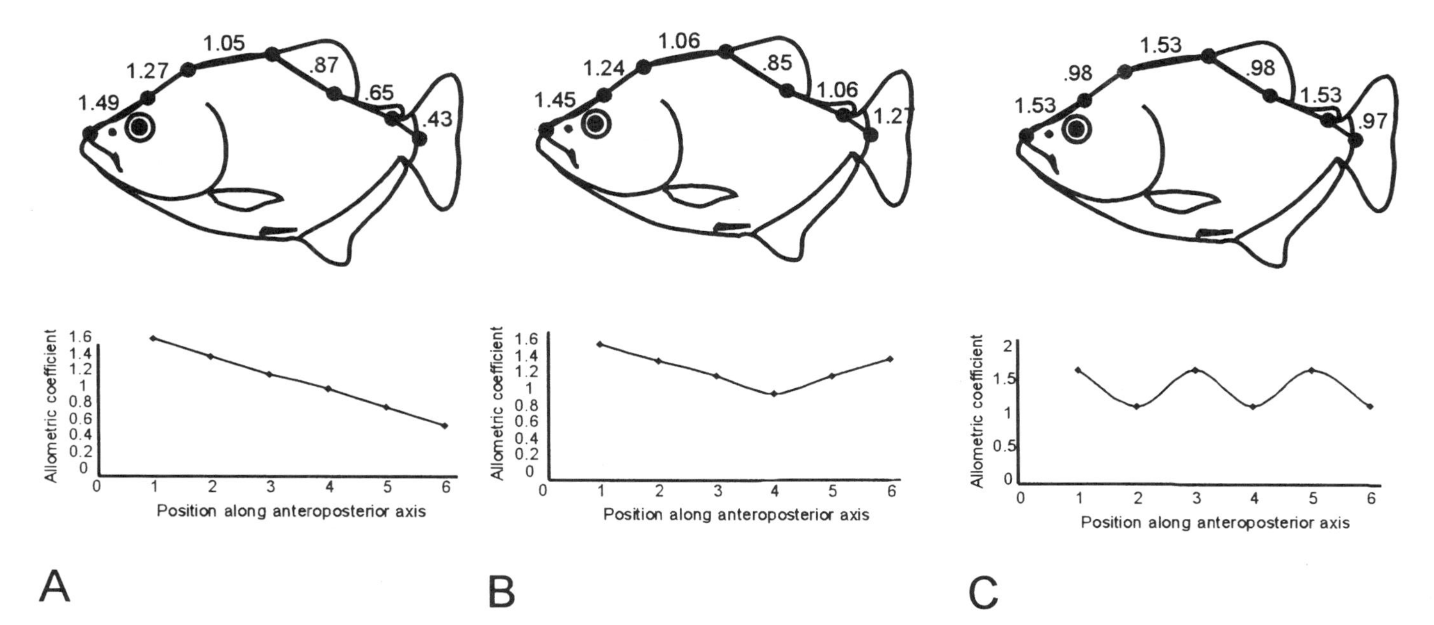

Figure 5.4. Growth profiles for measurements shown on the organism. The number above each measurement is the allometric coefficient, which is plotted on the *y* axis. Each measurement is numbered in sequence from most anterior to most posterior point on the organism; the ordinal number represents the position of the measurement in that sequence, which is plotted on the *x* axis. A: linear anteroposterior gradient. B: a U-shaped gradient. C: a two-peaked wave.

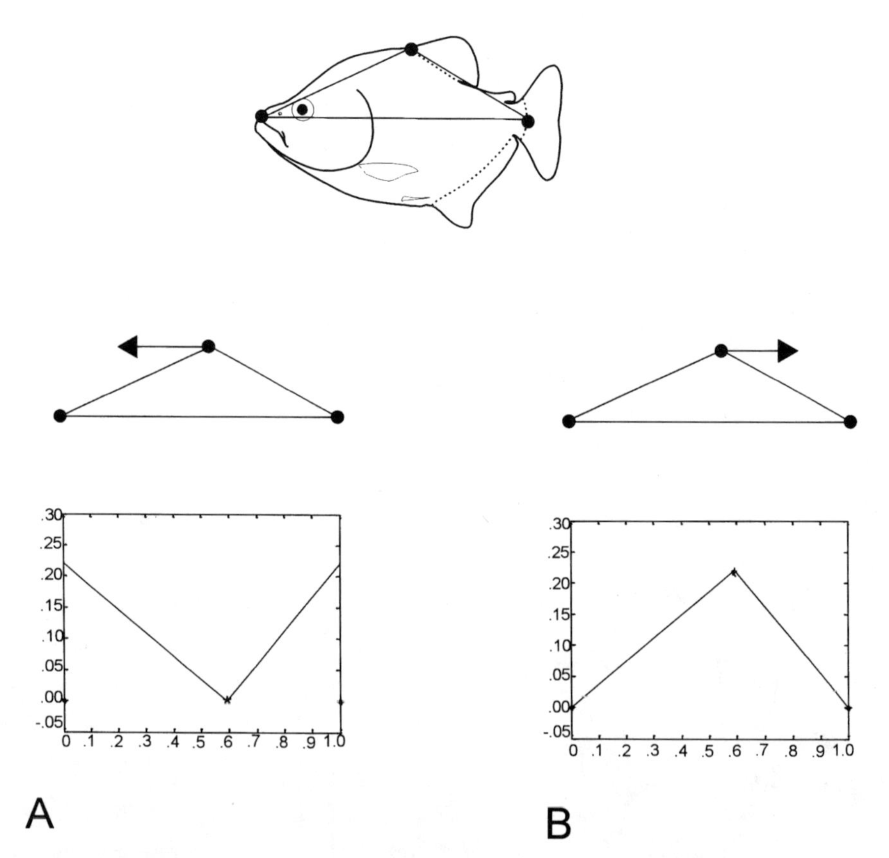

Figure 5.5. Growth profile for three shape coordinates. A: one moveable landmark is displaced toward the anterior fixed endpoint along the anteroposterior body axis; the allometric coefficient for the two fixed endpoints and the one moveable point are plotted on the *y* axis; the location of the landmark on the anteroposterior axis is plotted on the *x* axis. B: one moveable landmark is displaced toward the posterior endpoint along the anteroposterior body axis. Axes are as defined for A.

displaced posteriorly, the vector points to the right and the allometric coefficient is positive; the region between that landmark and the posterior end is shortened relative to the anterior region, that is, between the 0,0 endpoint and the moveable landmark (Fig. 5.5B). Thus, the slope of the first segment is positive, whereas it is negative over the second.

The same reasoning can be applied to any number of points, not just to a landmark between two fixed endpoints. For example, if two points are displaced towards each other, the distance between them shortens relative to the distance between each landmark and the endpoints (Fig. 5.6). The growth profile for this case has three segments, the first between the anterior endpoint (the 0,0 point) and the anterior moveable landmark, the second between the two

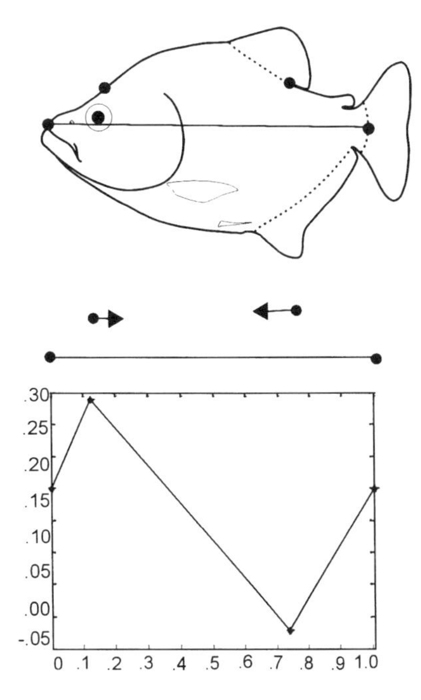

Figure 5.6. Growth profile for four shape coordinates; the two moveable landmarks are displaced in opposite directions along the anteroposterior body axis. Allometric coefficients for the two fixed endpoints and the two moveable points are plotted on the *y* axis; the location of the landmark on the anteroposterior axis is plotted on the *x* axis.

moveable landmarks, and the third segment is between the posterior moveable landmark and the posterior fixed point (the 1,0 point). The growth profile has a positive slope for the first segment, a negative slope for the second, and a positive slope for the third. Conversely, if two points are displaced towards the two fixed points (hence away from each other) the region between them lengthens relative to both ends.

The region between two landmarks can lengthen even if both are displaced in the same direction, as long as one is displaced to a greater extent than the other. For example, if both moveable landmarks are displaced to the right and the posterior of the two is displaced further, the region between them lengthens (Fig. 5.7). Because both vectors point to the right, their coefficients are positive. The slopes for the first two segments are positive, although they decrease over the second segment, indicating a decrease in relative growth rate. Over the final segment, the slope decreases further still, becoming negative (where the region between the posterior moveable landmark and the fixed endpoint shortens relative to the more anterior part of the body).

Given seven landmarks, we can detect the signature of a linear anteroposterior growth gradient (with posteriorly decreasing growth rates) in a growth profile of shape coordinates (Fig. 5.8A). All free landmarks are displaced posteriorly, and the more posterior regions are shortened relative to the more anterior ones. The slope of the growth profile for the dorsal landmarks

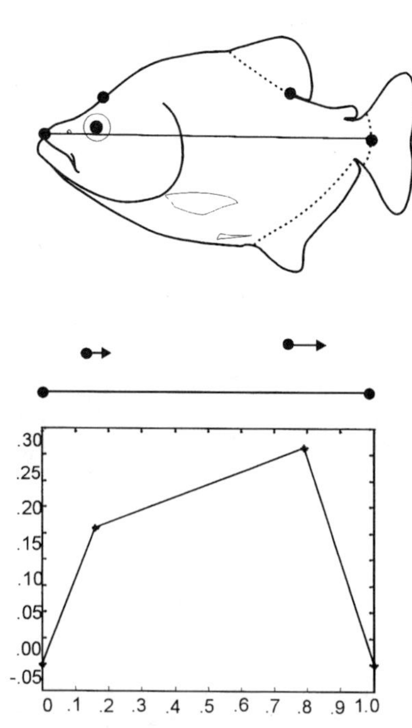

Figure 5.7. Growth profile for four shape coordinates; the two moveable landmarks are displaced by different amounts in the same direction along the anteroposterior axis. Allometric coefficients for the two fixed endpoints and the two moveable points are plotted on the *y* axis; the location of the landmark on the anteroposterior axis is plotted on the *x* axis.

(those numbered 1 though 7) is initially positive (where growth rates are high), but it decreases (indicating a deceleration of growth rates) until it becomes negative and further decreases toward the posterior end. We can also detect a U-shaped gradient, with high growth rates anteriorly and posteriorly, decreasing in the midbody (Fig. 5.8B). The slope is positive both anteriorly and posteriorly, indicating relative elongation of both regions. Between them, the slope is negative until it begins to rise posteriorly.

Interpretations of profiles for rates of deepening (dorsoventrally) are more straightforward. Vectors at landmarks orient vertically, either toward or away from the baseline. Vectors that point up from dorsal landmarks (Fig. 5.9) indicate deepening, as do downward displacements of ventral landmarks (Fig. 5.9). Thus, growth profiles with positive slopes for dorsal landmarks indicate deepening (Fig. 5.9A), as would negative slopes for ventral landmarks. The interpretation is complicated by the need to take into account the distance between each landmark and the baseline. If rates of deepening are uniform, the displacement of each landmark is proportional to its distance from the baseline. For example, a point twice as far from the baseline as another would be vertically displaced by twice the amount of the closer point. To interpret the growth profiles, it is necessary to examine the coefficients as a function of landmark locations on the dorsoventral axis; in this plot (Fig. 5.9B), the sequence of the

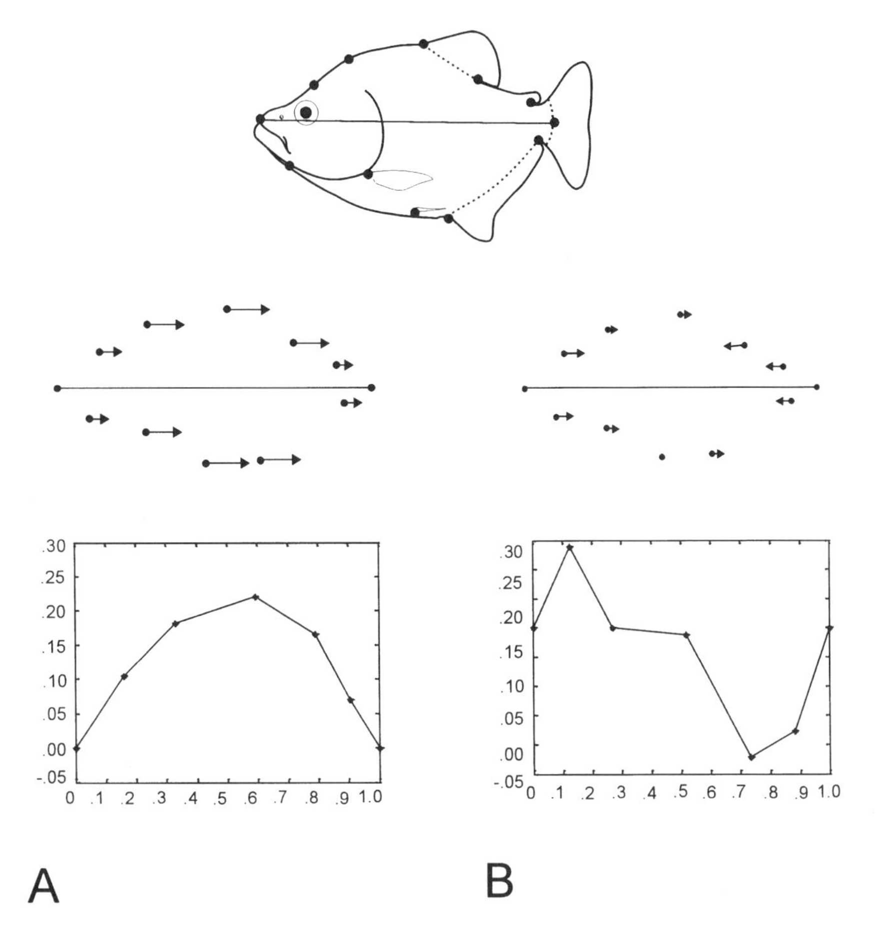

Figure 5.8. Growth profiles for seven shape coordinates. Allometric coefficients for the two fixed endpoints and the five moveable points are plotted on the *y* axis; the location of the landmark on the anteroposterior axis is plotted on the *x* axis. A: a liner anteroposterior gradient. B: a U-shaped gradient.

landmarks along the x axis does not reflect their anteroposterior position but rather their increasing vertical distance above the baseline. The coefficients are not strictly a function of that vertical distance; two landmarks (the third and sixth furthest from the baseline in the vertical direction) have lower than expected values according to a model of uniform growth rates.

Comparing Ontogenies of Shape

To compare ontogenetic allometries multivariately, we calculate the angle between pairs of allometric vectors. When two vectors point in the same direction,

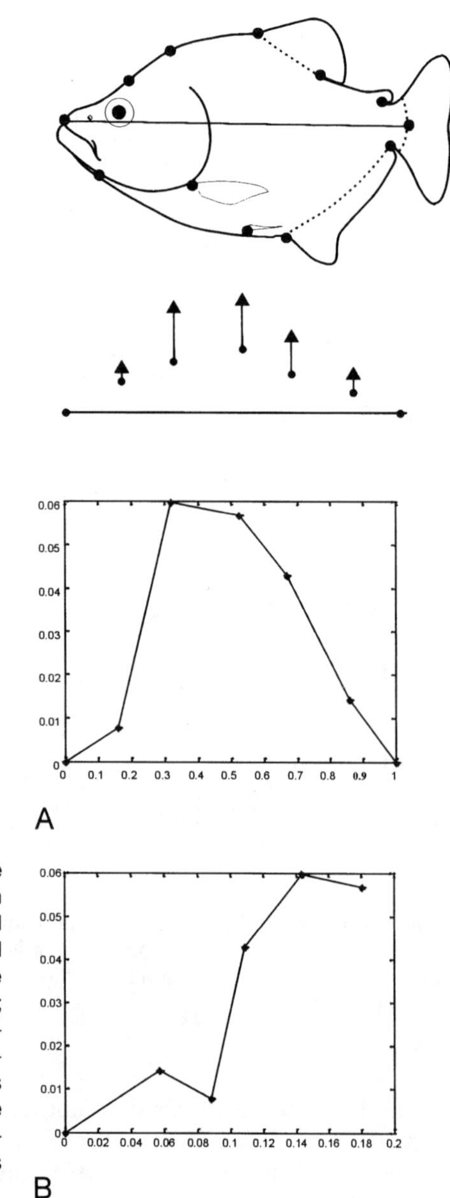

Figure 5.9. Growth profile for seven shape coordinates displaced by different amounts in the same direction, along the dorsoventral body axis. A: allometric coefficients for vertical displacements of two fixed endpoints and five moveable landmarks are plotted on the *y* axis; the location of the landmark on the anteroposterior axis is plotted on the *x* axis. B: allometric coefficients for vertical displacements of two fixed endpoints and five moveable landmarks are plotted on the *y* axis; the location of the landmark on the dorsoventral axis is plotted on the *x* axis.

the angle between them is 0.0°; the cosine of this angle is the vector correlation, R_v, which is 1.0 when the angle is 0.0°. To calculate the angle, vectors of allometric coefficients are normalized to unit length (removing the effects of differences in overall developmental rate) and the inner (dot) product is taken. This gives the observed angle between vectors, which may well be greater than 0.0° due to the effects of sampling from populations of variable ontogenetic allometries. To test this difference statistically, we obtain confidence intervals for the angles by resampling (Efron and Tibshirani, 1993). Specifically, we form bootstrap sets by bootstrapping residuals of the multivariate regressions. Each individual specimen in the original data set gives a set of residuals, which are bootstrapped as a complete set, preserving the covariance structure present in the residuals. The expected shape at each value of log-centroid size is estimated using the multivariate regression coefficients, and a set of residuals (drawn at random with replacement) is added to this value. To calculate angles that can be obtained by resampling within a single species, two vectors are produced per species, and the angle between them is calculated for each set of residuals. Should the interspecific angle exceed the 95% confidence interval of the within-species range, the interspecific difference is judged to be statistically significant. Because sample sizes vary across species, the analysis is done in terms of the distribution of bootstrapped data sets at comparable sample sizes. Resampling tests are performed using functions written in Matlab (Mathworks, 2000).

The angles between species are also compared with the angles between each allometric vector and vectors of randomly reshuffled allometric coefficients (Table 5.2). The object of this analysis is to determine whether interspecific correlations are greater than we would expect by chance. This randomization approach preserves the range of values found in data and also preserves the proportion of isometric, positively allometric and negatively allometric coefficients found in data. The vectors of coefficients are reshuffled randomly 400 times yielding 400 randomized vectors that are each compared with the data from one species. These 400 comparisons give a mean and upper and lower 95% confidence interval for comparison between observed and randomized vectors. When the angle between the two observed vectors is less than the value of the upper 95% confidence interval, species are judged to be more similar to each other than would be expected by chance. Conversely, when the angle exceeds the lower 95% confidence interval, species are judged to be less similar to each other than would be expected by chance.

RESULTS

P. denticulata

Allometric coefficients for traditional measurement (Fig. 5.10) suggest that growth rates are low both anteriorly and posteriorly, increasing toward the middle. The growth profile for dorsal measurements along the anteroposterior axis suggests a pattern more complex than a simple inverted U-shaped curve because there are two peaks: growth rates are low anteriorly, rising, then falling, then rising and falling again (Fig. 5.11A). However, the posterior peaks and troughs are higher than would be predicted by a model of a simple

Table 5.2. Angles and vector correlations (R_v) between empirical allometric vectors based on traditional morphometric variables and vectors obtained by randomly reshuffling the observed coefficients for each species (by bootstrapping, without replacement)

Species	Mean Angle (°)	95% Angle (°)	Mean R_v	95% R_v
		Traditional Measurements		
Pygopristis denticulata	8.85	7.34	0.9881	0.9918
S. elongatus	6.69	5.50	0.9932	0.9954
Serrasalmus gouldingi	11.01	9.11	0.9816	0.9874
S. manueli	8.31	6.93	0.9895	0.9927
S. spilopleura	6.33	5.25	0.9939	0.9958
Pygocentrus piraya	6.64	5.44	0.9933	0.9955
P. nattereri	7.16	5.90	0.9922	0.9947
P. cariba	6.33	5.19	0.9939	0.9959
		Shape Coordinates		
Pygopristis denticulata	77.30	63.45	0.2198	0.4469
S. elongatus	87.30	71.51	0.0472	0.3172
Serrasalmus gouldingi	90.37	71.99	−0.0065	0.3091
S. manueli	89.15	73.37	0.0149	0.2861
S. spilopleura	88.48	71.45	0.0266	0.3182
Pygocentrus piraya	80.14	64.80	0.1713	0.4258
P. nattereri	87.65	71.31	0.0411	0.3205
P. cariba	84.53	70.46	0.0953	0.3345
		Parameters of the Thin-Plate Spline		
Pygopristis denticulata	90.18	71.71	−0.0031	0.3140
S. elongatus	91.18	72.448	−0.0175	0.3009
Serrasalmus gouldingi	89.54	72.49	0.0080	0.2941
S. manueli	86.83	66.11	0.0553	0.4050
S. spilopleura	90.98	70.85	−0.0172	0.3280
Pygocentrus piraya	89.03	72.30	0.0170	0.3033
P. nattereri	89.75	68.24	0.0043	0.3706
P. cariba	89.27	70.22	0.0128	0.3380

Angles and correlations were calculated for each of 400 random shufflings. Given are the mean and upper 95% confidence intervals.

wave. Ventrally, the pattern more closely resembles a shallow U-shaped curve in that rates of growth gradually accelerate, reaching a peak anterior to the anal fin and then decreasing to a level maintained over the anal fin and caudal peduncle (Fig. 5.11B). Along the dorsoventral body axis, the growth profile also resembles a shallow inverted U-shaped curve, with rates increasing toward the middle of the body and falling posteriorly (Fig. 5.11C). However, the rate of

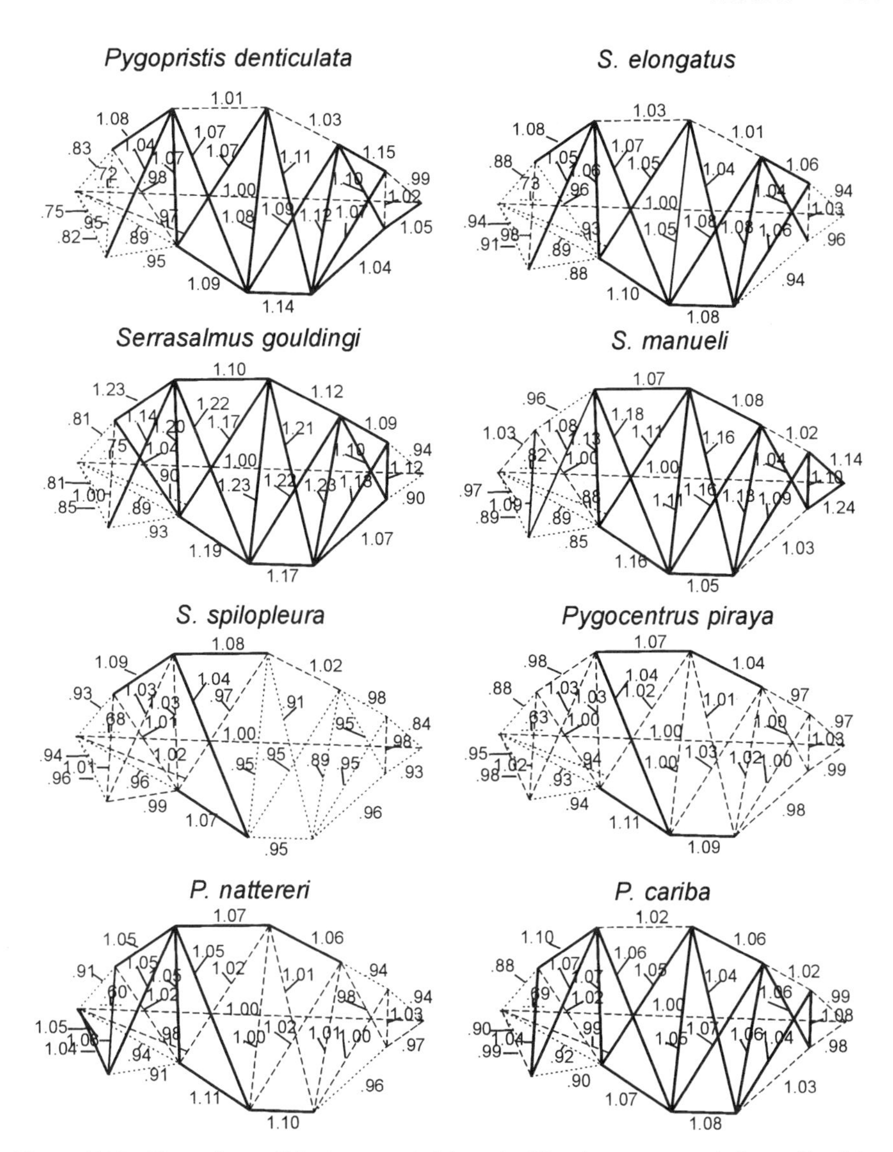

Figure 5.10. Allometric coefficients computed from traditional measurements by multivariate regression of all measurements on standard length. Solid lines indicate positively allometric measurements. Dashed lines indicate isometric measurements. Dotted lines indicate negatively allometric measurements.

deepening at the anterior base of the dorsal fin appears lower than anticipated by that model, and the posterior growth is higher than predicted by it, being substantially higher than cranial growth rates.

The shape coordinates confirm the impression of low cranial and posterior

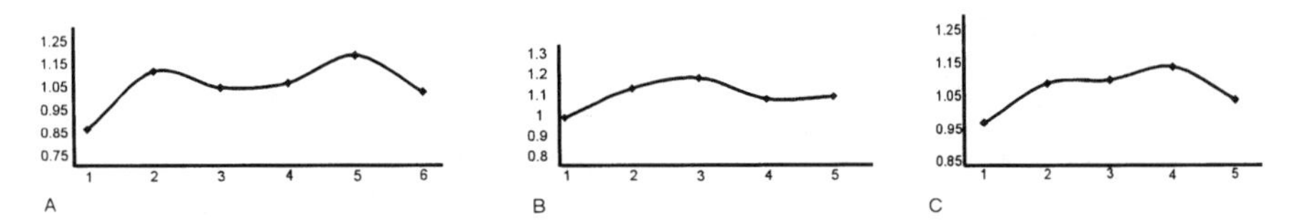

Figure 5.11. Growth profiles for traditional measurements of *P. denticulata*. A: dorsal measurements along the anteroposterior axis (measured between landmarks 1-2, 2-3, 3-4, 4-5, 5-6, and 6-7, plotted as a function of the order along the anteroposterior axis). B: ventral measurements along the anteroposterior axis (measured between landmarks 12-11, 11-10, 10-9, 9-8, and 8-7, plotted as a function of the order along the anteroposterior axis). C: measurements along dorsoventral axis (measured between landmarks 2-12, 3-11, 4-10, 5-9, and 6-8, plotted as a function of the order along the anteroposterior axis).

growth rates; vectors at the anterior moveable landmarks point toward the anterior fixed point, and vectors at posterior moveable landmarks point toward the posterior fixed point. This indicates a relative shortening of the head and peduncle (Fig. 5.12). The growth profiles for dorsal landmarks show that rates of growth along the anteroposterior axis are low in the anterior head (evident by the negative slope), rise substantially at the epiphyseal bar (where the slope changes from negative to very slightly positive), rise again in the region between the dorsal and adipose fins, and finally drop in the caudal peduncle (Fig. 5.13A). Ventrally, growth rates along the anteroposterior axis are low through the head, rising slightly from the jaw to the pectoral fin and increasing substantially in the region between the pectoral and anal fin, then decreasing to the point that the body is virtually isometric over the anal fin, and finally rising very slightly in the caudal peduncle (Fig. 5.13B). Both descriptions are largely consistent with those based on traditional measurements. Dorsally, both suggest a wave-like pattern, with one peak in the posterior head and a second between the dorsal and adipose fins. Ventrally, both suggest a shallow, inverted U-shaped curve with a peak located more posteriorly than anticipated and with a lesser than expected rate of deceleration posteriorly. The two descriptions differ regarding the growth rate of the dorsal posterior head relative to the anterior midbody, which may reflect the oblique orientation of the cranial measurements.

Along the dorsoventral axis dorsally, growth rates are low in the head, rising postcranially to a plateau maintained to the posterior base of the dorsal fin, dropping at the adipose fin, and dropping again (very slightly) over the caudal peduncle (Fig. 5.13C). From the growth profile, it is not evident that the rate accelerates at the posterior dorsal fin base, but this landmark is closer to the baseline than the two more anterior points; however, its rate of deepening is higher. Ventrally, deepening is indicated by negative slopes; rates of deepening are low in the head, increase in the (posterior) midbody, accelerate again at the anterior anal fin base, and then return to the level seen just anterior to that point (Fig. 5.13D). The acceleration at the anterior base of the anal fin is not

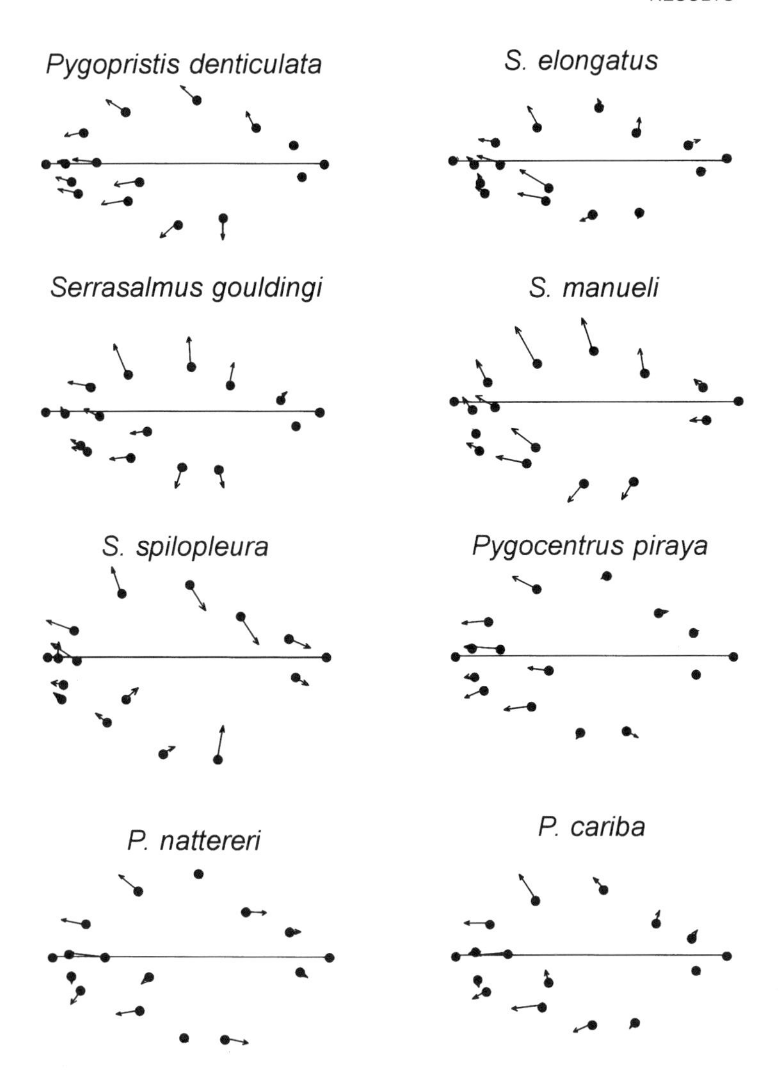

Figure 5.12. Ontogenetic shape change depicted by changes in the location of shape coordinates; the vectors indicate the amount and direction of shape at each landmark relative to two fixed endpoints (landmarks 1 and 7). These changes are calculated by multivariate regression of all moveable landmarks on the logarithm of centroid size.

obvious from the growth profile itself, but that landmark is virtually the same distance from the baseline as the one at the pelvic fin; however, its rate of deepening is considerably higher. Thus, dorsally, the pattern resembles an inverted U-shaped gradient with an elongated and fairly shallow peak, whereas ventrally, the pattern suggests an inverted U-shaped curve with a fairly sharp and posteriorly displaced peak. Taken together, these two curves correspond

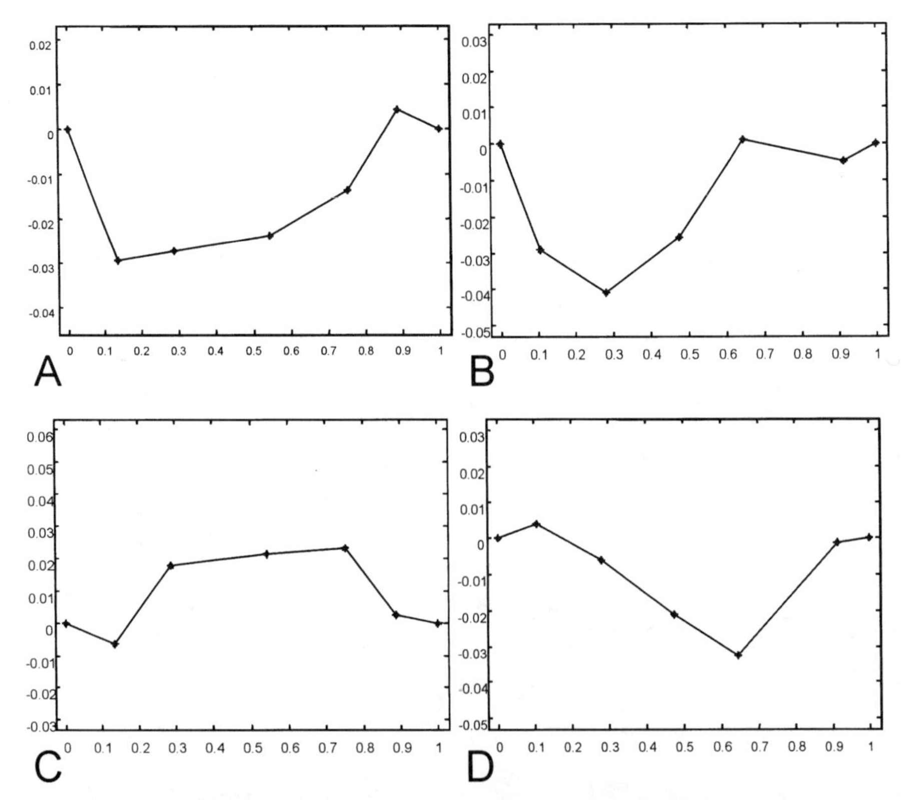

Figure 5.13. Growth profiles for shape coordinates of *P. denticulata*. A: anteroposterior component of peripheral dorsal coordinates (at landmarks 1–7). B: anteroposterior component of ventral peripheral coordinates (at landmarks 1, 11, 12, 10, 9, 8, and 7). C: dorsoventral component of peripheral dorsal coordinates. D: dorsoventral component of ventral peripheral coordinates.

to the pattern indicated by the traditional measurements. However, the pattern is more complex than can be revealed by traditional measurements because, in that representation, dorsal and ventral components of deepening are inseparable.

The depiction of growth by the thin-plate spline shows the relatively low rates of growth in the head and posterior body, and the grid suggests these regions contract (Fig. 5.14). They grow slowly relative to the region between them. It is also clear that the spatial distribution of rates of lengthening differs dorsally and ventrally, as indicated by vertical lines of the grid splaying outward or inward. Cranially, the diagram reveals abrupt changes in rates in the orbital regions; otherwise, the cranial growth rates change quite smoothly. Postcranially, there are too few landmarks to determine where growth rates accelerate or decelerate or whether they do so smoothly, but no abrupt changes are evident.

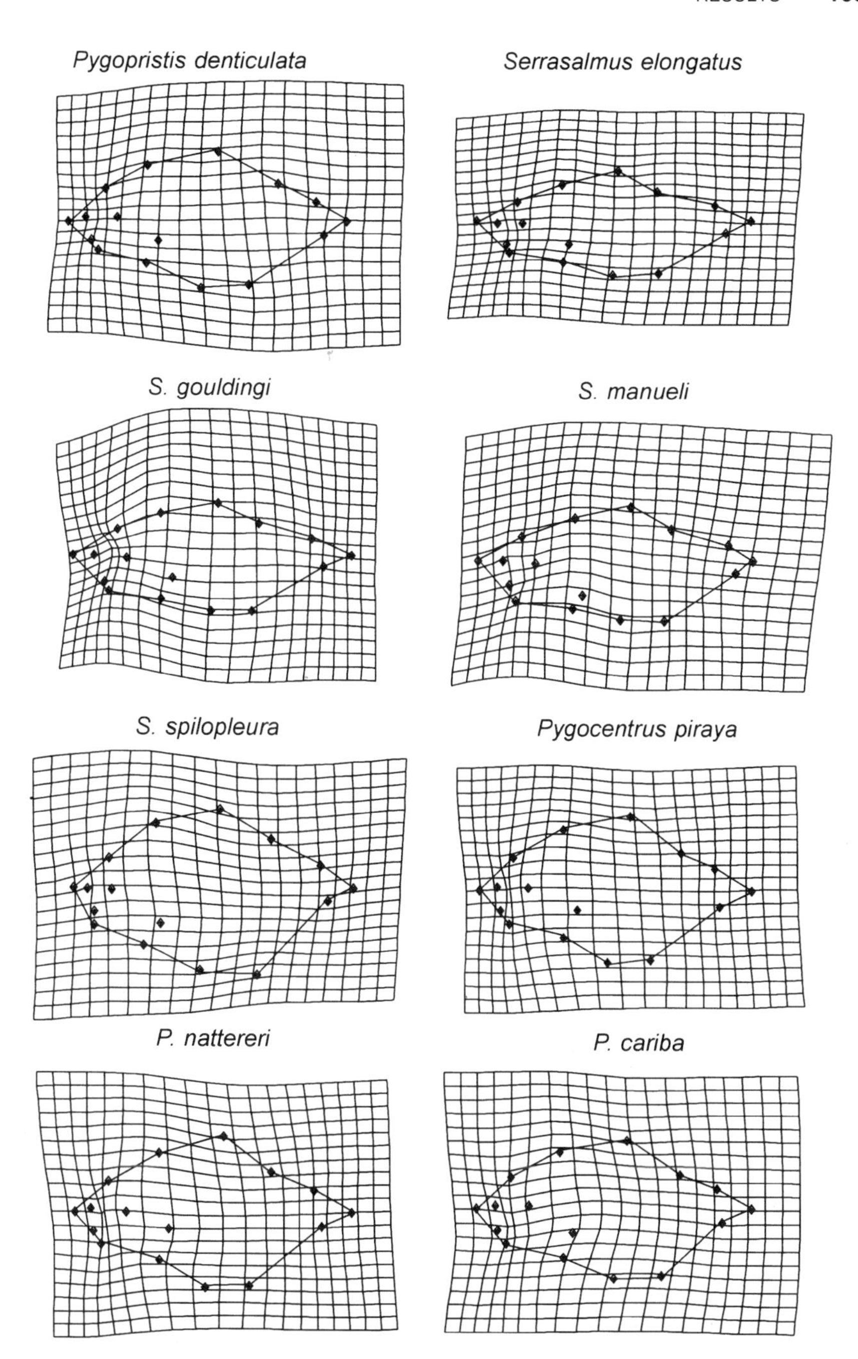

Figure 5.14. Ontogenetic shape change depicted as a deformation modeled by the thin-plate spline. These are calculated by multivariate regression of scores on all partial warps (including the uniform component) on the logarithm of centroid size. The outlines depict juvenile body shape, as represented by line segments connecting peripheral landmarks.

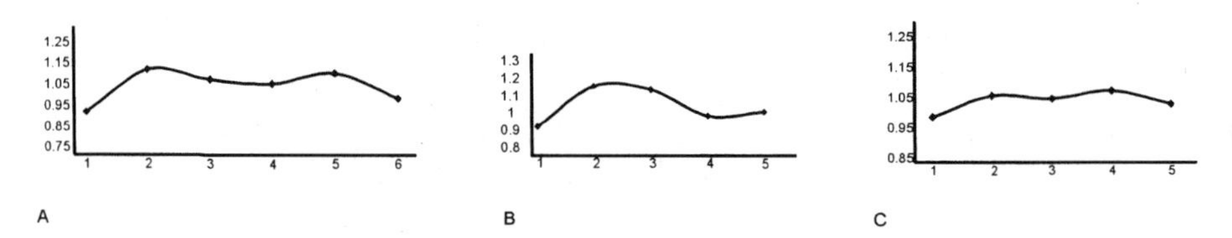

Figure 5.15. Growth profiles for traditional measurements of *S. elongatus*. A: dorsal measurements along the anteroposterior axis. B: ventral measurements along the anteroposterior axis. C: measurements along dorsoventral axis. Measurements as described in Fig. 5.11.

S. elongatus

S. elongatus resembles *P. denticulata* in that growth rates are generally lowest most anteriorly and posteriorly and rise toward the middle (Fig. 5.10). As in the case of *P. denticulata*, the growth profile for rates of lengthening dorsally suggest a two-peaked wave, with peaks in the posterior head and between the dorsal and adipose fins (Fig. 5.15A). Ventrally, the growth rates more closely resemble an inverted U-shaped curve with a peak extending over the entire midbody, differing from the model in that posterior growth rates do not descend to the level reached in the anterior head (Fig. 5.15B). Rates of growth along the dorsoventral axis are more nearly uniform, with very slight troughs in the middle (at the level of the anterior base of the dorsal fin) and at the two ends (Fig. 5.15C). Considering the very shallow body of the adult, it is interesting that postcranial body depth is positively allometric.

The shape coordinates show that little postcranial deepening occurs in the ontogeny of *S. elongatus*, but body shape changes nonetheless (Fig. 5.12). The growth profiles for dorsal shape coordinates show a low growth rate in the anterior head, which accelerates in the posterior head and maintains a nearly uniform rate to the adipose fin and then decelerates (Fig. 5.16A). However, this pattern does not strictly conform to an inverted U-shaped curve because of a subtle deceleration in the anterior midbody and a shallow secondary peak between dorsal and adipose fins. Ventrally, growth rates are low throughout the head, rising substantially posterior to the pectoral fin, and then decelerating over the anal fin and again in the caudal peduncle (Fig. 5.16B). Thus, the dorsal profile suggests a U-shaped curve, with a subtle indication of a wave-like pattern, whereas the ventral profile suggests an inverted U-shaped gradient. These results are consistent with those inferred from traditional measurements, although the deceleration over the anal fin is less striking in the shape coordinates, which again might reflect the oblique orientation of the anal fin measurement.

Along the dorsoventral axis, rates of deepening (dorsally) are low in the anterior head, rising substantially between the epiphyseal bar and supraoccipital process, and then falling postcranially except for an acceleration at the posterior base of the dorsal fin, after which they fall to the level seen in the head (Fig. 5.16C). This most closely resembles a wave-like pattern (if any) because

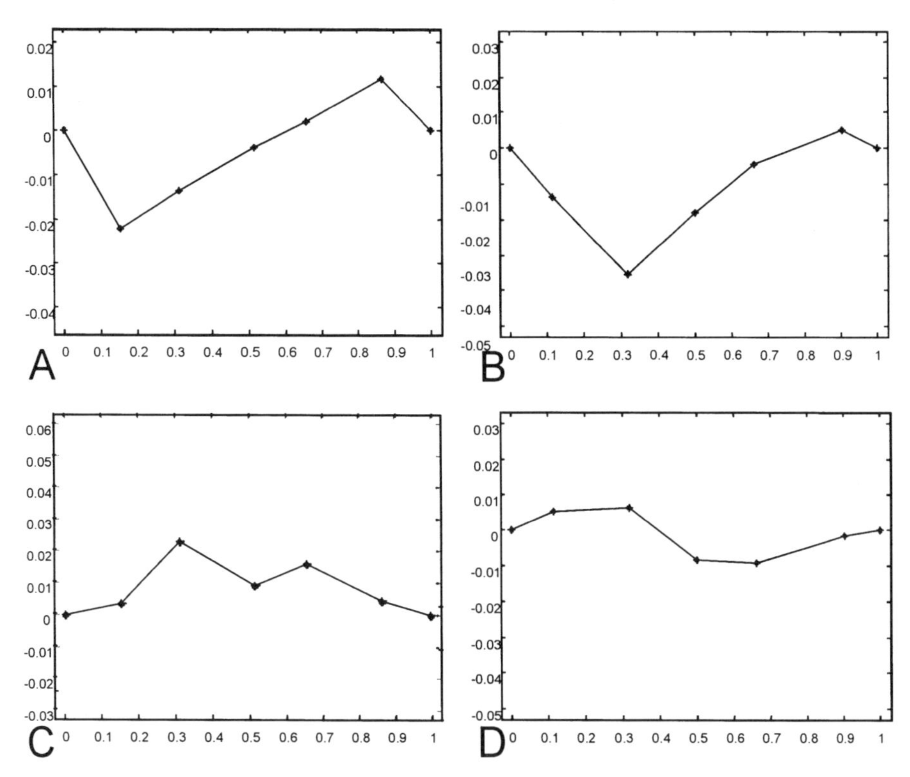

Figure 5.16. Growth profiles for shape coordinates of *S. elongatus*. A: anteroposterior component of peripheral dorsal coordinates. B: anteroposterior component of ventral peripheral coordinates. C: dorsoventral component of peripheral dorsal coordinates. D: dorsoventral component of ventral peripheral coordinates. Landmarks as described in Fig. 5.13.

it has two peaks, although the anterior peak is considerably higher. Ventrally, the growth rates are generally low, especially in the head—the ventral head shallows relative to body length (Fig. 5.16D). Postcranially, ventral growth rates increase and conform to a model of uniform growth rates—the postcranial ventral body is virtually isometric in depth. The pattern suggested by profiles of shape coordinates largely agrees with that detected with traditional measurements.

The depiction of growth by the thin-plate spline shows the wave-like pattern of relative elongation dorsally in contrast to the inverted U-shaped gradient ventrally (Fig. 5.14). In the posterior head and between the dorsal and adipose fins, relative growth rates are higher dorsally than ventrally, but the converse is true in the region between. In this species, unlike *P. denticulata*, rates of relative elongation do not decelerate dramatically over the caudal peduncle so there is no striking compression of the grid there. Rates of relative deepening show a somewhat more complex pattern, being higher dorsally than ventrally in the posterior head, so the dorsal head deepens relative to the ventral head. Ven-

Table 5.3. Angles (in degrees) and vector correlations (R_V) for the comparison between *S. elongatus* and *P. denticulata*, based on traditional morphometric measurements, shape coordinates, and parameters of the thin-plate spline

S. elongatus Compared With	Traditional		Shape Coordinates		Spline Parameters	
	Angle (°)	R_V	Angle (°)	R_V	Angle (°)	R_V
P. denticulata	3.1	0.999	29.4	0.871	31.1	0.856

trally, rates of relative deepening increase postcranially but decrease dorsally, and, in the posterior body (in the region of the posterior base of the dorsal fin), there is a second region with disparate dorsal and ventral relative rates.

Like the pattern seen in *P. denticulata*, that of *S. elongatus* does not fit a simple model of a global gradient. However, in *S. elongatus*, as in *P. denticulata*, the spatial pattern of growth rates seems far from haphazard. In both species, growth rates along the anteroposterior body axis follow a wave-like pattern dorsally (with more pronounced peaks and valleys in *P. denticulata* than in *S. elongatus*) and a U-shaped gradient ventrally, with higher than expected growth rates posteriorly. However, despite these similarities, the growth vectors of the two species differ significantly ($P < 0.05$), whether comparisons are based on traditional data or parameters of the thin-plate spline. The magnitude of the difference depends on the measurement scheme; the angles are larger in comparisons based on geometric data (Table 5.3). Regardless of the measurement scheme, the difference between the two species is less than that found in comparisons to vectors of randomized coefficients (Table 5.2).

The difference between growth vectors of *S. elongatus* and *P. denticulata* can be depicted as a deformation and also by vectors at individual landmarks (Fig. 5.17). In this comparison, the grid would be perfectly square if the two ontogenies were the same. The primary contrast between the two species is in their rates of relative deepening. Anteriorly, rates of relative deepening are higher dorsally but lower ventrally in *S. elongatus*; however, postcranially, rates of relative deepening are lower both dorsally and ventrally. There are also striking differences in the head, reflecting the disparate patterns of deepening dorsally and ventrally but also reflecting pronounced differences in rates of growth in the orbital region relative to the snout and postorbital head. In addition, between the pectoral and pelvic fins, *S. elongatus* has a higher rate of relative elongation, but that rate is considerably lower between the pelvic and anal fins and slightly lower more posteriorly, except in the caudal peduncle, where the disparity between species increases.

S. gouldingi

Growth rates of *S. gouldingi* seem to follow an inverted U-shaped pattern, with the peak in the posterior head (Fig. 5.10). The profile for dorsal traditional measurements along the anteroposterior axis suggests this inverted U-shaped

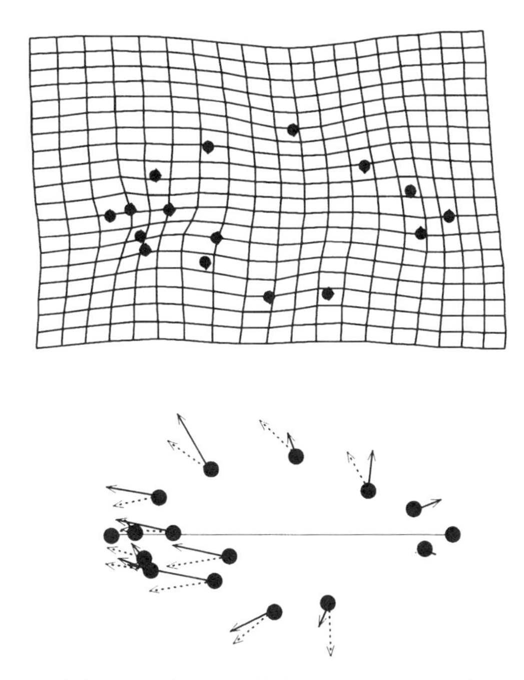

Figure 5.17. Differences between ontogenetic trajectories of allometric coefficients of *P. denticulata* and *S. elongatus*, represented by the thin-plate spline (top) and by shape coordinates to a common baseline (bottom). Ontogenetic trajectories are normalized to unit length, removing effects of differences in overall developmental rate.

curve, although the data deviate from expectations because the peak is more anteriorly located than expected and the posterior growth rates are unexpectedly high (Fig. 5.18A). Ventrally, growth rates also seem to follow an inverted U-shaped curve, again with a more anteriorly than expected peak (Fig. 5.18B). Along the dorsoventral body axis, growth rates are low cranially, but there is a marked acceleration of rates of deepening in the posterior head, with notable deceleration postcranially (Fig. 5.18C).

The shape coordinates show the striking ontogenetic changes in body depth in this species (Fig. 5.12). Along the anteroposterior axis dorsally, the growth

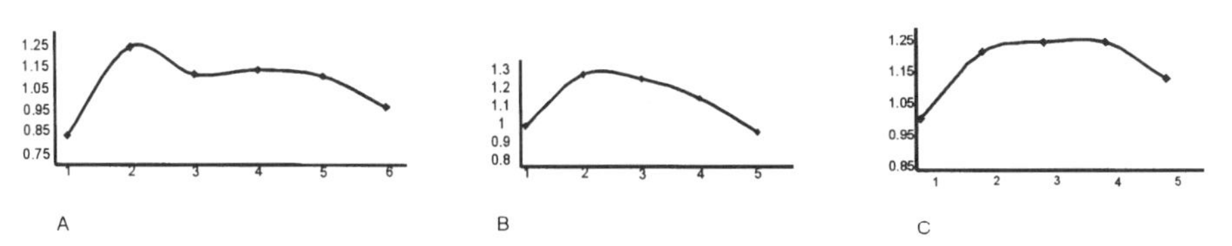

Figure 5.18. Growth profiles for traditional measurements of *S. gouldingi*. A: dorsal measurements along the anteroposterior axis. B: ventral measurements along the anteroposterior axis. C: measurements along dorsoventral axis. Measurements as described in Fig. 5.11.

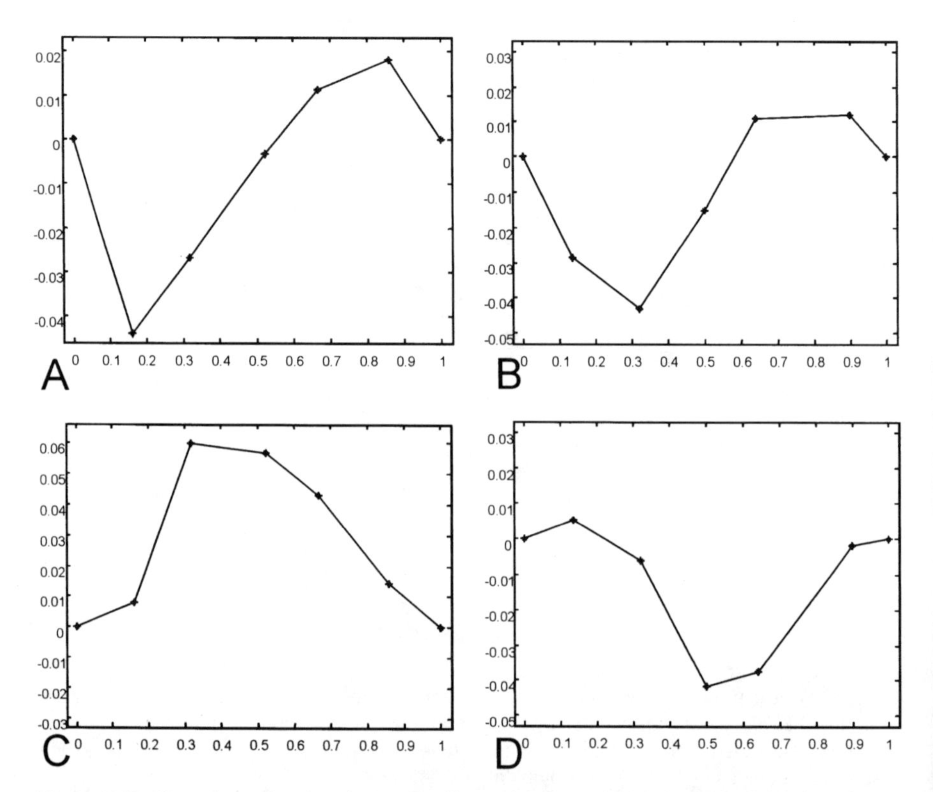

Figure 5.19. Growth profiles for shape coordinates of *S. gouldingi.* A: anteroposterior component of peripheral dorsal coordinates. B: anteroposterior component of ventral peripheral coordinates. C: dorsoventral component of peripheral dorsal coordinates. D: dorsoventral component of ventral peripheral coordinates. Landmarks as described in Fig. 5.13.

profile for shape coordinates suggests that rates closely match the hypothetical inverted U-shaped curve, deviating from it primarily in that the peak is anteriorly displaced relative to the model (Fig. 5.19A). The acceleration in the posterior head is greater anticipated by the model, and rates do not subsequently increase as they would if they were reaching a peak in the middle. Ventrally, growth rates along the anteroposterior body axis resemble an inverted U-shaped gradient in that the rates gradually increase towards the middle of the body and subsequently decrease (Fig. 5.19B). However, the pattern deviates from the model in that the peak extends from the pectoral to the anterior anal fin base; therefore, it is located more anteriorly than expected and extends further than anticipated. Also, the deceleration over the anal fin is more abrupt than postulated by a model of a smooth gradient. The patterns described by shape coordinates closely match those inferred from traditional measurements, with a few exceptions that probably reflect the orientation of the traditional measurements relative to body axes.

Along the dorsoventral body axis, rates of deepening dorsally are low in the posterior head, rise substantially in the posterior head, and then drop at the anterior base of the dorsal fin, rising again posteriorly and remaining virtually uniform more caudally until dropping slightly over the caudal peduncle (Fig. 5.19C). This is not fully evident in the plot because the landmark at the dorsal fin is furthest from the baseline and thus should be the furthest displaced but is not. This pattern is most consistent with a wave-like model, if any. Ventrally, rates of deepening are very low in the head and nearly uniform postcranially, except they are higher than anticipated in the caudal peduncle (Fig. 5.19D).

The pronounced relative deepening of the postcranial body and the relative elongation of the region of the postcranial head and midbody relative to the more anterior head and caudal peduncle are both evident in the depiction of ontogeny as a deformation by the thin-plate spline (Fig. 5.14). In the midbody, the vertical lines of the grid splay outward ventrally (in the region between pelvic and anal fins), suggesting that the growth rate along the anteroposterior axis is higher ventrally than dorsally. Within the head, the dorsal region of the snout elongates relative to both the eye and the more ventral part of the snout, and the dorsal head deepens relative to the suborbital head. Also, the posterior head shows a far more pronounced lengthening dorsally than ventrally. Caudally, the deceleration in growth rates along the anteroposterior axis is considerable but seems to occur quite smoothly (to the extent that this can be inferred from the sparse sampling of the posterior body).

Considering the dramatic difference in rates of postcranial deepening between *S. gouldingi* and the other two species, it is not surprising that *S. gouldingi* differs significantly from both ($P < 0.05$). The magnitude of difference depends on the measurement scheme (Table 5.4), but the angle is invariably significantly greater than 0.0° and also invariably smaller than found in comparisons between *S. gouldingi* and vectors of randomly reshuffled coefficients (Table 5.2).

S. gouldingi and *P. denticulata* differ greatly in rates of deepening. The rates postcranially are generally higher in *S. gouldingi* dorsally but lower ventrally (Fig. 5.20A). As a result, it appears that the dorsal body is deepening more in relation to the ventral body in that species. Another striking difference is in rates of relative elongation both cranially and in the posterior body (posterior

Table 5.4. Angles (in degrees) and vector correlations (R_V) for the comparisons between *S. gouldingi, S. elongatus,* and *P. denticulata,* based on traditional morphometric measurements, shape coordinates, and parameters of the thin-plate spline

S. gouldingi	Traditional		Shape Coordinates		Spline Parameters	
Compared With	Angle (°)	R_V	Angle (°)	R_V	Angle (°)	R_V
P. denticulata	3.6	0.998	32.9	0.840	32.7	0.842
S. elongatus	3.9	0.998	30.4	0.863	38.8	0.779

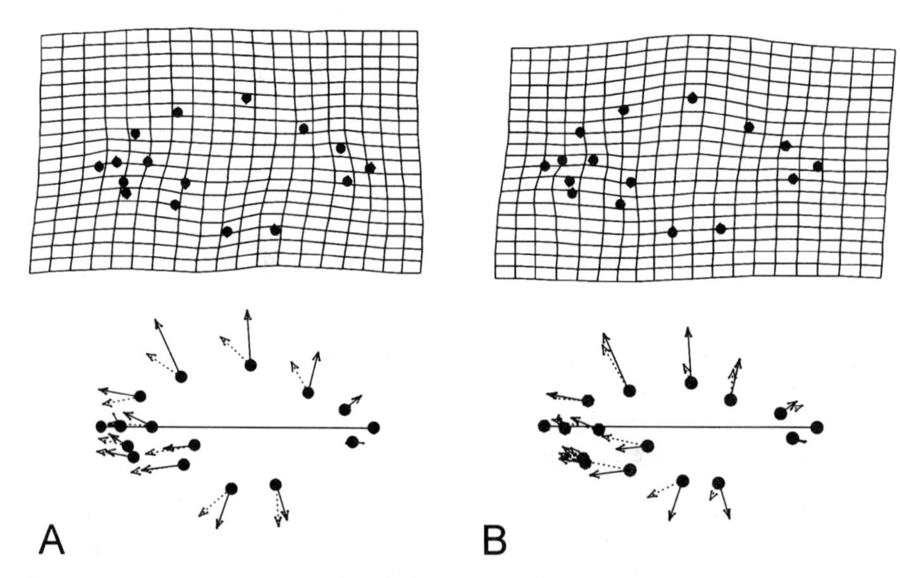

Figure 5.20. Differences between ontogenetic trajectories of allometric coefficients; A: *P. denticulata* vs. *S. gouldingi*. B: *S. elongatus* vs. *S. gouldingi*. Differences are represented by the thin-plate spline (top) and by shape coordinates to a common baseline (bottom). Ontogenetic trajectories are normalized to unit length, removing effects of differences in overall developmental rate.

to the dorsal fin dorsally and to the pelvic fin ventrally). Cranially, *S. gouldingi* undergoes a greater relative elongation of the snout and posterior dorsal head; however, ventrally, there is a lesser relative elongation in the orbital and postorbital region. However, its rate of relative elongation exceeds that of *P. denticulata* in the most posterior region of the postorbital head to anterior midbody. Posteriorly, the contrasts are just as complex. Dorsally, between the dorsal and adipose fins, rates of relative elongation are far lower in *S. gouldingi*, although they are considerably higher ventrally. Slightly more posteriorly, rates of relative elongation are lower in *S. gouldingi*; the disparity between species increases from point to point posteriorly.

S. gouldingi and *S. elongatus* differ most obviously in rate of relative deepening (Fig. 5.20B). In the anterior midbody, the rate of relative deepening is far higher in *S. gouldingi*; however, in some regions, it is lower, particularly in the supraorbital region and also in the dorsal posterior body (Fig. 5.20B). The most pronounced difference in rates of relative elongation is found in the head and anterior midbody. Cranially, the disparity between species has a complex pattern. Rates of relative elongation in the ventral snout are comparatively low in *S. gouldingi*, but they exceed those of *S. elongatus* in the orbital and ventral postorbital region to the opercular landmark, where they seem to decrease in *S. gouldingi* compared with *S. elongatus*. Postcranially, rates of relative elongation in the anterior midbody ventrally are lower in *S. gouldingi* compared with *S. elongatus*, as they are between dorsal and adipose fins.

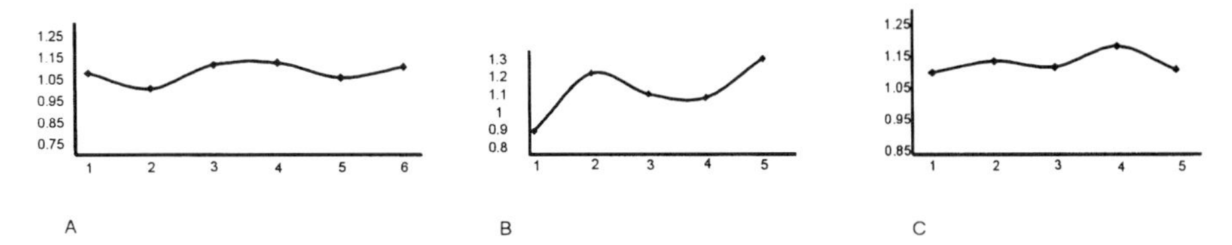

Figure 5.21. Growth profiles for traditional measurements of *S. manueli.* A: dorsal measurements along the anteroposterior axis. B: ventral measurements along the anteroposterior axis. C: measurements along dorsoventral axis. Measurements as described in Fig. 5.11.

S. manueli

The spatial pattern of growth rates in *S. manueli* is distinctive in that the highest relative growth rate is found in the caudal peduncle. Also, the relative growth rate of the anterior head exceeds that of the posterior head (Fig. 5.10). The growth profile for dorsal measurements suggests a wave-like pattern along the anteroposterior axis, with two troughs, each located where the peaks rates are found in *P. denticulata* and *S. elongatus*, that is, in the posterior head and between the dorsal and adipose fins (Fig. 5.21A). Ventrally, the curve resembles an anteriorly displaced inverted U-shaped curve, except for the precipitous rise in growth rate posteriorly (Fig. 5.21B). Along the dorsoventral body axis, relative growth rates are high over the whole body, accelerating in the region of the dorsal fin and then decreasing (but only slightly) in the caudal peduncle (Fig. 5.21C). The relative rates are almost uniform, except for a localized increase in the region of the dorsal fin.

The shape coordinates confirm the distinctness of growth in *S. manueli* (Fig. 5.12). The growth profiles along the anteroposterior axis, dorsally, suggest low rates of relative elongation in the head, even more so posteriorly than anteriorly, with a substantial postcranial acceleration, until notably decreasing in the region between dorsal and adipose fins, which is succeeded by a precipitous acceleration (Fig. 5.22A). The dorsal pattern resembles a two-peaked wave that is nearly an inverse of the one seen in *P. denticulata* and *S. elongatus*. Ventrally, relative growth rates are low (and uniform) through the head, rising substantially postcranially, but decelerating between the pelvic and anal fins and even more so over the anal fin, and then rising to a remarkable degree in the caudal peduncle (Fig. 5.22B). Thus, the ventral measurements also suggest a two-peaked wave. Both patterns are largely consistent with the pattern seen in the traditional measurements, except that the decrease in relative growth rates over the anal fin is not evident with traditional measurements probably because of the positive allometry of body depth in that region.

Along the dorsoventral body axis, rates of relative deepening are fairly uniform except for a deceleration at the anterior base of the dorsal fin and then a return to the level comparable to that found in the more anterior body, followed by a modest drop in the caudal peduncle (Fig. 5.22C). Ventrally, the profile is also fairly uniform, except for the low relative rate of deepening of the

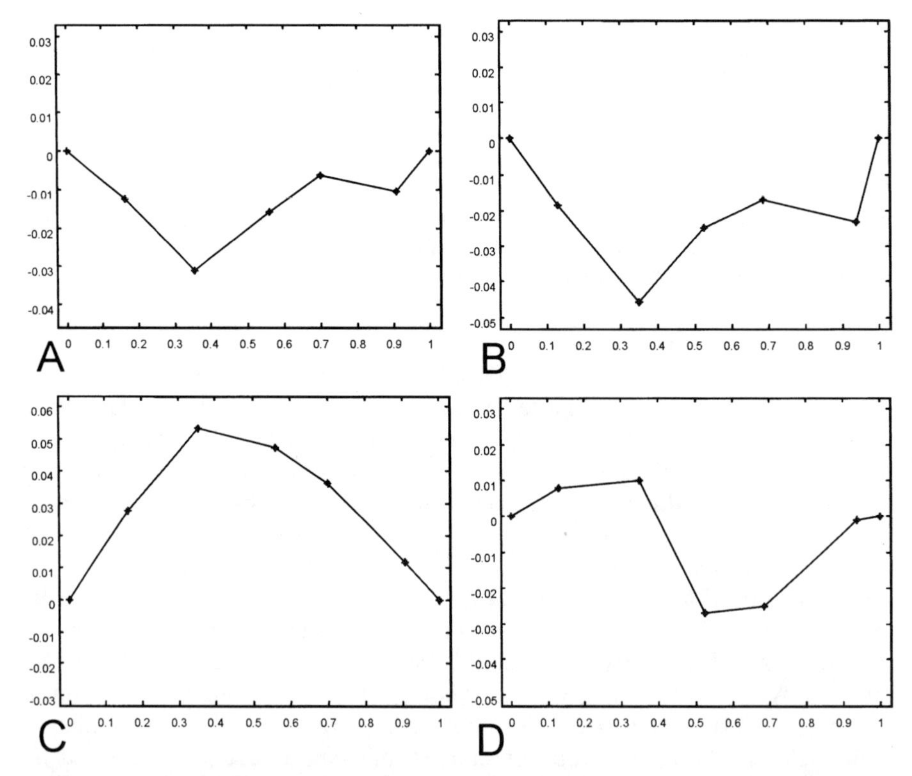

Figure 5.22. Growth profiles for shape coordinates of *S. manueli.* A: anteroposterior component of peripheral dorsal coordinates. B: anteroposterior component of ventral peripheral coordinates. C: dorsoventral component of peripheral dorsal coordinates. D: dorsoventral component of ventral peripheral coordinates. Landmarks as described in Fig. 5.13.

head, more so anteriorly than posteriorly, and the deceleration over the anal fin, after which it accelerates slightly (Fig. 5.22D). In general, the rates of deepening are higher dorsally than ventrally. The pattern is similar to that described by the traditional measurements, which show a largely uniform pattern of relative deepening, except for decelerations at the anterior base of the dorsal fin and the caudal peduncle.

The depiction of ontogenetic shape change as a deformation shows the wave-like pattern of relative elongation along the anteroposterior axis. This is particularly notable posteriorly where decelerating relative growth rates lead to a visible contraction of the region between dorsal and adipose fins, succeeded by a striking expansion posteriorly (Fig. 5.14). The pronounced relative deepening of the body is also visually striking. The pattern of relative deepening is more complex than could be seen in the growth profiles because there are localized changes within the head that cannot be seen in peripheral landmarks. Although the growth profiles suggest a fairly uniform rate of growth in the

Table 5.5. Angles (in degrees) and vector correlations (R_V) for the comparison between *S. manueli*, *P. denticulata*, *S. elongatus*, and *S. gouldingi*, based on traditional morphometric measurements, shape coordinates, and parameters of the thin-plate spline

S. manueli	Traditional		Shape Coordinates		Spline Parameters	
Compared With	Angle (°)	R_V	Angle (°)	R_V	Angle (°)	R_V
P. denticulata	4.8	0.996	39.0	0.771	45.7	0.698
S. elongatus	4.3	0.997	34.8	0.821	45.9	0.696
S. gouldingi	6.0	0.994	34.2	0.827	34.9	0.820

head, the additional landmarks show a nonuniform pattern: the snout lengthens more rapidly than the eye, and rates of elongation accelerate in the postorbital region, to a degree higher than those found in the snout (especially ventrally). Also, the anterior head is relatively deepened ventrally; however, in the orbital region, it is relatively deepened dorsally more than ventrally. In the postorbital region, the rates of relative deepening increase gradually along the dorsoventral axis, accelerating abruptly below the posterior point on the operculum.

Not surprisingly, *S. manueli* differs from *P. denticulata* and its closer relatives in its ontogenetic allometries (Table 5.5). On the bases of all three measurement schemes, *S. manueli* differs significantly from all three species ($P < 0.05$). Nevertheless, the differences are smaller than those found in comparisons between the growth vector of *S. manueli* and vectors of randomly reshuffled coefficients (Table 5.2).

Comparisons with the other three species show numerous distinctive features of *S. manueli* (Fig. 5.23). Compared with *P. denticulata*, the most striking contrast is the lower rates of relative elongation in the midbody of *S. manueli*

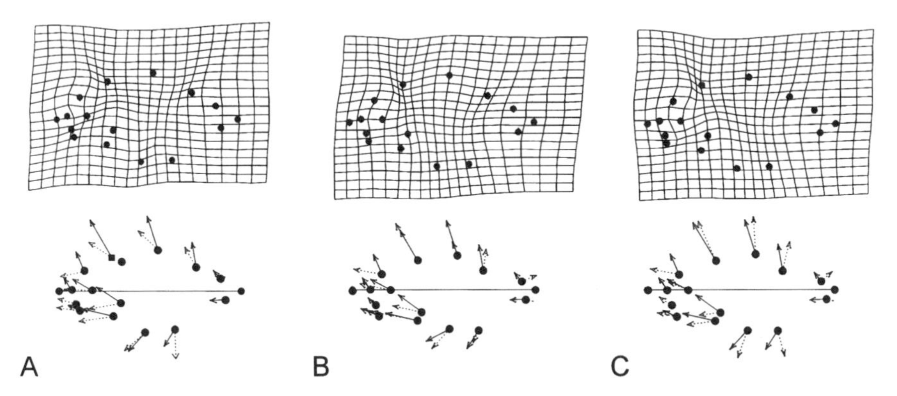

Figure 5.23. Differences between ontogenetic trajectories of allometric coefficients. A: *P. denticulata* vs. *S. manueli*. B: *S. elongatus* vs. *S. manueli*. C: *S. gouldingi* vs. *S. manueli*. Differences are represented by the thin-plate spline (top) and by shape coordinates to a common baseline (bottom). Ontogenetic trajectories are normalized to unit length, removing effects of differences in overall developmental rate.

(Fig. 5.23A). This pattern is more pronounced dorsally than ventrally, except between pelvic and anal fins, where relative growth rates accelerate in *P. denticulata* but decelerate in *S. manueli*. Dorsally, the difference is subtle; in *P. denticulata*, rates of relative elongation are nearly uniform postcranially to the region between dorsal and adipose fins, where they accelerate. In *S. manueli*, rates of relative elongation decelerate near the anterior base of the dorsal fin and then accelerate over the fin. There are also two striking differences in the head. One is the comparatively higher rates of deepening dorsally relative to ventrally in *S. manueli*, a contrast that increases just inferior to the eye; the other is the higher rates of elongation of the anterior head relative to the posterior head.

Compared with *P. denticulata*, *S. manueli* is distinctive for its very low rates of relative elongation in the midbody, its exceptionally high rate of relative elongation posteriorly, and its high rate of cranial deepening dorsally compared with ventrally (Fig. 5.23A). Some of these same features distinguish *S. manueli* from the other two *Serrasalmus*; compared with them, *S. manueli* has strikingly high rates of relative elongation in the postorbital region and posterior body, as well as more localized differences in rates of relative deepening cranially, especially in the sub- and postorbital head (Fig. 5.23, B and C).

S. spilopleura

The spatial pattern of growth rates in *S. spilopleura* resembles that seen in the other species (with the exception of *S. manueli*) in that relative growth rates are higher in the midbody than head and posterior body (Fig. 5.10). However, in this species alone, almost all measurements posterior to the dorsal fin are negatively allometric, suggesting a shortening and shallowing of the posterior body relative to both overall body length and head. The growth profile for dorsal measurements along the anteroposterior axis resembles a shallow inverted U-shaped curve with an anteriorly located peak (Fig. 5.24A). However, there may also be a very modest secondary peak in the region between dorsal and adipose fins. Ventrally, the growth rates suggest an inverted U-shaped curve except for the unexpectedly anterior location of the peak and the uniformity of growth rates posterior to the pelvic fin (Fig. 5.24B). Along the dorsoventral body axis, growth rates are highest anteriorly, although, at their maximum, they are iso-

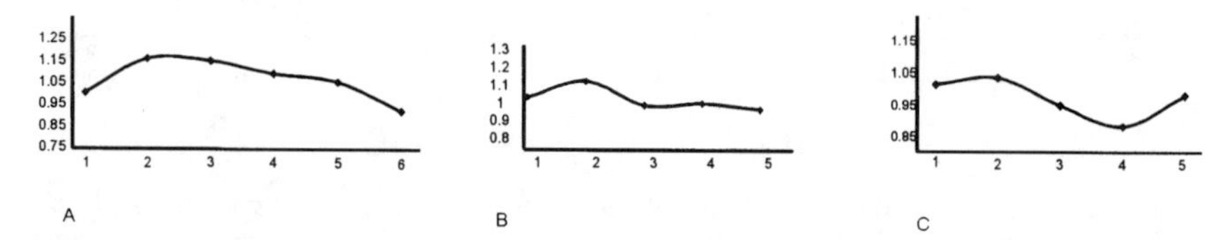

Figure 5.24. Growth profiles for traditional measurements of *S. spilopleura*. A: dorsal measurements along the anteroposterior axis. B: ventral measurements along the anteroposterior axis. C: measurements along dorsoventral axis. Measurements as described in Fig. 5.11.

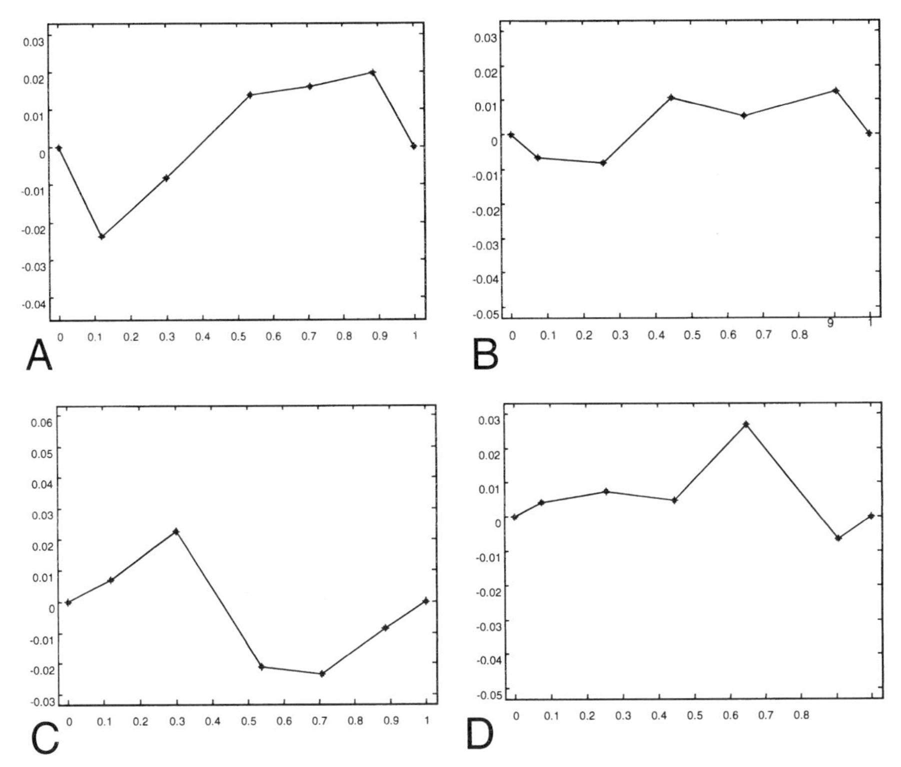

Figure 5.25. Growth profiles for shape coordinates of *S. spilopleura*. A: anteroposterior component of peripheral dorsal coordinates. B: anteroposterior component of ventral peripheral coordinates. C: dorsoventral component of peripheral dorsal coordinates. D: dorsoventral component of ventral peripheral coordinates. Landmarks as described in Fig. 5.13.

metric (Fig. 5.24C). Posteriorly, growth rates are negatively isometric and become more negatively allometric at the posterior base of the dorsal fin, rising to isometry in the caudal peduncle.

The shape coordinates show a relative deepening of the posterior head dorsally, whereas the midbody and posterior body undergo relative shallowing (Fig. 5.12). The growth profile along the anteroposterior axis dorsally reveals the familiar pattern, an inverted U-shaped gradient, with a peak extending from the posterior head to dorsal fin (Fig. 5.25A). Ventrally, the profile suggests a complex wave-like pattern, with low rates of relative elongation cranially, increasing slightly in the posterior head, then accelerating markedly in the anterior midbody, dropping in the region between pelvic and anal fins, increasing to a secondary peak over the anal fin, and finally decelerating over the caudal peduncle (Fig. 5.25B). Dorsally, the pattern is consistent with that seen in traditional measurements, but the complex ventral pattern is not. Whereas the shape coordinates suggest an acceleration over the anal fin and a deceleration over the caudal peduncle, the traditional measurements suggest more uniform rates. In this case, the discrepancy may be due to the pronounced negative allometry of body depth posteriorly, which would reduce the allometric coefficient for the oblique measurement over the anal fin.

Along the dorsoventral axis, rates of relative deepening dorsally are uniform through the head, decreasing between the posterior head and dorsal fin, then falling precipitously at the anterior base of the dorsal fin, decelerating further over the dorsal fin, and then rising over the caudal peduncle (Fig. 5.25C). Ventrally, rates of relative deepening are generally low, accelerating slightly at the pelvic fin and caudal peduncle (Fig. 5.25D). These patterns are largely consistent with those inferred from traditional measurements, except that the postcranial deceleration is more pronounced in the shape coordinates.

The depiction of growth as a deformation shows the considerable deepening of the posterior head relative to the postcranial body and the inverted U-shaped gradient in rates of elongation, with its anteriorly located peak (Fig. 5.14). As evident from the relative heights of the rectangles along the dorsoventral axis, relative growth rates are especially low in the anterior midbody dorsally and posterior midbody ventrally. The inclusion of the interior cranial landmarks reveals the elongation of the snout relative to the eye and also the elongation of the ventral postorbital head relative to the eye and snout. In addition, the ventral head deepens relative to the dorsal head. This complexity is not evident in the peripheral landmarks or in profiles of individual components taken separately.

Not surprisingly, *S. spilopleura* differs significantly from *P. denticulata* and the other *Serrasalmus* ($P < 0.05$), and the angles are remarkably large (Table 5.6); they are so great that we cannot reject the null hypothesis of uncorrelated growth vectors (Table 5.2).

The contrasts between *S. spilopleura* and *P. denticulata* and *S. elongatus* are difficult to summarize succinctly because they vary from region to region (Fig. 5.26). *S. spilopleura* is distinctive for its low relative rates of postcranial deepening, and its cranial rates of relative elongation are higher dorsally, especially posteriorly. One striking difference can be interpreted in light of the growth profiles. In both *P. denticulata* and *S. elongatus*, dorsal rates of elongation fol-

Table 5.6. Angles (in degrees) and vector correlations (R_v) for the comparison between *S. spilopleura*, *P. denticulata*, *S. elongatus*, *S. gouldingi*, and *S. manueli*, based on traditional morphometric measurements, shape coordinates, and parameters of the thin-plate spline

S. spilopleura	Traditional		Shape Coordinates		Spline Parameters	
Compared With	Angle (°)	R_v	Angle (°)	R_v	Angle (°)	R_v
P. denticulata	6.2	0.994	93.5	−0.06	107.5	−0.301
S. elongatus	4.3	0.997	75.3	0.254	85.2	0.084
S. gouldingi	7.0	0.993	84.6	0.094	101.0	−0.191
S. manueli	6.8	0.993	92.4	−0.042	108.1	−0.311

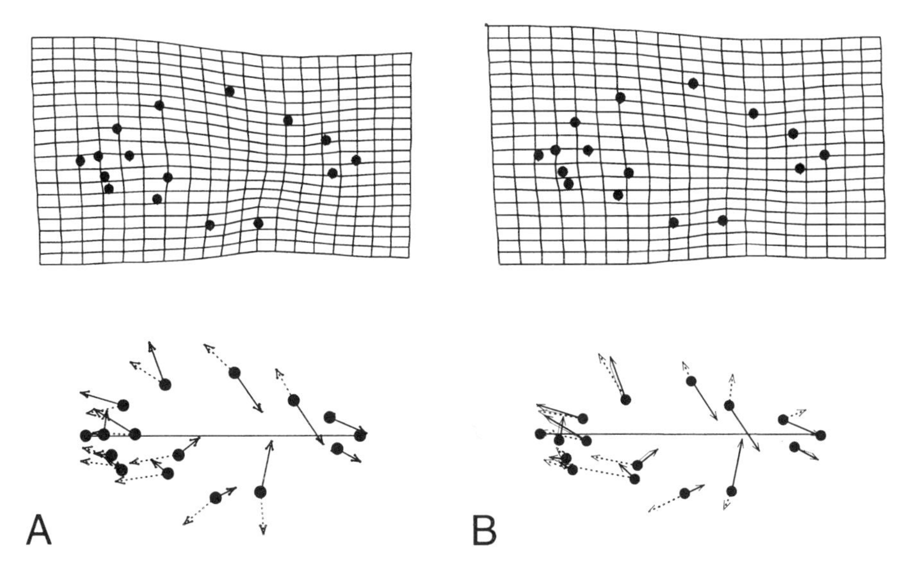

Figure 5.26. Differences between ontogenetic trajectories of allometric coefficients. A: *P. denticulata* vs. *S. spilopleura*. B: *S. elongatus* vs. *S. spilopleura*. Differences are represented by the thin-plate spline (top) and by shape coordinates to a common baseline (bottom). Ontogenetic trajectories are normalized to unit length, removing effects of differences in overall developmental rate.

low a wave-like pattern, whereas the ventral rates follow an inverted U-shaped curve; however, that pattern is reversed in *S. spilopleura*. Thus, the relative rates of elongation are higher in *S. spilopleura* where its peak coincides with the troughs of the other species but lower in *S. spilopleura* where its troughs coincide with the peaks of the other species. The disparity between *S. spilopleura* and *S. elongatus* is less marked than that between *S. spilopleura* and *P. denticulata* because *S. elongatus* has a less pronounced wave-like pattern dorsally than does *P. denticulata*; in this, it more closely resembles *S. spilopleura* with its inverted U-shaped pattern. Also, the relative deepening near the anterior base of the dorsal and anal fins is less marked in *S. elongatus* and in this too it resembles *S. spilopleura*.

P. piraya

The relative growth rates of *P. piraya* are generally either isometric or negatively allometric relative to body length, as is the case for *S. spilopleura* (Fig. 5.10). However, in *P. piraya*, the head and caudal peduncle are negatively allometric with respect to the midbody, whereas the posterior head is the most positively allometric region of *S. spilopleura*. Also, the expansion of the midbody relative to the ends encompasses a larger region in *P. piraya*, extending posteriorly over the dorsal fin. The growth profiles for traditional measurements confirm the impression of a smooth increase in relative growth rates

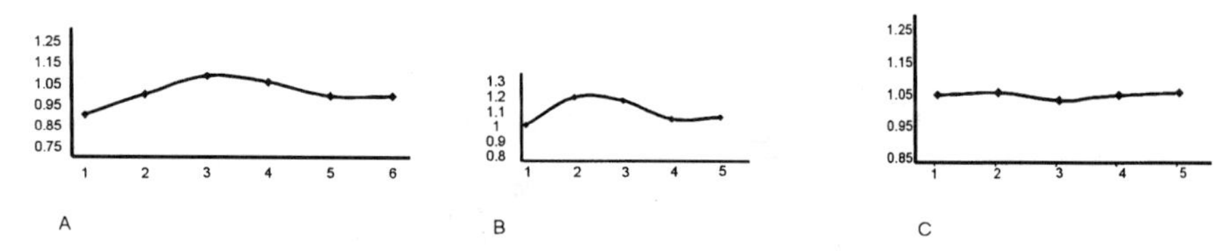

Figure 5.27. Growth profiles for traditional measurements of *P. piraya*. A: dorsal measurements along the anteroposterior axis. B: ventral measurements along the anteroposterior axis. C: measurements along dorsoventral axis. Measurements as described in Fig. 5.11.

toward the middle of the body, succeeded by a decrease, differing from a hypothetical inverted U-shaped curve only in that the posterior relative growth rates are not as low as expected (Fig. 5.27A). Ventrally, there is a more pronounced U-shaped pattern, with an anteriorly located peak and a posteriorly extended trough, which (like that seen in the curve for dorsal measurements) does not descend to the low level found in the head (Fig. 5.27B). Relative growth rates along the dorsoventral axis increase to their maximum at the supraoccipital process, subsequently declining and then (perhaps) slightly increasing over the caudal peduncle (Fig. 5.27C).

The shape coordinates also suggest a marked expansion of the midbody relative to the head and posterior body, with modest rates of relative deepening except in the posterior head (Fig. 5.12). Dorsally, the profile for rates of relative elongation resembles an inverted U-shaped gradient, except that the relative rates rise more abruptly than expected at the supraoccipital process and remain high to the posterior base of the dorsal fin. Posteriorly, they are also higher than expected from the model (Fig. 5.28A). Ventrally, the pattern also resembles an inverted U-shaped gradient except that the peak extends a bit more posteriorly than expected, resembling the pattern dorsally, and the posterior relative rates do not decrease to the low level predicted by the model, again resembling the pattern dorsally (Fig. 5.28B). Rates of relative deepening dorsally are low except in the posterior head; elsewhere, the dorsal body is virtually unmodified in relative depth (Fig. 5.28C). Ventrally, little relative deepening occurs anywhere except in the posterior head and the most posterior body; elsewhere, rates of relative deepening are low and uniform (Fig. 5.28D). This description is largely consistent with the pattern described by the traditional measurements in that relative deepening is most pronounced in the posterior head.

The depiction of ontogeny by the thin-plate spline shows expansion of the midbody relative to the cranial and caudal ends resulting from the U-shaped pattern in rates of relative elongation (Fig. 5.14). It also shows the substantial increase in relative depth of the posterior head due to the increase in rates or relative deepening there. In general, accelerations and decelerations in relative growth rates seem to be fairly smooth, aside from the abrupt deceleration in the orbital region and the exceptionally rapid acceleration in rate of relative elongation postorbitally.

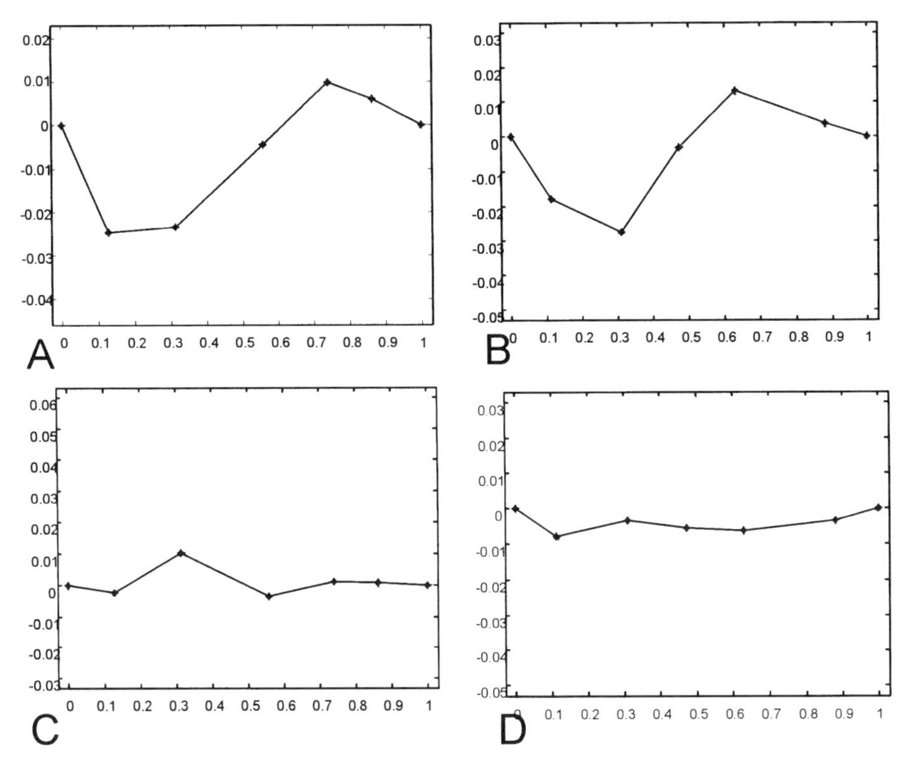

Figure 5.28. Growth profiles for shape coordinates of *P. piraya*. A: anteroposterior component of peripheral dorsal coordinates. B: anteroposterior component of ventral peripheral coordinates. C: dorsoventral component of peripheral dorsal coordinates. D: dorsoventral component of ventral peripheral coordinates. Landmarks as described in Fig. 5.13.

P. piraya differs significantly from *P. denticulata* and the three *Serrasalmus* ($P < 0.05$). On the basis of traditional measurements, the differences seem modest, but they are far from subtle in comparisons based on geometric data (Table 5.7). On the basis of the parameters of the thin-plate spline, *P. piraya* is apparently no more similar to *S. spilopleura*, *S. gouldingi*, and *S. manueli* than to a vector of randomized coefficients (Table 5.7).

Compared with *S. spilopleura* (accepted here as the basal species of the clade including *P. piraya*), the major difference is in the pattern of relative deepening (Fig. 5.29A). Both species deepen their heads relative to the postcranial body; in *P. piraya*, this results from high rates of relative deepening in the head (especially ventrally), and, in *S. spilopleura*, this results from remarkably low rates of relative deepening in the posterior body. In addition, these two species differ in relative growth rates along the anteroposterior axis; *P. piraya* has higher rates of relative elongation in the dorsal posterior head and anterior midbody. Ventrally, the pattern is more complex because of the wave-like pattern in *S. spilopleura* vs. the inverted U-shaped gradient in *P. piraya*.

Table 5.7. Angles (in degrees) and vector correlations (R_v) for the comparison between *P. piraya*, *S. spilopleura*, *P. denticulata*, *S. elongatus*, *S. gouldingi*, and *S. manueli*, based on traditional morphometric measurements, shape coordinates, and parameters of the thin-plate spline

P. piraya Compared With	Traditional		Shape Coordinates		Spline Parameters	
	Angle (°)	R_v	Angle (°)	R_v	Angle (°)	R_v
S. spilopleura	3.2	0.998	65.2	0.420	60.3	0.496
P. denticulata	4.4	0.997	38.7	0.780	59.3	0.511
S. elongatus	2.5	0.999	36.7	0.802	47.9	0.670
S. gouldingi	5.4	0.996	47.3	0.678	65.0	0.423
S. manueli	5.0	0.996	57.8	0.533	73.3	0.287

Compared with *P. denticulata* (the sister to the other piranhas in this study), *P. piraya* undergoes lesser relative deepening postcranially (Fig. 5.29B). These two species also differ in growth rates along the anteroposterior axis. Although relative growth rates (dorsally) generally follow an inverted U-shaped curve in both species, the anterior trough is higher in *P. piraya* and the dorsal peak is located more posteriorly. Over the midbody, relative growth rates are generally higher in *P. piraya*, but, ventrally, *P. piraya* does not have the rapid acceleration anterior to the anal fin that occurs in *P. denticulata*. In addition, there are subtle differences within the head, most notably, the higher rates of relative deepening suborbitally in *P. piraya*.

Compared with *S. elongatus*, the species most similar in its growth pattern to *P. piraya*, the differences are largely in rates of relative deepening; these are lower in *P. piraya* both cranially and posteriorly (Fig. 5.29C). These species also differ in relative rates of elongation; the contrasts between them resemble

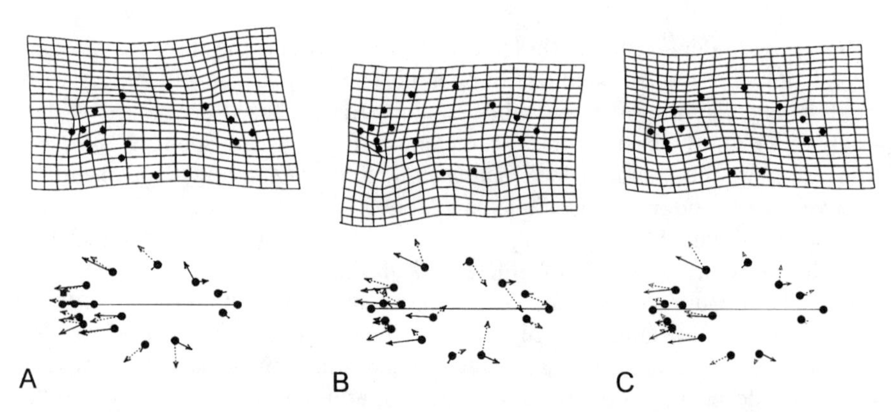

Figure 5.29. Differences between ontogenetic trajectories of allometric coefficients. A: *P. denticulata* vs. *P. piraya*. B: *S. spilopleura* vs. *P. piraya*. C: *S. elongatus* vs. *P. piraya*. Differences are represented by the thin-plate spline (top) and by shape coordinates to a common baseline (bottom). Ontogenetic trajectories are normalized to unit length, removing effects of differences in overall developmental rate.

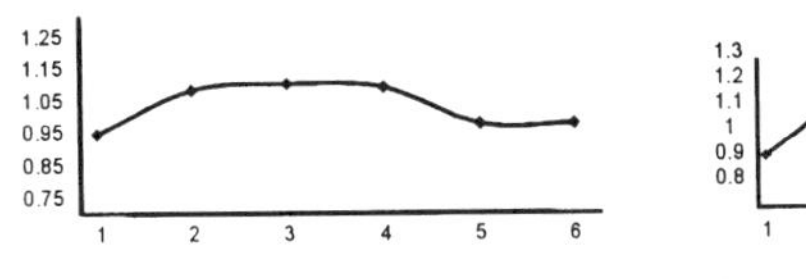

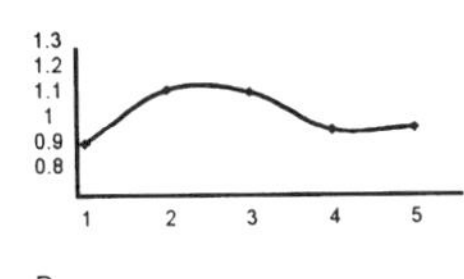

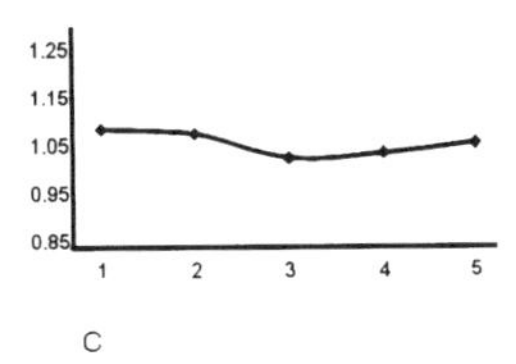

Figure 5.30. Growth profiles for traditional measurements of *P. nattereri.* A: dorsal measurements along the anteroposterior axis. B: ventral measurements along the anteroposterior axis. C: measurements along dorsoventral axis. Measurements as described in Fig. 5.11.

those already described in the comparison between *P. piraya* and *P. denticulata*. However, there are additional differences between *P. piraya* and *P. elongatus*, especially in the pattern of cranial deepening. In *P. piraya*, rates of relative deepening and elongation are higher ventrally than dorsally, although both decrease most ventrally.

P. nattereri

The spatial pattern of relative growth rates in *P. nattereri* closely resembles that of *P. piraya*, although cranial relative growth rates are higher (Fig. 5.10). The growth profiles for traditional measurements show virtually the same pattern described for *P. piraya*, except for the higher coefficients cranially (Fig. 5.30).

The shape coordinates show that *P. nattereri*, like *P. piraya*, develops a deep head in relation to the postcranial body and that the midbody elongates relative to the cranial and caudal ends, perhaps to a slightly greater extent in *P. nattereri* (Fig. 5.12). Not surprisingly, the growth profiles for shape coordinates resemble those of *P. piraya*, with a few minor differences, the most notable being the greater acceleration in rates of relative elongation at the epiphyseal bar (Fig. 5.31A). Generally, the rates of relative elongation dorsally accelerate and decelerate to a slightly greater degree in *P. nattereri* than in *P. piraya*. Ventrally, the pattern is virtually identical to the one described for *P. piraya* (Fig. 5.31B). Along the dorsoventral body axis, the patterns are also virtually the same in these two species (Figs. 5.31, C and D).

As depicted by the thin-plate spline, relative growth rates in *P. nattereri* are especially high in the postorbital region (Fig. 5.14). Rates of relative elongation decline quite notably in the posterior body. Cranial growth rates have a fairly complex spatial distribution; the snout elongates relative to the eye, as does the postorbital region (to an even greater degree, especially ventrally), and relative growth rates in the orbital region are higher ventrally than dorsally.

Statistically, the growth vector of *P. nattereri* cannot be distinguished from that of *P. piraya* by either traditional or geometric data ($P > 0.05$). Not surprisingly, the angles between *P. nattereri* and other species are similar in magnitude to those found between *P. piraya* and the other species, although *P. nattereri* is more similar to *S. spilopleura* than to *S. manueli*, whereas the reverse is shown for *P. piraya* (Table 5.8). However, because the distinction

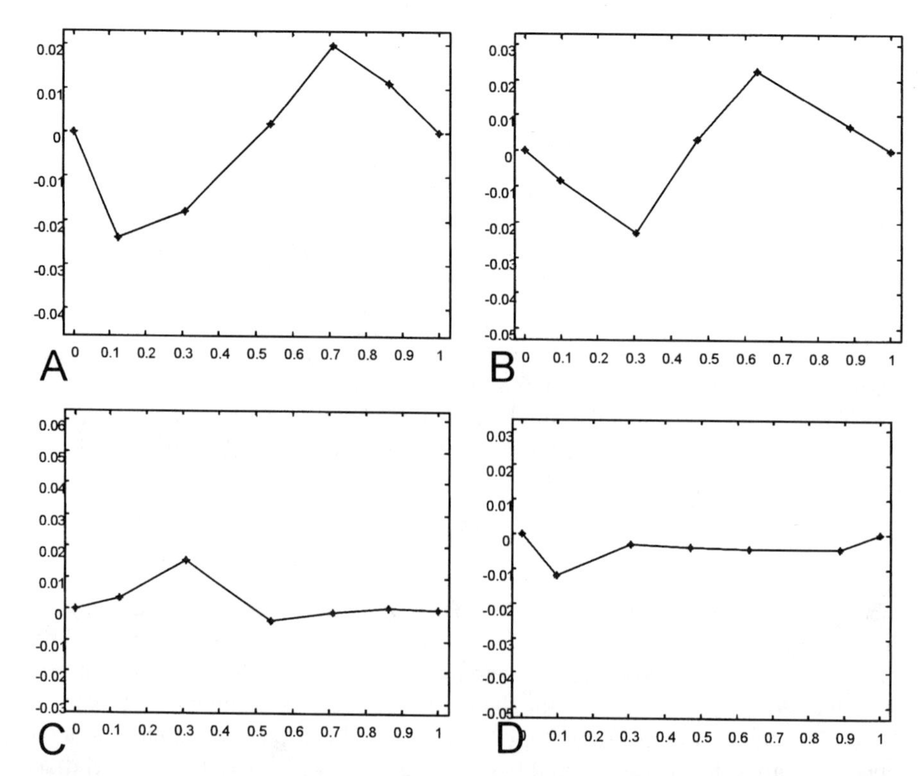

Figure 5.31. Growth profiles for shape coordinates of *P. nattereri*. A: anteroposterior component of peripheral dorsal coordinates. B: anteroposterior component of ventral peripheral coordinates. C: dorsoventral component of peripheral dorsal coordinates. D: dorsoventral component of ventral peripheral coordinates. Landmarks as described in Fig 5.13.

Table 5.8. Angles (in degrees) and vector correlations (R_v) for the comparison between *P. nattereri*, *P. piraya*, *S. spilopleura*, *P. denticulata*, *S. elongatus*, *S. gouldingi*, and *S. manueli*, based on traditional morphometric measurements, shape coordinates, and parameters of the thin-plate spline

P. natterei	Traditional		Shape Coordinates		Spline Parameters	
Compared With	Angle (°)	R_v	Angle (°)	R_v	Angle (°)	R_v
P. piraya	1.6	1.000	26.3	0.896	22.7	0.923
S. spilopleura	3.1	0.999	56.2	0.556	53.8	0.591
P. denticulata	5.8	0.995	58.3	0.526	70.6	0.332
S. elongatus	3.4	0.998	50.2	0.640	54.7	0.578
S. gouldingi	6.3	0.994	53.9	0.589	68.1	0.373
S. manueli	5.7	0.995	69.5	0.350	76.5	0.233

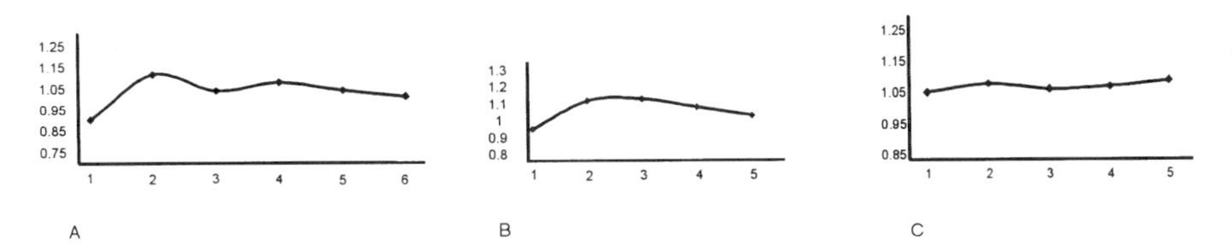

Figure 5.32. Growth profiles for traditional measurements of *P. cariba*. A: dorsal measurements along the anteroposterior axis. B: ventral measurements along the anteroposterior axis. C: Measurements along dorsoventral axis. Measurements as described in Fig. 5.11.

between *P. nattereri* and *P. piraya* cannot be established statistically, we cannot assume that these discrepancies are meaningful. For that reason, we presume that the contrasts between the ontogenetic vectors of *P. nattereri* and other species follow those already described for *P. piraya* and therefore need not be detailed herein.

P. cariba

The spatial pattern of relative growth rates in *P. cariba* resembles that familiar from other the piranhas—they are generally lowest anteriorly and posteriorly (Fig. 5.10). Like the other two species of *Pygocentrus*, in *P. cariba*, the highest rates of relative deepening are found in the posterior head, although *P. cariba* is positively allometric in depth over the whole body. Growth profiles for traditional measurements along the anteroposterior axis dorsally suggest a wave-like pattern, with relative growth rates being lowest anteriorly, peaking in the posterior head, dropping slightly, rising slightly at the dorsal fin, and finally decreasing slightly to the end of the caudal peduncle (Fig. 5.32A). Ventrally, rates of relative lengthening follow an inverted U-shaped curve, being low anteriorly, rising to a peak in the anterior midbody, and then smoothly decelerating over the anal fin and caudal peduncle (Fig. 5.32B). However, relative growth rates posteriorly are higher than expected from a model of a U-shaped gradient, and the peak is unexpectedly anterior in location; moreover, the deceleration is slightly greater than anticipated posterior to the head dorsally. Dorsoventrally, relative growth rates are low and virtually uniform (Fig. 5.32C).

The shape coordinates also suggest a pronounced lengthening of the postorbital region relative both to the anterior head and to the anterior midbody (Fig. 5.12). In addition, the posterior head deepens dorsally. The growth profile for dorsal coordinates shows that rates of relative elongation are low in the anterior head, increase substantially at the epiphyseal bar, and then decrease through the midbody until accelerating over the dorsal fin, after which they drop progressively, if not to the low level found in the anterior head (Fig. 5.33A). Ventrally, the rates of relative elongation are low in the head, increase substantially at the pectoral fin, and maintain an almost uniform rate over the midbody, dropping slightly in the anal fin then more substantially in the caudal

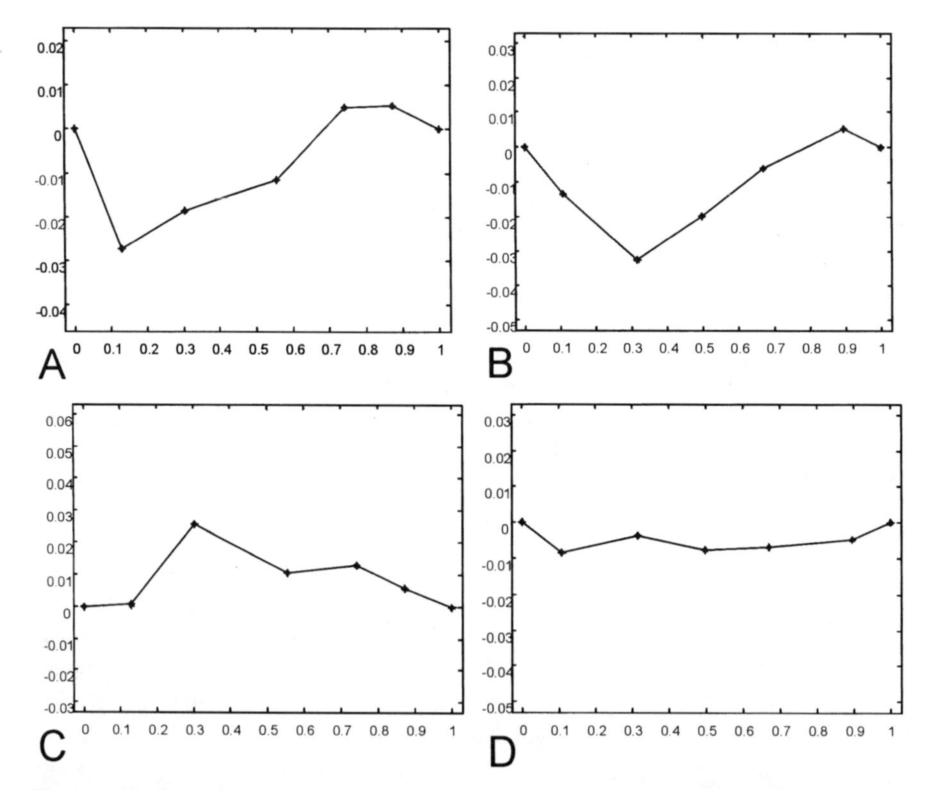

Figure 5.33. Growth profiles for shape coordinates of *P. cariba*. A: anteroposterior component of peripheral dorsal coordinates. B: anteroposterior component of ventral peripheral coordinates. C: dorsoventral component of peripheral dorsal coordinates. D: dorsoventral component of ventral peripheral coordinates. Landmarks as described in Fig. 5.13.

peduncle but not to the low level seen in anterior head (Fig. 5.33B). Thus, dorsally, the growth profile suggests a wave-like pattern with peaks in the posterior head and at the dorsal fin and with troughs located in the anterior head, anterior midbody, and most caudally. In contrast, the ventral profile suggests an inverted U-shaped gradient, except that the relative rates of elongation rise too abruptly and are too uniform to the anal fin. Both patterns are similar to those inferred from traditional measurements, except that dorsal rates of relative elongation seem less uniform in the mid- and posterior body in the analysis based on shape coordinates.

Along the dorsoventral axis, dorsal rates of relative deepening are generally low and nearly uniform except for decelerations in the anterior head and near the anterior base of the dorsal fin (Fig. 5.33C). Ventrally, rates of relative deepening are also low and nearly uniform except for a very slight acceleration in the anterior head and caudal peduncle (Fig. 5.33D). Both dorsal and ventral patterns suggest a subtle wave-like pattern, taking into account that the ampli-

Table 5.9. Angles (in degrees) and vector correlations (R_V) for the comparison between *P. cariba*, *P. nattereri*, *P. piraya*, *S. spilopleura*, *P. denticulata*, *S. elongatus*, *S. gouldingi*, and *S. manueli*, based on traditional morphometric measurements, shape coordinates, and parameters of the thin-plate spline

P. cariba	Traditional		Shape Coordinates		Spline Parameters	
Compared With	Angle (°)	R_V	Angle (°)	R_V	Angle (°)	R_V
P. nattereri	2.2	0.999	44.2	0.717	50.6	0.635
P. piraya	1.9	0.999	36.4	0.805	52.0	0.616
S. spilopleura	3.7	0.998	74.1	0.274	76.0	0.241
P. denticulata	4.7	0.997	39.6	0.771	58.4	0.524
S. elongatus	2.2	0.999	30.0	0.866	36.7	0.802
S. gouldingi	5.4	0.996	41.0	0.755	53.6	0.593
S. manueli	4.9	0.997	43.0	0.731	51.6	0.621

tudes of the peaks and troughs are barely detectable. The pattern differs from that inferred from traditional measurements only in that the shape coordinates suggest slight departures from uniformity.

In the depiction of growth as a deformation, there is the marked elongation of the postorbital head relative to the midbody seen in the other species but not a shortening of the caudal body relative to the midbody (Fig. 5.14). Within the head, the orbital region shows a local deceleration in relative growth rates, especially ventrally. Anteriorly, relative rates of lengthening seem comparatively high anteriorly, dropping near the epiphyseal bar (dorsally) and the jaw (ventrally), and then accelerating in the postorbital region, an acceleration that seems to occur slightly more posteriorly in the ventral head.

On the basis of traditional measurements, the growth vector of *P. cariba* does not differ significantly from the growth vectors of the other two *Pygocentrus* or of *S. spilopleura* ($P > 0.05$). However, the differences are statistically significant in comparisons based on geometric data ($P < 0.05$). In those geometric comparisons, *P. cariba* is most similar to *S. elongatus* (Table 5.9). *P. cariba* is so different from the most basal member of this clade, *S. spilopleura*, that their resemblance is no greater than we would expect by chance (Table 5.2).

Compared with *P. piraya*, the basal species of *Pygocentrus*, *P. cariba* has lower rates of midbody elongation, generally higher rates of relative elongation postorbitally, and a lower rate of relative cranial deepening ventrally (Fig. 5.34A). Compared with *S. elongatus*, the species most similar to *P. cariba* (in comparisons based on geometric data), the major difference is also the low rate of midbody elongation in *P. cariba* (Fig. 5.34B). In this comparison, the region of higher rates extends more posteriorly and is more ventrally located than in the comparison to *P. piraya*. An additional difference is in cranial growth rates. Generally, *P. cariba* has higher rates of relative elongation in the suborbital and postorbital regions, and its rates of relative deepening are higher ventrally than dorsally.

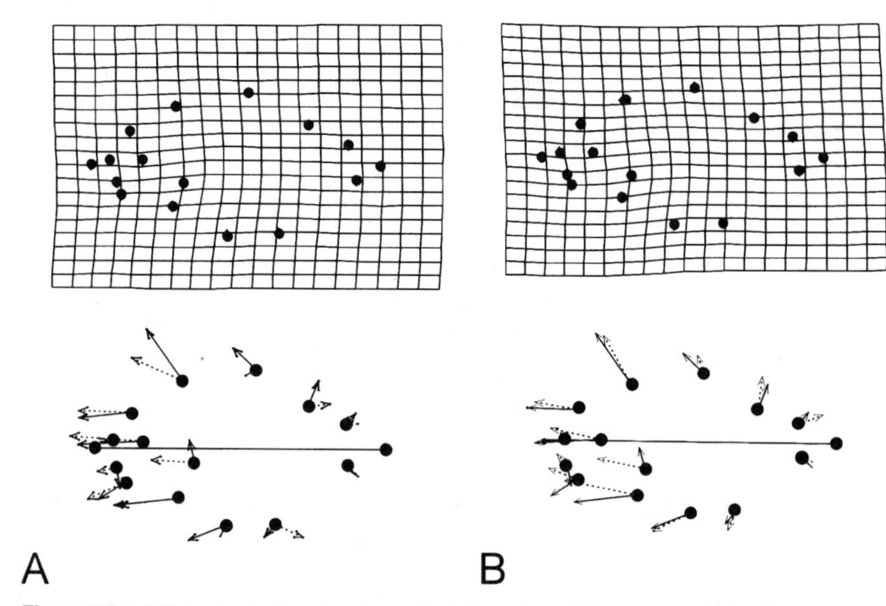

Figure 5.34. Differences between ontogenetic trajectories of allometric coefficients. A: *P. piraya* vs. *P. cariba*. B: *S. elongatus* vs. *P. cariba*. Differences are represented by the thin-plate spline (top) and by shape coordinates to a common baseline (bottom). Ontogenetic trajectories are normalized to unit length, removing effects of differences in overall developmental rate.

DISCUSSION

The Spatial Complexity of Juvenile Growth

The spatial patterns of juvenile growth are moderately complex in these eight species of piranhas. They are clearly not as simple as anticipated by the ideal models, all of which postulate that growth rates are a function of the position of a point along the anteroposterior axis. Contrary to these predictions, cranial growth rates cannot be predicted from caudal growth rates, as they would be were rates in both regions a function of a global growth gradient. Nor can dorsal growth rates be predicted from ventral growth rates, as they would be were rates along a single axis purely a function of position along that axis. Nor can rates of lengthening along the anteroposterior axis be predicted from rates of deepening along the dorsoventral axis, as they would be were growth a one-dimensional system. Clearly, the models are simplistic; none of them, or any other one-dimensional model, predicts the spatial distribution of growth rates. A two-dimensional model would also be insufficient. A successful two-dimensional model would predict rates of growth (in both anteroposterior and dorsoventral directions) based on the coordinates of any point along those two axes, but anteroposterior and dorsoventral growth rates do not seem predictable from a single model.

Growth rates show a complex spatial pattern that is not entirely predictable; however, some regularities can be found, even if no generalization holds for all species. One such regularity is that growth rates along the anteroposterior axis dorsally are usually lowest in the anterior head, accelerating in the posterior head, then either maintaining that high rate over the midbody or else decelerating then rising again, and finally decelerating (to a rate higher than that of the anterior head). The rates suggest either an inverted U-shaped gradient or a wave-like pattern. Another is that, along that same axis ventrally, growth rates are lowest in the head and highest near the pectoral fin. Growth rates along the dorsoventral axis also reveal some consistencies; specifically, they are usually higher dorsally than ventrally and also higher postcranially than cranially. Finally, in most species, cranial growth is more complex than postcranial growth, exhibiting several regions of localized accelerations or decelerations in various directions. These consistent patterns suggest order and regularity, if not a simple one-dimensional, global gradient.

Cranial growth rates seem especially complex, in part due to the localized deceleration in the orbital region. But it is not as if this is the sole departure from an otherwise smooth pattern, except perhaps in *P. denticulata*. In the other species, the pattern is not otherwise smooth. Rather, cranial growth rates change quite abruptly and to some extent unpredictably between closely spaced points. It may be that the complexity of cranial growth rates is an artifact of our measurement scheme because the head is sampled densely, whereas the postcranium is sampled at relatively few points, primarily where fins insert. Obviously, we cannot detect localized changes between closely spaced points when points are not closely spaced. However, the greater complexity of cranial growth might instead have a biological explanation—the greater osteological complexity of the skull. Not only are there many bones in the skull but also (and more importantly) they vary in orientation. For that reason, growth can occur in a variety of directions, and growth rates can vary in all of them. To distinguish between these two explanations, studies of closely spaced postcranial landmarks are necessary. Unfortunately, in these teleosts, the only structures available for such a study (vertebral centra, hemal and neural spines) do not in most cases meet the criterion of homology. For example, it is arguable whether the eighth centrum in one specimen corresponds to the eighth in another—the number of vertebrae in teleosts can differ between species and it often varies intraspecifically. Landmarks at those structures would not be homologous if the structures are not. Nevertheless, sampling these points would enable testing the hypothesis of spatially integrated postcranial growth.

The Diversity of Juvenile Growth

These eight species do not share a single pattern of growth retained from an ancestral ontogenetic allometry; thus, few generalizations apply to all species. With the exception of *P. nattereri* and *P. piraya*, no two species share the same spatial patterning. The average interspecific correlation (based on geometric data) is 0.477 (Tables 5.3–5.9), intermediate between entirely uncorrelated and perfectly correlated (Table 5.2). That value is much lower than the average correlation based on traditional data, $R_V = 0.997$, but that value also indicates near intermediacy between entirely uncorrelated ($R_V = .991$) and perfectly correlated ($R_V = 1.0$). Given that these species are related, we expect some similarity among them; what we find is far less similarity than expected were the evolution of growth constrained.

The diversity of growth is evident in the number of distinctive growth profiles and becomes even more evident when comparisons incorporate the interior landmarks, that is, those of the jaw, eye, and posterior opercular point. Counting just along the anteroposterior axis, there are four distinct growth patterns or possibly five, depending on whether *S. elongatus* is considered to have one intermediate between *P. denticulata* and *S. gouldingi* or is classified as unique. Even if *S. elongatus* is considered similar to *P. denticulata* in having a somewhat wave-like pattern of anteroposterior growth rates dorsally and an inverted U-shaped gradient ventrally, these species differ dramatically in rates of deepening. Except in the dorsal head, those rates are generally lower in *S. elongatus*. The differences are even more pronounced between *S. manueli* and *S. gouldingi*, especially cranially and ventroposteriorly, where *S. manueli* has strikingly higher rates of elongation over the anal fin and caudal peduncle. The differences between species are not only statistically significant, they are usually far from subtle.

The lack of constraint on spatial patterns of growth might seem unremarkable, especially in light of our previous study of piranhas (Zelditch et al., 2000), except for the large literature suggesting that ontogenetic allometries are highly conserved. Many studies conclude that late stages of growth evolve solely by extending or truncating a conserved ontogenetic vector, and this conclusion seems well supported by the high vector correlations often reported between species (e.g., Shea, 1992; Voss and Marcus, 1992; Smith, 1998). Had we relied solely on traditional measurements and had we taken the high correlations at face value, we would have concluded that juvenile ontogenetic allometries of piranhas are conserved. However, those high correlations are not as impressive as they seem in light of the correlations yielded by comparisons to randomized vectors. The generalization that allometric vectors are typically conservative badly needs a critical reanalysis.

Relating Spatial Complexity to Evolutionary Diversity

The diversity of juvenile growth is consistent with expectations of structuralist theories relating complexity to diversity (e.g., Liem, 1973, Vermeij, 1973, Lauder, 1981, Schaeffer and Lauder, 1986). However, a complex spatial distribution of growth rates does not necessarily mean that growth is regulated by numerous independent factors. Spatial complexity does not equate to developmental dissociability. Obviously, dissociability is more likely when spatial patterns are complex than when they are one-dimensional. A one-dimensional global growth gradient cannot be subdivided into units, and, as a one-dimensional whole, it can only change in one dimension. Because larval growth has been interpreted as one-dimensional (at least along the anteroposterior axis), we would expect larval growth to be highly constrained, certainly more than juvenile growth. That is why it seems paradoxical to say that juvenile growth is both more spatially complex and more conservative than larval growth. Our analysis indicates that the paradox is based on the faulty assumption of conservatism—juvenile growth may be at least moderately complex, but it is not historically conserved. However, the paradox may also be based on a second faulty assumption—that larval growth is simple in comparison to juvenile growth.

Larval growth may seem so simple because studies detecting global anteroposterior gradients in larvae examined measurements along only one axis and at only one transect along it, for example, along the sagittal axis (Fuiman, 1983; Gisbert, 1999; Koumoundouros et al., 1999). Had we analyzed only one dimension at one level, for example, analyzing only ventral growth rates along the anteroposterior axis, we might have also concluded that juvenile growth is simple, since that dimension often conforms to a global, inverted U-shaped gradient. Lacking profiles for dorsal growth rates along that axis, as well as interior points not analyzed by those profiles, we would not have found the complexity of juvenile growth patterns. Complexity of spatial structure is not likely to emerge from data limited to growth profiles of peripheral measurements at one level along one axis. Until larvae are measured more comprehensively, we cannot say that their patterns of growth are especially simple.

Such measurements of larvae are needed to assess the impression that larvae are more evolutionarily flexible than juveniles (Klingenberg, 1998). This generalization is based on a wealth of comparative studies, but most of these infer larval lability indirectly, that is, by finding that that intercepts of allometric equations differ. Such findings suggest divergence at a very small size, which is a questionable basis for the inference that larval growth is evolutionarily more flexible because very subtle differences in larval allometric coefficients could have a large impact on form, considering how much shape change occurs during larval growth. At present, it is difficult to say whether larval allometries are any more labile than juvenile ones because few studies directly compare larval allometries and also compare juvenile allometries. From the few studies that do, it does not appear that younger stages of growth differ more than older ones. For example, in a comparison among sculpins, 50% of the comparisons among larval growth vectors revealed significant positive correlations between species, whereas none of the comparisons among juvenile growth vectors did (analyses based on data published by Strauss and Fuiman, 1985). Even less divergence is found in comparisons between two species of anchovies. Comparisons between species at two stages of growth yield significant but slightly decreasing positive correlations when interspecific correlations are considered in context of correlations with randomized growth vectors (analyses based on data published by Sarpédonti et al., 2000).

CONCLUSIONS

The diversity of juvenile growth patterns in these piranhas is clearly inconsistent with the generalization that this phase of growth is typically conservative, evolving primarily or solely by extensions/truncations (Klingenberg, 1998). Were this conservatism an a priori expectation rather than a conclusion from data, it would seem wildly implausible. It is not grounded in either evolutionary theory or developmental biology. Certainly, it is possible to imagine conditions under which development could constrain variance to that degree, making constrained evolutionary change a plausible expectation. But those conditions, even if imaginable, are unrealistic. For example, we might imagine a situation in which growth is a perfectly integrated, one-dimensional system, as would be the case were growth rates determined by a global morphogenetic gradient. Under that condition, variance (hence evolutionary change) would be restricted to one dimension. Or we could imagine a case in which growth is spatially complex, but the complex system is so highly integrated by epigenetic factors that it is effectively one-dimensional. Under that condition, variance (hence evolutionary change) would also be restricted to one dimension. Neither of these conceivable models is plausible, even if development may seem one-dimensional, as is the case when it is described by a geometrically one-dimensional figure—a line representing an ontogenetic trajectory. But individuals vary around that line, which simply describes the ontogenetic change of an average individual. And it is the dimensionality of that variation, not the dimensionality of the ontogenetic trajectory itself, that is relevant to evolutionary predictions. Developmental biology provides no reason for assuming that variance is one-dimensional, much less that its single dimension is along the ontogenetic trajectory.

Unfortunately, developmental biology contains little theory that might enable more realistic predictions. From a growing body of empirical evidence, it is clear that spatial patterns of development do evolve and that they change in numerous ways, including changes in the location of inductive interactions (e.g., Brylski and Hall, 1988), in positional information (e.g., Lovejoy et al., 1999), and, as implicated herein, in the spatial distribution of growth fields (see also Lieberman, 2000). In the context of current trends in the study of evolution and development, the evolution of spatial patterning may seem less interesting than heterochrony because there is no implication of intrinsic constraints and no suggestion that novelties are inexplicable by conventional theories. However, those are not good reasons for dismissing them. These transformations in developmental spatial patterns may be responsible for much of the diversity of morphology and they well exemplify how complex developmental systems can be modified. Ignoring them because they are plausible and compatible with conventional theory would be unfortunate and would only retard our understanding of the evolution of development.

ACKNOWLEDGMENTS

We thank Donald Swiderski for his assistance with several aspects of this project.

REFERENCES

Bookstein FL (1986): Size and shape spaces for landmark data in two dimensions. Stat Sci 1: 181–242.

Bookstein FL (1989): Principal warps: thin-plate splines and the decomposition of deformations. IEEE Trans Pattern Anal Machine Intelligence 11: 567–585.

Bookstein FL (1991): *Morphometric Tools for Landmark Data: Geometry and Biology*. Cambridge: Cambridge University Press.

Brylski P, Hall BK (1988): Ontogeny of a macroevolutionary phenotype: the external cheek pouches of Geomyoid rodents. Evolution 42: 391–395.

Efron B, Tibshirani RJ (1993): *An Introduction to the Bootstrap*. New York: Chapman and Hall.

Fink WL (1993): Revision of the piranha gnus *Pygocentrus* (Teleostei, Characiformes). Copeia 1993: 665–687.

Fink WL, Machado-Allison A (1992): Three new species of piranhas from Brazil and Venezuela (Teleostei: Characiformes). Ichthyol Explorations of Freshwaters, 3: 55–71.

Fuiman LA (1983): Growth gradients in fish larvae. Journal of Fish Biology 23: 117–123.

Gisbert E (1999): Early development and allometric growth patterns in Siberian sturgeon and their ecological significance. J Fish Biol 54: 852–862.

Goulding M (1980): *The Fishes and the Forest: Explorations in Amazonian Natural History*. Los Angeles, CA: University of California Press.

Huxley JS (1932): *Problems of Relative Growth*. London: MacVeagh.

Klingenberg CP (1998): Heterochrony and allometry: the analysis of evolutionary change in ontogeny. Biol Rev 72: 79–123.

Koumoundouros G, Divanach P, Kentouri M (1999): Ontogeny and allometric plasticity of *Dentex dentex* (Osteichthyes: Sparidae) in rearing conditions. Mar Biol 135: 561–572.

Lauder GV (1981): Form and function: structural analysis in evolution morphology. Paleobiology 7: 430–442.

Lieberman DE (2000): Ontogeny, homology and phylogeny in the hominid craniofacial skeleton: the problem of the browridge. In: O'Higgins, P, Cohn, M (eds), *Development, Growth and Evolution: Implications for the Study of the Hominid Skeleton*. San Diego, CA: Academic Press, p. 85–122.

Liem KF (1973): Evolutionary strategies and morphological innovations: cichlid pharyngeal jaws. Syst Zool 22: 425–441.

Lovejoy CO, Cohn MJ, White TD (1999): Morphological analysis of the mammalian postcranium: a developmental perspective. Proc Natl Acad Sci USA 96: 13247–13252.

Machado-Allison A (1985): Studies on the subfamily Serrasalminae. Part III: on the generic status and phylogenetic relationships of the genera *Pygopristis*, *Pygocentrus*, *Pristobrycon*, and *Serrasalmus* (Teleostei-Characidae-Serrasalminae). Acta Biol Venez 12: 19–42.

Mathworks (2000): MATLAB, v. 6.0.0.88.

Orti G, Petry P, Porto JIR, Jegu M, Meyer A (1996): Patterns of nucleotide change in mitochondrial ribosomal RNA genes and the phylogeny of piranhas. J Mol Evol 42: 169–182.

Reis RE, Zelditch ML, Fink ML (1998): Ontogenetic allometry of body shape in the neotropical catfish *Callichthys* (Teleostei: Siluriformes). Copeia 1998: 177–182.

Sarpédonti V, Ponton D, Ching CV (2000): Description and ontogeny of young *Stolephorus baganensis* and *Thryssa kammalensis*, two Engraulididae from Peninsular Malaysia. J Fish Biol 56: 460–1476.

Schaeffer SA, Lauder GV (1986): Historical transformation of functional design: evolutionary morphology of feeding mechanisms in loricariod catfishes. Syst Zool 35: 489–508.

Shea BT (1992): Ontogenetic scaling of skeletal proportions in the talapoin monkey. J Hum Evol 23: 283–307.

Smith J (1998): Allometric influence on phenotypic variation in the Song Sparrow (*Melospiza melodia*). Zool J Linn Soc 122: 427–454.

Strauss RE, Fuiman LA (1985): Quantitative comparisons of body form and allometry in larval and adult Pacific sculpins (Teleostei: Cottidae). Canadian Journal of Zoology 63: 1582–1589.

Vermeij GJ (1973): Adaptation, versatility and evolution. Syst Zool 22: 466–477.

Voss RS, Marcus LF (1992): Morphological evolution in muroid rodents. 2. Craniometric factor divergence in 7 neotropical genera, with experimental results from *Zygodontomys*. Evolution 46: 1918–1934.

Zelditch ML, Fink WL (1995): Allometry and developmental integration of body growth in a piranha, *Pygocentrus nattereri* (Teleostei, Ostariophysi). J Morphol 223: 341–355.

Zelditch ML, Fink WL (1996): Heterochrony and heterotopy: stability and innovation in the evolution of form. Paleobiology 22: 241–254.

Zelditch ML, Sheets DH, Fink WL (2000): Spatiotemporal reorganization of growth rates in the evolution of ontogeny. Evolution 54: 1363–1371.

6

SPATIAL AND TEMPORAL GROWTH PATTERNS IN THE PHENOTYPICALLY VARIABLE *LITTORINA SAXATILIS*: SURPRISING PATTERNS EMERGE FROM CHAOS

Robert Guralnick

CU Museum and Department of EPO Biology, University of Colorado, Boulder, Colorado

James Kurpius

Museum of Paleontology and Department of Integrative Biology, University of California, Berkeley, California

INTRODUCTION

Understanding the causes and consequences of intraspecific phenotypic variation has been an enduring theme for ecologists and evolutionists. Although nature has provided a near-infinite variety of test cases, few systems have been as well studied as the shells of *Littorina saxatilis* for questions regarding the evolution of phenotypic variation (see review by Reid, 1996). *Littorina saxatilis*

Beyond Heterochrony: The Evolution of Development, Edited by Miriam Leah Zelditch
ISBN 0-471-37973-5

has been used as model system so often because members of the taxon are widespread in the Atlantic Ocean (northern latitudes), numerous, and easy to collect on rocky shores. Most importantly, the shells are highly variable in size and shape, the most variable of any *Littorina*. For example, snails that live in high-wave energy environments tend to be more squat and have larger apertures than those in more sheltered areas. This squat shape has been hypothesized to be an adaptation to avoid dislodgement in high-wave energy environments compared with the relatively more elongate and smaller-apertured littorines found in sheltered environments, where adaptations to heat stress and higher rates of predation are presumed. Figure 6.1 shows examples of *Littorina saxatilis* shells from several habitats.

Given this morphological variability associated with habitat, the focus of the research on *Littorina saxatilis* has generally been of two kinds: (1) determining how shell traits vary with environment and the heritability of those traits (e.g., Hull et al., 1999) and (2) studies of genetic or adult morphologic divergence as evidence for gene flow or lack thereof (e.g., Gosling et al., 1998). These studies attempt to understand the ultimate causations of morphological variation in different microhabitats (shape and size differences), either through environmentally mediated phenotypic plasticity or selection on heritable shell traits.

In this chapter, we approach the question of the causes of phenotypic variation by first trying to quantify and understand the ontogenetic paths individual snails from different habitats take to reach different adult shell shapes and how genetic and environmental effects relate to ontogeny. The shells of *L. saxatilis* provide an almost ideal system to study variation and changes in ontogeny and its relation to environmental vs. genetic effects. The first advantage is that littorine shells, like all shells, preserve a record of their morphological development through time due to the nature of how they grow. Shells grow by spiral accretion—later growth is added onto earlier growth as the mantle of the snail facilitates deposition of calcium carbonate. Spirality refers to the manner of growth, with the point of new growth appearing to rotate around the y axis while translating downward along the z axis and outward along the x axis.

Brian Hayes (1995) gives an excellent summary of this property of shells in an *American Scientist* article: "[if] you could unroll a shell and flatten it out into a rectangle, the pattern would form a space-time diagram, with position in space measured along one axis (say from left to right) and sequence in time recorded along the other axis (from top to bottom)." This means that a large amount of ontogenetic information can be gleaned from one individual. This ontogenetic record cannot only be read in individual shells but also compared between shells of the same species in different environmental contexts. Comparisons of ontogeny are crucial to pinpointing when, where, and how changes have occurred within and between individuals and populations.

A second advantage is that *L. saxatilis* broods its young, often up until the deposition of the second whorl (one whorl = 360 degrees of shell growth) (Reid, 1996). Thus, determining when ontogenetic changes are manifested can support an important argument about the relative importance of environmental and genetic effects. If ontogenetic change happens early in development, before

Figure 6.1. Examples of the shell shapes in top view and side view of different ecotypes of *Littorina saxatilis*, following Reid (1996). A: sheltered/brackish. B: wave exposed. C: moderate. D: barnacle.

embryos exit the brood chamber, then this change is occurring before many physical factors are directly acting on individuals. In this situation, shape differences are most likely either (1) genetic, where shape is predetermined by decent, or (2) due to the indirect effect of environmental cues as mediated through the mother, plasticity. If change in shape is happening later in ontogeny, then there may be both genetic and direct and indirect environmental effects. Thus, the study of ontogeny can be used to examine the importance of genetic and direct environmental effects.

Despite the many advantages for synthesizing ontogenetic and morphological variation, using *L. saxatilis* shells has a potentially major disadvantage. This species has traditionally been categorized as an indeterminate grower (Reid, 1996). Indeterminate growth complicates ontogenetic comparisons because the end points of ontogeny are not recognizable for individual shells. However, there is evidence that littorines do not deposit much shell after reaching sexual maturity and have some features like shell thickening that accompany adulthood (Reid, 1996). Therefore, it is unclear whether littorines are indeterminate growers or whether the end of growth is recognizable and similar between individuals.

Lack of a distinguishable stopping point complicates ontogenetic comparisons, but the problem is, in practice, easy enough to overcome if the starting point of adult growth or the protoconch/teleoconch boundary, which demarcates metamorphosis, is unambiguous in all shells. In that case, as Gould (1969) suggested, individual post-metamorphic whorls can be used as the units of comparison as long as comparable units are measured. The second problem in *L. saxatilis* (and virtually all intertidal gastropods) is that the earliest whorls (protoconch) are seldom preserved in adult shells; thus, neither the beginning nor the end points of growth are known with precision.

Why is the shell apex (top) not well preserved in adult littorines? The apex is the smallest part of the shell [<800 μm in diameter (Reid, 1996)] and the oldest. Therefore, over the lifetime of the animal, shell material is eroded and damaged in the intertidal zone. Wave action, among other abiotic factors, contributes most heavily to these processes.

The problem with not knowing the beginning or ending point in ontogeny is illustrated in Figure 6.2. Imagine that we are privy (i.e., through a lab growth experiment) to the full ontogeny of both shells in Figure 6.2A and therefore know that they have exactly the same growth parameters but different stopping points in ontogeny, with one growing more quickly or longer than the other. Now wipe that knowledge of the full ontogeny from your mind and imagine that all we have are the final shells. How would we align the two shells for comparison if there was wear on the apex and we therefore lost any record of metamorphosis at the protoconch/teleoconch boundry? We might align the two shells to their ending points as in Figure 6.2B. However, if we were comparing any units of the shells for ontogenetic analysis, our measurements would be of noncomparable units—we would be assuming that points 1a and 2a on one shell are homologous to points 1 and 2 on the other when in fact the comparable units are points 1b and 2b. Ultimately, comparisons of ontogenies must be of comparable units—otherwise, even measurements of relative timing are meaningless. As discussed below, one outcome of this chapter is a method that allows valid comparisons of indeterminate growing gastropods.

Given the advantages and potential disadvantages of using *L. saxatilis*, we expand and improve on the methods initially described in Johnston et al. (1991) for comparing shell shape and size and attempt to answer the following four questions:

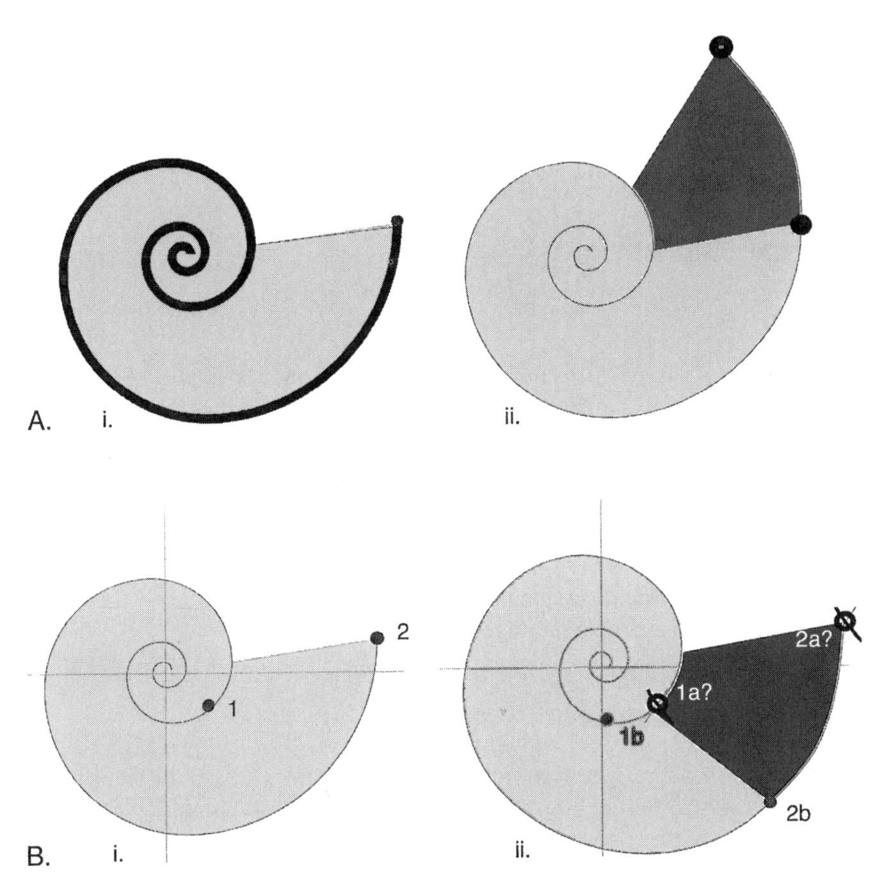

Figure 6.2. Shell growth in two different, isometrically growing and almost identical shells showing the problem of choosing homologous points. A: shells are aligned correctly and the shell labeled ii has grown ~45 degrees more than the one labeled i. B: assuming that the beginning point is gone, shells may be aligned incorrectly and points assumed homologous are actually not. The homologues to the points in Bi are 1b and 2b not 1a and 2a in Bii.

1. Can morphometric analyses of adult *Littorina saxatilis* shells be used to test whether valid comparisons of worn, indeterminate growing snails can be made? Can this method be extended to other recent and fossil shells?
2. Given different populations from various habitats, are there clear populational or habitat differences in starting and stopping points in growth, or, more generally, are there growth differences at all?
3. What are the ontogenetic trends in growth for different populations living in various habitats? That is, how are habitat and populational differences manifested in shell growth? How are adult shapes built during ontogeny? Are they built by major shifts in early or late ontogeny? Are there clear

points where the shapes diverge from a common trajectory? Do the ontogenetic patterns resemble heterochrony or are they more heterotopic?

4. How does ontogenetic change relate to plasticity and genetically predetermined change?

MATERIAL AND METHODS

Materials

Seven different lots of *Littorina saxatilis* shells from the British Museum of Natural History were sampled for this study. These lots represent four different qualitative ecotypes, as defined by Reid (1996), and are from seven different localities throughout the United States and Europe. From each lot, we sampled between four and six specimens ($n = 33$). Table 6.1 shows the number of individuals subsampled from each lot, habitat designations, ecotype designations, and collection localities. We sampled shells that appeared to have the most complete whorl information, including the least amount of wear on the apex of the shell and no discernible damage to other parts of the shell.

Introduction to Methods

We use a number of different morphometric methods below to answer different kinds of questions. Before addressing each analysis in more detail, we present a brief overview for our choice of methods and how those methods are appropriate for the questions we are addressing. One of our main goals is to determine whether segments of the shell can be delimited and compared between individuals in the same species. If they can be compared, then we also want to examine shape and size differences between these units at smaller and larger morphological spatial scales (e.g., change in segments of the shell vs. changes across the whole shell). Geometric morphometric methods provide an appropriate toolkit to examine these issues because they allow holistic yet statistical analysis of shape at many spatial scales. In particular, we use Relative Warp Analysis (RWA; it is the geometric analog to a Principal Components Analysis) because our main interest is in major axes of shape variation and how these axes relate to the alignment (or misalignment) of shell segments. For more information on RWA and its use on shells, see Chapter 1.

Part of our argument about the utility of RWA in this particular case requires ancillary information about ontogenetic allometry. In particular, allometric changes during ontogeny must be smaller than changes due to misalignment of shells (see below for more detail). We use multivariate allometry analysis to test for shape changes due to increasing size of whorls (see Chapter 4 for a more detailed description of shape/size allometry methods). Because our

Table 6.1. Sample sizes, habitat, locality and number of observable whorls for each of the lots examined

Lot No.	No. of Specimens	Habitat	Ecotype	Locality	Estimated Whorl No.
1	5	Sheltered	Sheltered/brackish	Muscongus Island off Round Pond, Maine, USA	6
2	4	Fairly sheltered	Moderate	Flatey, Breidafjordur, Iceland	6
3	5	Very exposed	Wave	Cabo Silleiro, Baiona, Galicia, Spain	5
4	6	Sheltered estuary	Sheltered/brackish	Scotland	5
5	4	Semi-exposed	Moderate	Anglesey, Wales, UK	5
6	5	Marsh	Sheltered/brackish	Norfolk, England	5
7	4	Barnacle	Barnacle and/or wave	Achill Island, Ireland	3

Total number = 33.

interest is in the magnitude of overall allometric change, we do not focus on significance of univariate measures like partial warps but instead visually present allometric change as output from multivariate analyses conducted in Regress12 (Mathworks, 2000).

Initial Placement of Homologous Points

The first step in any geometric morphometric study is to choose easily identified points that evenly sample across the form of interest. In new morphometrics, these points are usually referred to as landmarks. The procedure for placing landmarks depends on the goals of the study. In this study, we want to choose points on the shell that lie along the spiral and allow comparative descriptions of putatively homologous sections of the spiral across all our samples. Our goals are similar to those of Ackerly (1989) who argued that, although it would be ideal to analyze spiral growth continuously, it is in practice necessary to compare many discrete (and presumably homologous) points between shells. Our view on landmark choice is also similar to that of Stone (1998) who followed many other students of the gastropod shell and took a utilitarian approach toward the placement of comparable points on shells. He used the aperture centroid (throughout ontogeny at 180-degree intervals) as a landmark for evolutionary-developmental analysis.

Specimens sampled from the seven lots were photographed in apical view using a computerized image capture system. All specimens were mounted with the columella (traditionally considered the axis of coiling) oriented perpendicular to the field of view and with the aperture located in as nearly an identical position as possible between specimens. The apical views in Figure 6.1 show the general configuration, with the aperture of the shell oriented toward the bottom right. A digital camera with a Nikkor 50-mm lens was mounted on a light stand, and digital photographs of shells with scale bars were captured for further processing. All images were transformed to the same scale using Photoshop, and, once scaled, the digital photographs of each shell were ready for further processing and placement of landmarks.

Stone (1998) sampled along each whorl at every 180 degrees. This sampling method allows for small-scale changes to perhaps go unrecognized between those 180-degree points. To account for small scale shape change and to more completely sample the shell, we placed landmarks along whorl boundaries at 30-degree increments. By definition, each whorl has 12 landmarks and the spacing between landmarks increases as the whorl size increases. To accurately place the landmarks onto the images, we built an image with a composite set of axes spaced 30 degrees from one another using Adobe Illustrator. Images with overlying axes were imported into TPSDIG (Rohlf, 1998), and landmarks were placed on the shells at the junction of the axes and whorl boundaries (Fig. 6.3). We chose to start sampling all shells at the same position along the first visible whorl as shown in Figure 6.3.

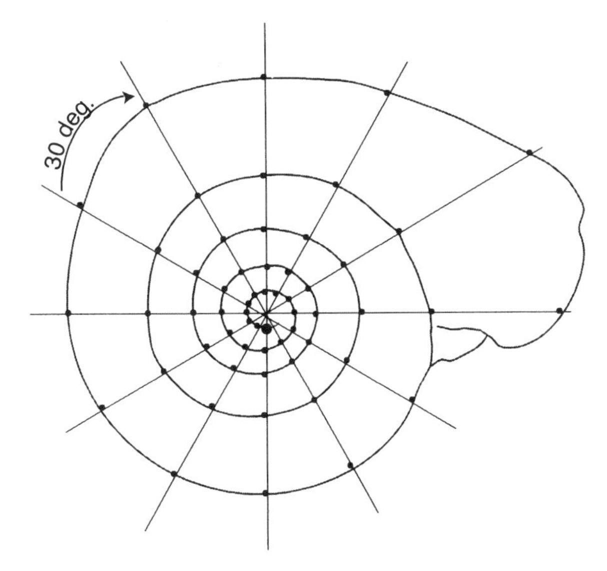

Figure 6.3. Placement of landmarks on the shell at 30-degree increments. First, a 30-degree spaced grid is placed at the centroid of the visible first whorl, and then points are placed at the intersection of the grid and whorl divisions.

Testing Growth Patterns—A Rationale for Sampling

One of the main questions in this chapter is whether a supposedly indeterminate growing snail like *L. saxatilis* can be used to make meaningful comparisons of ontogenetic trajectories. One way to determine whether comparable units of ontogeny are being used is to test for what we call misalignment. Misalignment would be a case where presumed whorl segments were assumed comparable but are not. The seeds of using morphometrics to test for misalignment come from Johnston et al. (1991). They determined ontogenetic trajectories by placing landmarks at varices—relatively evenly spaced sculpturing on the shell whorls—and numbering the landmarks from the beginning of the shell to the end. They then sampled landmarks 1–9, then 2–10, then 3–11, and onward, until sampling of the whole shell was complete. Finally, they used this nine-landmark data set for morphometric analysis. Thus, they followed the track of ontogeny through time.

When their data were ordinated using RWA, in almost all cases the track of one whorl in relative warp space was circular. Why? The explanation depends on the dynamics of spiral growth. Imagine that shell growth is purely isometric, with no shell growth allometries (Fig. 6.4, A and B). In the case of spiral growth, isometry means that each whorl of the shell has the same shape as the other whorls, and only size is increasing. Allometry means that the shape of

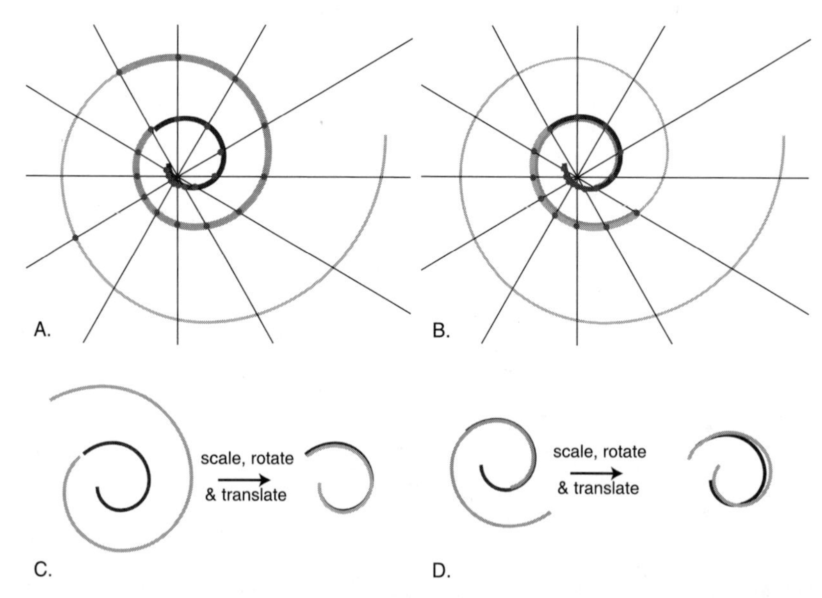

Figure 6.4. A and C: view of two different whorls, one light gray and one dark, from the same isometrically growing shell. If the two whorls are isolated, scaled, rotated, and translated to each other, they have the same shape. B and D: view of one whorl from the beginning of growth (dark gray) and another whorl offset from the first by 180 degrees. When these whorls are isolated, scaled, and rotated, they have different shapes.

individual whorls varies with size. Now, imagine comparing the shape of two different whorls, as shown in Figure 6.4A. Given isometry, these shapes once scaled, rotated, and translated are identical (Fig. 6.4C). Now imagine examining the first whorl and then comparing it with a whorl offset not by 360 degrees but instead by 180 degrees, as in Figure 6.4B. If we attempt to fit these shapes together as best as possible by scaling, rotating, and translating, they still look quite different (Fig. 6.4D) compared with the previous example (Fig. 6.4C). In fact, if the sections of the whorl analyzed were offset by 180 degrees, for example, when ordinated using relative warps, they showed up on the relative warp axes offset by 180 degrees. Therefore, if spiral growth is isometric during ontogeny, segments at the same starting points on different whorls of the same shell should look more similar once scaled, rotated, and translated than at different starting points within whorls. To reiterate, the same starting points on different whorls look the same, whereas different starting points on the same whorl look different.

This argument can be extended to the case of comparisons *between* shells. Imagine two shells that have the same growth parameters except that one shell has a different ending point compared with the other, as in Figure 6.2. Also, assume that both have worn apices. If the shells are misaligned, the shape of what are assumed to be the same whorls between the shells should be different.

Sampling—180 and 90 Degree Data Sets

The sampling of ontogeny by Johnston et al. (1991) was extremely dense in order to follow the track of ontogeny, but it created a huge amount of data for each shell. In this study, we also subsample our landmarks but less densely, since even visualizing the patterns can be difficult with so much data. We constructed two different data sets, each one composed of 12-landmark shell whorl units (1 whorl = 360 degrees around the spiral). One data set was constructed by sampling at 180-degree offsets so that every sixth landmark from the beginning of each whorl unit becomes the starting point of the next whorl unit. Thus we sampled landmarks 1–12, 7–18, 13–24, 19–30, and so forth, until each shell was fully sampled (Fig. 6.5). Sampling was completed when there were no longer enough landmarks to construct complete 12-landmark whorl units; therefore, some landmarks at the end of the shell might not get sampled in the analysis. We refer to this data set as the "180° data set" from here on out. The other data set was sampled more densely at 90-degree offsets—thus, we constructed a larger data set sampled at landmarks 1–12, 4–15, 7–18, 10–21, 13–24, and so forth. The 360-degree whorl units were sampled so that they are offset from each other by 90 degrees. This data set was constructed after an initial analysis using the 180° data set to provide a finer-scale analysis of shape change. We refer to it as the "90° data set." Based on the results of the initial 180° data set, we also removed landmarks from two lots in the 90° data set (discussed further in RESULTS).

Analysis of Ontogenetic Allometry

In our example above showing how spiral growth can be used to detect whether shells are aligned correctly, we assumed isometry and similar growth parameters. Allometry during shell growth complicates matters slightly, in that it is difficult to partition changes due to potential misalignment vs. changes due to growth allometries. For example, we assume that the shape of nonoffset whorls (e.g., landmarks 1–12 and 13–24) should be almost identical, but allometric change may make those shapes look different. If allometric change is severe enough, it could swamp out the signal of shape similarities and differences inherent in the spiral growth trajectories. To tell whether allometric change is too great to use the method, we tested within-individual allometries on a shell-by-shell basis. We then determined whether the magnitude of allometry compared with shape differences due to whorl offsets. We refer to within-individual allometry as ontogenetic allometry, since we are testing whether shape is changing with size during the growth of a single individual by comparing whorls within an individual.

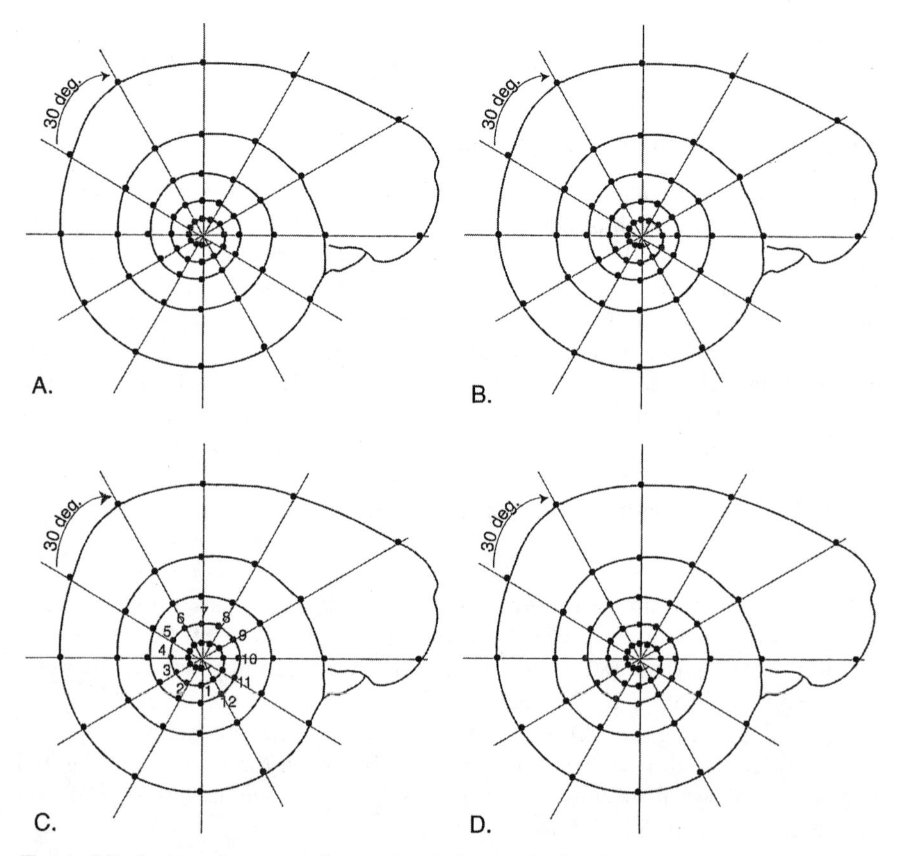

Figure 6.5. A view of our sampling protocol showing landmarks on one whorl (whorl = 360 degrees = 12 landmarks), as in A, incrementing three landmarks, or 90 degrees of growth, and then sampling another 12 landmarks, as in B. This continues through C and D and only stops when there are <12 landmarks to sample. Numeric labels on C reflect the numbering of landmarks, discussed in the text.

To determine whether there were ontogenetic allometries, one or two shells were selected haphazardly from each lot. Because we are interested in ontogenetic allometry (variation in an individual shell), each shell was tested separately. Also, instead of using the offset data—360-degree whorls sampled at 90 or 180° increments—we ignored the 90, 180, and 270-degree offsets and just used nonoverlapping whorl units: landmarks 1–12, 13–24, and so forth. Using nonoverlapping whorl units is equivalent to comparing the size and shape of different whorls in one shell to each other. The initial shape data was converted to Procrustes distances using CoordGen, a program written in Matlab (Mathworks, 2000). That data file was then used in Regress12—also written in Matlab (Mathworks, 2000)—which calculates a regression of partial warp and uniform deformation scores on log-centroid size. We chose to log transform centroid size because the size differences between early and late whorls were more than an order of magnitude different. The reference used to calculate overall shape change is the consensus of the two smallest whorls. The final outputs from Regress12 are a total measure of shape differences as vector displacements or Cartesian deformations (e.g., see Fig. 6.7, A and C).

Analysis of the Whorl Offset Data Sets

Each group of 12 landmarks, making up a 360-degree rotation of the shell, is considered a separate unit in the analysis. For the two data sets, we performed an initial least-squared Procrustes fit of all whorl units of all the shells and then used a principal components analysis (PCA) of the residual variation left over after the Procrustes fit to summarize the landmark data. This is equivalent to an unweighted RWA, and we used TPSRELW (Rohlf, 1998) to perform the analyses. Scores of the specimens on the first two relative warp axes were then imported into Systat 8.0 (1998) to generate and display scatterplots of the groupings.

For tracking of data, units can be grouped by lots (1–7 in Table 6.1) and by whorl offsets. We use the term "whorl offset" often in the rest of this chapter and want to clarify its meaning here. Whorl offset refers to the position of the beginning landmark of each whorl unit relative to the absolute first landmark selected when the landmarks were originally placed on the shell. Thus, the whorl offset can either be 0 or 180 for the 180° data set or 0, 90, 180, or 270 degrees for the 90° data set. A 0-degree offset means that the whorl being sampled starts at the same point as the first landmark from the first whorl. To understand individual ontogenetic changes vs. group changes, the data were also grouped by individual and relative ontogenetic stage—from earliest growth (the first whorl) to the latest (e.g., see Fig. 6.10).

Models of Growth

Given isometry, if growth stops at the same point during ontogeny and the starting point is already chosen to be the same for each specimen, then our expectation would be that the shapes of whorl units at similar positions along the spiral (say, for example, all the whorl units at 180 degrees from the beginning point) should form a cluster together on the principal component axes because the shape of the whorl at those points is most similar (Fig. 6.6A). If, however, growth does not have similar ending points, the pattern should be random (Fig. 6.6B) because our assumed comparable units are really not, and shapes at similar whorl positions are free to vary. It is possible that the actual patterns fall somewhere between Figure 6.6A and 6.6B. Perhaps some shells have the same stopping points along the whorl and others do not.

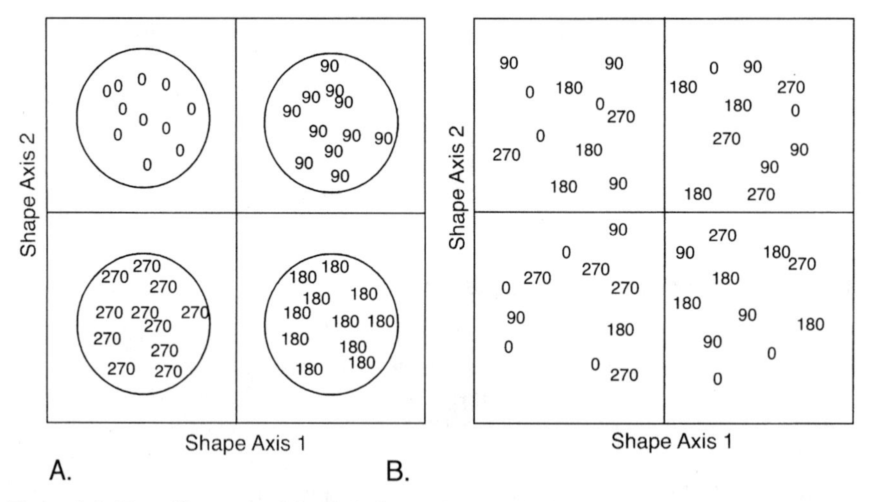

Figure 6.6. Two different models of shell growth derivable from spiral growth and the morphometric techniques. Both panels show a shape space on which whorls offset by 90 degrees are ordinated. A: ordination shows whorl offsets from all whorls on all shells, and each 90-degree offset clumps in a separate part of the shape space. This is the model for growth with the same stopping points. B: each whorl offset is randomly distributed in the shape space and is the expected pattern for shells with different stopping points in ontogeny.

Apical Wear and Correcting for Whorl Loss

Another important piece of information is total number of whorls. For example, lots 1 and 2 have six visible whorls, whereas lots 3–6 have five whorls and lot 7 has three whorls. For lots 3–6, it may be that this is a real difference in whorl number or it is possible that one more whorl has been worn off. If lots 3–6 have lost one more whorl compared with lots 1 and 2, then the first visible whorl in those lots is actually comparable to the second whorl in lots 1 and 2. Thus, there are two possible assumptions: one in which all lots (with the exception of lot 7) have the same number of whorls (six) but one more whorl is worn off in lots 3–6 and one in which the lots have different numbers of whorls. How can we test which is more likely?

A simple way to address this question is to compare sizes of whorls for each lot because whorls increase in size throughout accretionary growth. If the first visible whorl in lots 3–6 is really the second visible whorl and the second visible whorl is really the third visible, and so forth, then the sizes of the second visible whorl in lots 3–6 should be more similar to third visible whorl in lots 1 and 2 and the overall measured variation would be smaller for all whorls. We can look at the overall variation across all lots under the assumption of apical wear and compare it with the overall variation assuming the loss of a whorl. It is possible that variation in centroid size is similar under both assumptions, and we cannot distinguish between the two. Alternatively, centroid size variation may be much smaller under the assumption of apical wear compared with

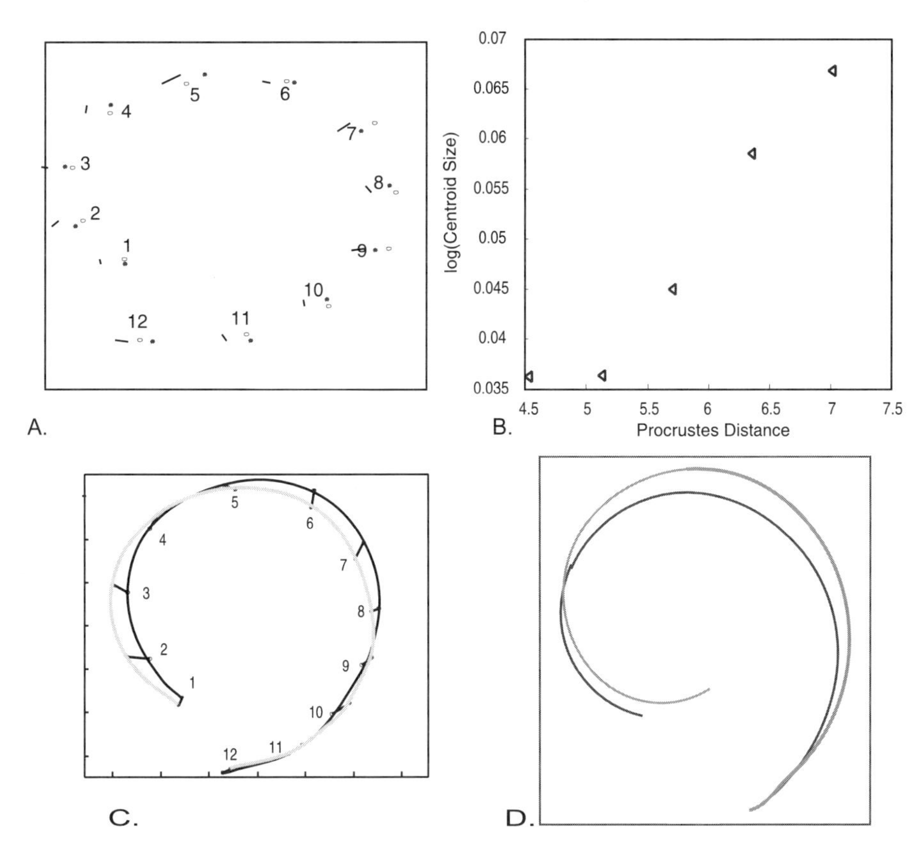

Figure 6.7. A: overall allometric shape change magnitude, as shown by vector deformations for a shell from lot 5. B: graph of Procrustes distance vs. log (centroid size) from the same lot 5 shell. C: overall allometric shape change magnitude for the shell with the most ontogenetic allometry, the second shell from lot 1. D: magnified version of Figure 6.4D, showing the magnitude of shape differences due to whorl offset.

actual shorter growth trajectories or vice versa. If we find a pattern of smaller variation under one assumption or the other, the one with the significantly lower variation over all the whorls would be the more parsimonious solution and therefore preferred.

RESULTS

Ontogenetic Allometries

The main question we want to address regarding ontogenetic allometry is whether the magnitude of shape change due to allometry is greater than the magnitude of change due to whorl offset. To examine magnitude of allometry, we use Procrustes distance as our metric and compare those distances with the log transform of centroid size.

Results from our analysis suggest that, for many specimens, there is a significant relationship between Procrustes distance and log-transformed centroid size (Fig. 6.7B for a representative plot of shape vs. size). The magnitude of this allometry is measured by comparing Procrustes distance between the consensus of the two smallest vs. the largest whorls. This distance can be represented as vector displacements at each landmark. For each of our haphazardly sampled specimens, we created vector displacement graphs to understand the magnitude of shape change with size. In Figure 6.7C, we plot that shape difference for the one shell (specimen 2, lot 1) that has the most allometric differences between early and late whorls. An example of the magnitude of change due to whorl offset, magnified from Figure 6.4D, is shown in Figure 6.7D. Comparisons of Figure 6.7C and 6.7D suggest that, although the magnitude of shape change due to allometry is noticeable, it is not as great as the magnitude due to whorl offset. Given that this is the single largest significant allometry, we argue that the shape differences due to misalignment will not be swamped out by ontogenetic allometries; therefore, we were able to proceed with further analyses of whorl shape.

Morphometric Analysis of the 180° Data Set

A full analysis of the 180° data set produces a PCA in which the first two axes, RW1 and RW2, explain 56.55% and 16.30% of the overall variation. In order to avoid information overload when presenting scatterplots of the Relative Warp Analysis, we focus on pairwise comparisons of lots instead of examining all the lots together (Fig. 6.8). That is, the RW1 and RW2 axes and overall patterns are based on all the samples, but only the whorl units from two lots are shown in each graph. Not all of the possible lot groupings are shown in Figure 6.8 (lots 4 and 5 are left out), and the complete results are summarized in Table 6.2.

The most important finding is that patterns for the 180° data set are not random but that, in all lots, the position of the 0- and 180-degree offset whorl units are in distinctly different parts of the shape space (Fig. 6.8, A and D). That is, within each lot, the specimens always appear to have separate groups of 0-degree whorl offsets and 180-degree whorl offsets. However, two lots (6 and 7 in Table 6.2), the sheltered/brackish and barnacle ecotypes, have a very different but consistent pattern compared with the other lots (Fig. 6.8, B and C). In each shell from these lots, the beginning point is offset exactly 90 degrees from the beginning points in lots 1–5 on the relative warp ordinations. Thus, it appears that either there has been 90 degrees of extra growth or 270 degrees of less growth compared with the other lots. We interpreted this as a potential case of misalignment.

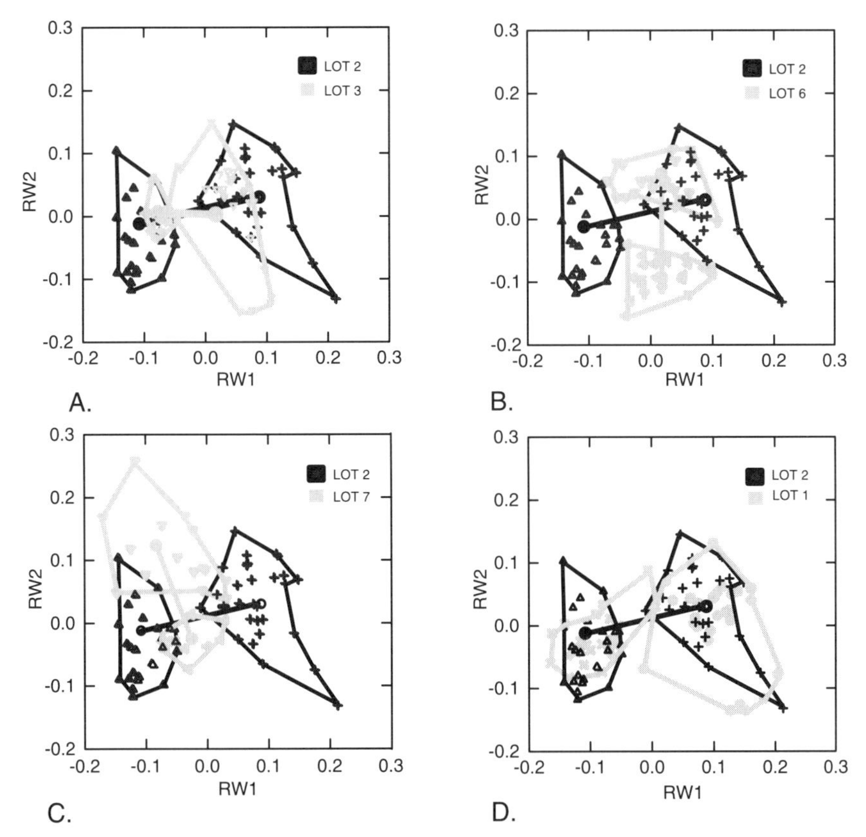

Figure 6.8. Ordinations of the whorl segments in the shape space for the 180-degree data set. Four panels show comparisons of lot 2 to other lots. The data points are faintly visible; more visible are the hulls, showing the overall range of variation. A: comparison of lot 3 to lot 2, showing that lot 3 clusters more toward the midline than lot 2. B: lot 2 compared with lot 6, showing that lot 6 is offset by 90 degrees compared with lot 2. C: lot 2 compared with lot 7, showing a similar 90-degree offset in lot 7 and that lot 7 is shifted more toward the left in the shape space. D: lots 2 and 1 are very similar in shape.

Table 6.2. Position of 180-degree whorl offsets when ordinated on relative warp axis 1 and 2 for each lot

Lot No.	Habitat	Whorl Offset on PCA Axes (degrees)
1	Sheltered	0 and 180
2	Fairly sheltered	0 and 180
3	Very exposed	0 and 180
4	Sheltered estuary	0 and 180
5	Semi-exposed	0 and 180
6	Marsh	90 and 270
7	Barnacle	90 and 270

Morphometric Analysis of the 90° Data Set

Because the results of the 180° data set suggested that two lots, 6 and 7, were not properly aligned, we refined our analysis. We first added more whorl offsets and then "realigned" the specimens from lots 6 and 7 by removing three landmarks and finally recompared these lots with the other five. The other possibility would have been to add an extra 270 degrees of growth, but, in this case, adding data is much more difficult to justify than removing data. Once the landmarks were removed from lots 6 and 7, we performed a RWA using the 90° data set. The first relative warp explained 50.44% of the total variation, and the second explained 38.01%. The first two relative warps summarize 88.45% of the overall variation; therefore, we focus on those two axes below.

As in the 180° data set, it is difficult to show all the patterns simultaneously. Thus, we again focus on comparison between two lots to show the patterns more clearly (Fig. 6.9). Again, the plots represent summaries of all the data, some that are not shown. In this case, we use lot 2 as a standard for comparison and compare it to lots 1, 3, 6, and 7 in four different graphs (Fig. 6.9). When lot 6 and lot 7 are reanalyzed with the equivalent of 90 degrees of growth removed (equal to three landmarks), they are each in the same phase as the other lots (Fig. 6.9C). This is most evident for lot 6. As is clearly visible from the graphs, each different whorl segment, as predicted, is relatively separate from the others, especially for lots 1, 2, 3, and 6 (Fig. 6.9, A–C). There is more overlap between different whorl segments in the barnacle ecotype lot 7 (Fig. 6.9D).

Interestingly, the analysis clearly shows different kinds of patterns for different lots. Some lots have very similar overall ordinations—for example, lots 1 and 2 have significant overlap for all whorl offsets and generally similar means and variation (Fig. 6.9A). When compared with lot 2, lot 3 has clearly reduced variation, especially along the extremes of the axes; therefore, the mean also looks shifted toward the origin (Fig. 6.9B).

The barnacle ecotype (lot 7) has a unique and different shape ordination compared with other lots (Fig. 6.9D); its whole growth pattern has been shifted toward the negative portion of RW1 and slightly toward the positive portion of RW2. This shift represents a shape difference that appears to affect all whorls at all points during growth.

Visualizations of Shape Change

Shape differences along the relative warp axes for the 90° data set can be represented by Cartesian deformations. Imagine that a square grid is placed over all the specimens and that individual squares that make up the grid can expand or contract. The consensus configuration, determined by the Procrustes fit, is used as a standard for comparison, and all specimens have some deformation of their shapes from the standard (grid with all right angles), and their overlying grid, away from the consensus. Figure 6.10 shows Cartesian deformations for shapes in the different quadrants of the graph of relative warps 1 and 2. These

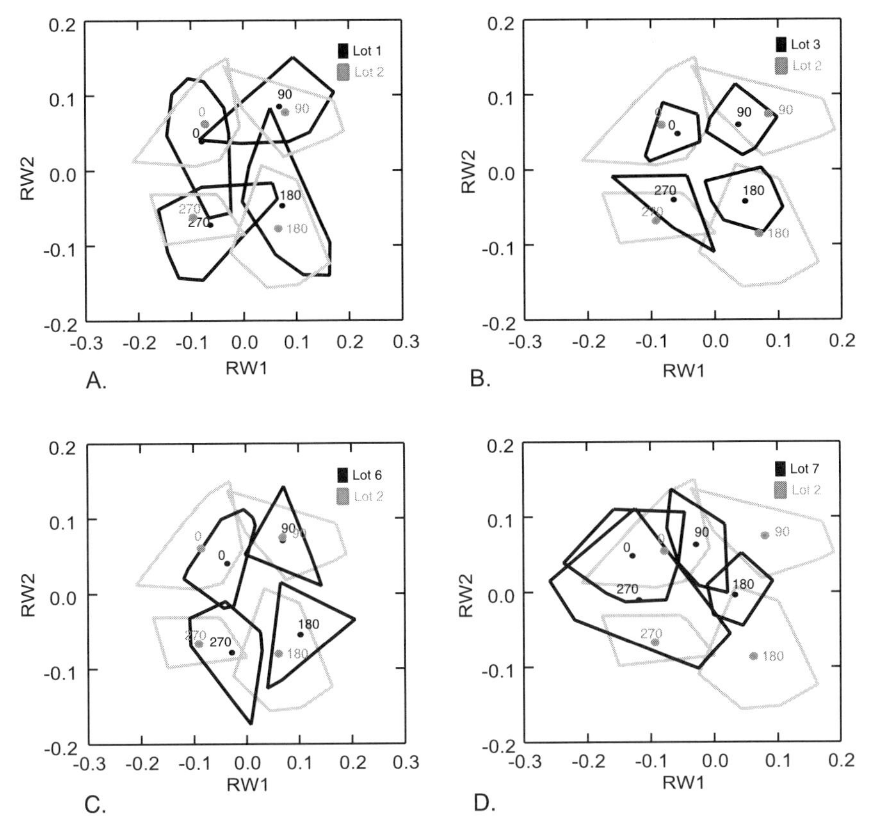

Figure 6.9. Ordinations of the whorl segments in the shape space for the 90-degree data set with removal of three landmarks from lots 6 and 7. Four panels show comparisons of lot 2 to others. Only the range of variation is shown as hulls around the scatter; data points themselves have been omitted. A: comparison of the similar lots 1 and 2. Note how each whorl offset has a distinct cluster in the shape space. B: lot 2 compared with lot 3. Here, it is even clearer than in Figure 6.4A that lot 3 has less variation and mean shapes closer to the origin. C: lot 2 compared with lot 6. Note that, when 90 degrees of growth is removed, the whorls now sync and their positions in the shape space are similar. D: lot 2 compared with lot 7. Here, it is clearer that lot 7 has a dramatically shifted position in the shape space leftward and slighty upward, representing expanded whorls.

quadrants correlate with the 90° data set. For orientation, the space between landmarks 1 and 12 was at one time the former position of the aperture during spiral accretionary growth. We refer to this as the "aperture" when discussing shape.

When examining Figure 6.10, in the 0-degree quadrant of the shape space, the whorls are relatively enlarged near the aperture and compressed near landmarks 3–6. At the 180-degree offset, whorls show an opposite pattern of contraction at the aperture, especially at landmarks 1 and 12, and expansion at

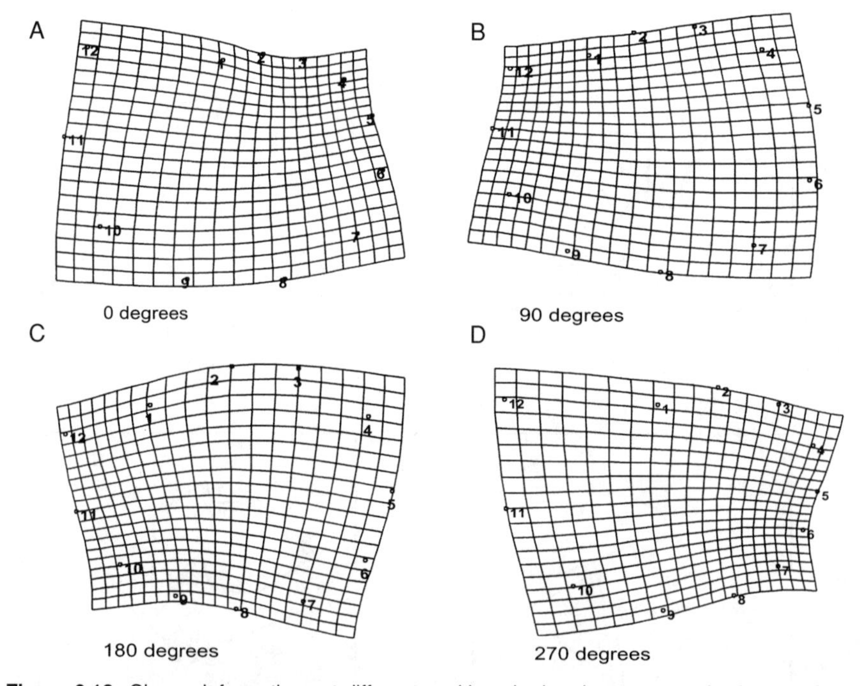

Figure 6.10. Shape deformations at different positions in the shape space. A: shape deformation at 0 degrees, representing contraction near the beginning of whorl (landmarks 1–5) and expansion near the end (landmarks 8–12). B: shape deformation at 90-degree whorl offsets, representing expansion near landmarks 5–8 and contraction near landmarks 9–12. C: shape deformation at 180-degree whorl offsets representing expansion near landmarks 2–6 and contraction near landmarks 8–12. D: shape deformation at 270-degree whorl offsets, representing expansion near landmarks 1 and 9–12 and contraction near landmarks 2–8.

landmarks 3–7. This pattern with 180-degree offset is strikingly similar to Figure 6.4D, which also shows shape differences for 180-degree whorl offsets—in a perfectly isometrically growing spiral. Thus, the shape differences from the analysis are consistent with expectations based on the model of shape change.

Whorl offsets at 90 and at 270 degrees show opposite patterns of expansion and contraction and are focused more at landmarks 2–5 and 8–11. At 90-degree whorl offsets, the shape of the whorl is inflated outward at landmarks 2–5 and contracted inward at landmarks 8–11. Landmarks 1 and 12 are directed toward each other, leading to a narrowing of the aperture that gives the "oblong" look of the shell from an apical view (see Fig. 6.1). In contrast, at 270-degree whorl offsets, the aperture and later portions of the whorl, especially regions between landmarks 9 and 11, are expanded. Initial whorl segments are greatly contracted, especially areas between landmarks 2 and 5. Regions between landmarks 1 and 12 are expanded, because those landmarks shift in opposite directions, leading to a wider aperture.

Some lots, like 3 and 7, have different ordinations, either in mean position (lot 7) or variance (lot 3) compared with the others, and these differences can be examined as shape changes in the Cartesian deformations. All the data for lot 7, for example, fall more toward the negative end of the RW1 axis and along the positive end of the RW2 axis compared with the other lots (Fig. 6.9). Taken together, this suggests that all have more expansion near the end of the whorls and less expansion in the beginning, near landmarks 3–7. This suggests an overall pattern of a higher whorl expansion rate that appears in the first whorl and is propagated in subsequent whorls.

The wave ecotype from Spain (lot 3) has the lowest overall variation in all whorls and at all points along the whorl, but it is hard to separate ontogenetic variation from individual specimen variation. To examine these separately, we plotted ontogenetic change for each specimen in lot 1 and lot 3 to compare the variations (Fig. 6.11). Each graph in Figure 6.11 represents the ontogenetic change in the plane of RW1 and RW2 for one specimen, and the graphs are separated by lot (Fig. 6.11A shows lot 1 and Fig. 6.11B shows lot 3). A proviso about such ontogenetic plots is necessary. Individual ontogenies may not lie in the same plane defined by the major axes of variation (RW1 and RW2) as the full sample. However, our main interest is within and between lot variation in those ontogenies, and this variation should likely be concordant with the first two warp axes.

The graphs shown in Figure 6.11 suggest that the relatively smaller amount of overall variation in lot 3 is driven by all the specimens having very similar, tighter trajectories. Lot 1, on the other hand, has some specimens with more tightly coiled trajectories (the first and third graph in Fig. 6.11A) and some with much wider or larger trajectories (the second and fourth graph in Fig. 6.11A). Wider trajectories reflect greater shape changes along the entire trajectory, and shell shapes tend to have high expansion and contraction at different points, leading to shells that from top view appear to be more rounded, as in Figure 6.1A. Shells from very exposed areas appear to have smaller contraction and expansion rates and have shells that are more oval shaped, as in Fig. 6.1D.

Analysis of Size and Shell Whorl Numbers

Shells in lots 1 and 2 have six whorls, whereas shells in lots 3–6 have five. Lot 7 is unique in having only three whorls per shell. Disregarding lot 7 for the moment, we want to test whether lots 1–6 had the same number of whorls but differential wear. To test this question, we examined variation in the centroid size of the whorl under the assumptions that lots 3–6 really do have five visible whorls and that the first visible whorls in lots 3–6 are equivalent to whorl 1 in lots 1 and 2. We also examined variation in centroid size under the assumption that the first whorl has been worn away and the first visible whorl in lots 3–6 is instead equivalent to the second whorl in lots 1 and 2, and so forth.

Figure 6.12 shows box-and-whisker plots for the two assumptions, with the top boxes-and-whiskers (shown in grey) representing the assumption of more

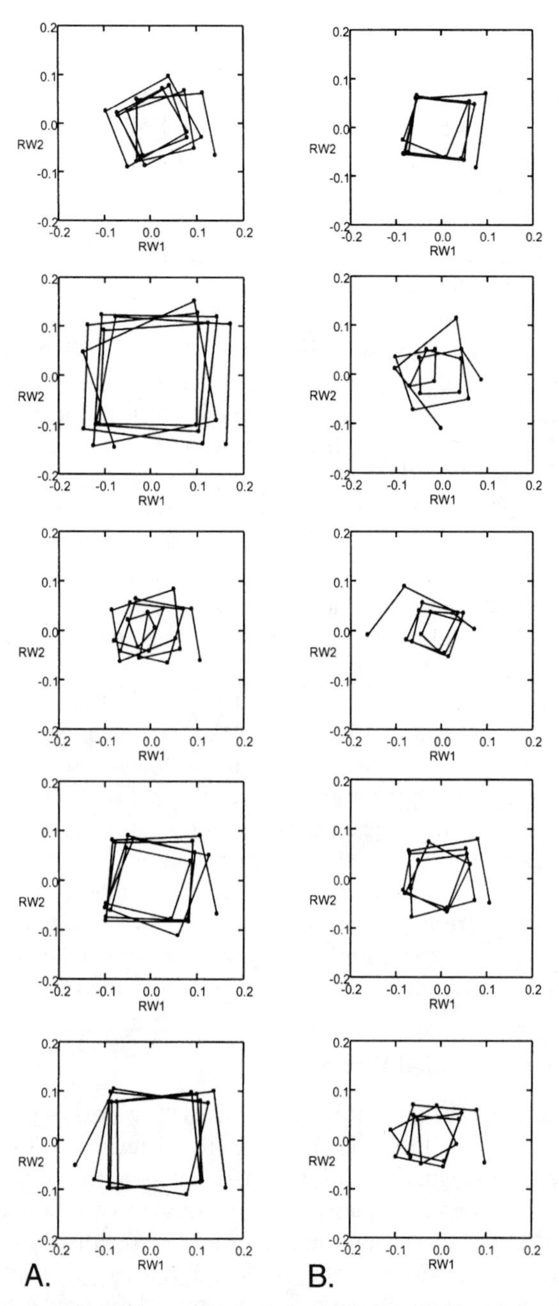

Figure 6.11. Ontogenetic shape change during growth along RW1 and RW2 for individuals within lots 1 and 3. A: ontogenetic trajectories for the five individuals sampled in lot 1. Note the variation in the width of the trajectories. B: ontogenetic trajectories for the five individuals sampled in lot 3. Note the lack of variation in the width of the trajectories.

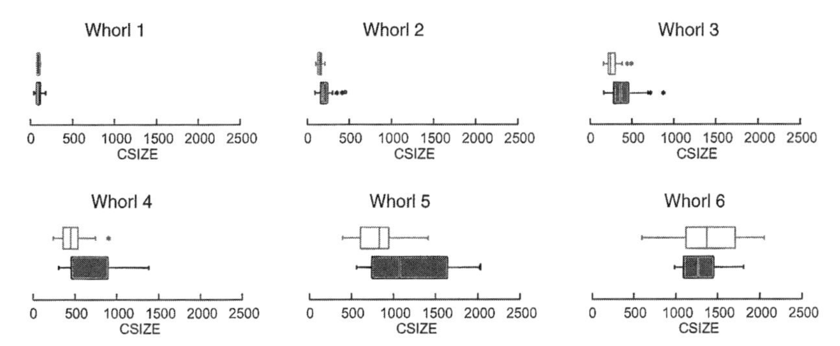

Figure 6.12. The top box-and-whisker plots (unfilled) show variation under the assumption that the first visible whorl is actually the second whorl and the second whorl is the third, and so forth, for lots 3–7. Bottom plots (filled dark grey) show variation under the assumption that the first whorls in lots 3–7 are the same as those in lots 1–2. Variation for most whorls is lower in the top plots.

apical wear and the bottom ones (shown in black) representing the assumption of less wear. In whorls 1–5, variation is reduced under the assumption that the first whorl in lots 3–6 was removed. The last whorl variation increases, but this is due to the addition of many more samples. In the top box-and-whisker plots, only lots 1 and 2 have six whorls, whereas, in the bottom plots, lots 3–6 are also assumed to have six whorls (Fig. 6.12). Shells in lot 6 are much smaller than lots 1 and 2; therefore, the last whorl is also smaller and the range of variation increases for that one whorl. In the same respect, the whorl 1 variation decrease in the top box-and-whisker plot may be due to the removal of the first whorl in lots 3–6. However, the message from whorl 2–5 is very clear. There is much less variation in centroid size if we assume that lots 3–6 have lost a whorl due to apical wear. Thus, lots 3–6, at least, appear to also have originally had six whorls like lots 1 and 2.

DISCUSSION

With all the interest in *Littorina saxatilis* as a model system for understanding the causes and consequences of morphological variation (Grahame and Mill, 1993; Grahame et al., 1995; Rolán-Alvarez and Johannesson, 1996; Johannesson et al., 1997, to name a few), it is surprising that the ontogeny of the shell has not been more rigorously examined, especially since ontogeny is the obvious proximate cause of these differences, whether the ultimate causation is an environmentally induced change or a genetic change (or both). Rolán-Alvarez and Johannesson (1996) examined embryonic shells from the brood pouch of *L. saxatilis* to understand morphological differences in early shell ontogeny between two different morphs from the same shoreline at two localities, and other workers have compared juvenile and parent morphologies from the same localities (Grahame and Mill, 1993). However, more broad comparisons of ontogeny have not been attempted.

Given the title of this volume, it is appropriate to discuss broad patterns of ontogenetic change by attempting to go beyond a simple heterochronic explanation. It is perhaps even more appropriate to look beyond heterochrony in snail shells because the rebirth of heterochrony and the ties between ontogeny and phylogeny as a vibrant research topic can be traced back to Gould's (1969) work on the shells of *Cerion*, an indeterminately growing snail. If the rebirth of heterochrony started with snail shells, it is only fitting that a look beyond heterochrony to other evolutionary developmental models also happens with shells. Heterochrony in snail shells has been documented by many workers (Gould, 1969; Tissot, 1988; Laurin and Garcia-Joral, 1990). However, these have been based in many cases on allometric heterochronies (Laurin and Garcia-Joral, 1990; Lieberman et al., 1994), and it is unclear whether these patterns are driven by methods as opposed to true timing differences (Zelditch, 1995). Analysis of partial warps provides one way to test for both spatial and temporal changes in ontogeny (Guralnick and Lindberg, 1999; Zelditch et al., 2000).

Looking at our data set, what is perhaps most amazing is the diversity of the sometimes subtle patterns of ontogenetic change within the species. At least one major ontogenetic shift is not that subtle. In the barnacle ecotype, all the shapes of the whorls have been shifted toward the negative portion of relative warp 1. That is, each whorl, from the first onward, has greater expansion in all portions of the whorl, leading to overall higher whorl expansion rate compared with any other lot. Interestingly, this shift toward greater expansion seems most pronounced at the start of the ontogeny, as can be seen in plots of individual trajectories along RW1 and RW2 in Figure 6.13. Note how the starting point of ontogeny (labeled start) is located in the most negative region along RW1. As ontogeny proceeds, this trend toward expansion along RW1 and RW2, compared with other lots, continues but is weaker. Because the shape of the barnacle ecotype's first whorl is radically different from the shape of the first whorl of the other lots, it would be inappropriate to call shape differences heterochronic. Heterochronic patterns assume a similar starting point for ontogenetic shape trajectories (as plotted along the axes of shape change represented by RW1 and RW2), which in one taxon could shorten or lengthen compared with the other. In the case of the barnacle ecotype, the ontogenetic trajectories start out highly displaced compared with the other lots and then become less displaced but still different during ontogeny. The shape patterns perhaps better fit a heterotopic pattern in which not timing but the spatial aspects of spiral growth have been directly affected. In this case, the spatial aspects of growth affected appear to be an overall greater whorl expansion at every point during growth.

Although the shape patterns suggest a heterotopic explanation, the number of whorls in the barnacle ecotype suggests a heterochronic pattern. This ecotype appears to deposit fewer whorls than in other lots, and this truncation of

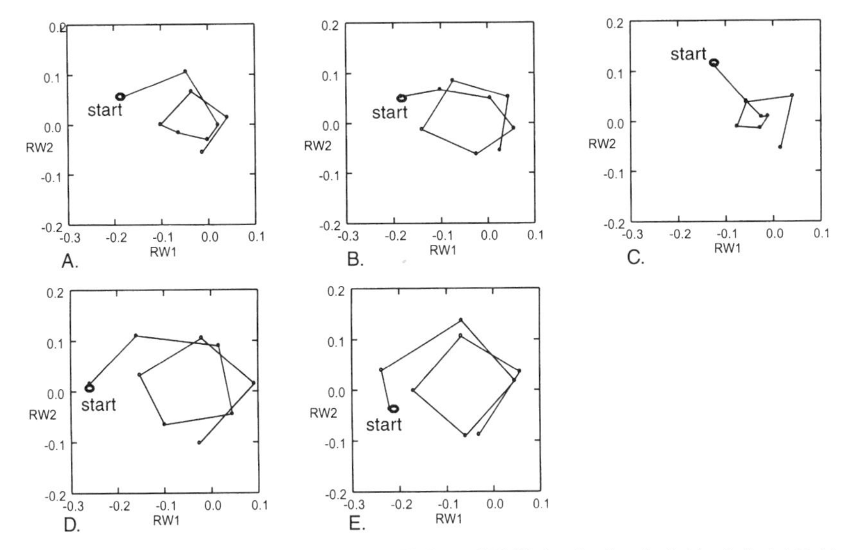

Figure 6.13. Ontogenetic trajectories along RW1 and RW2 for the five individuals in lot 7. Note that the beginning of the trajectories (labeled with circle and the word "start") always starts in the most negative portion of RW1 shape space, which represents the most whorl expansion.

growth can be labeled heterochronic. It is difficult to decide which kind of heterochrony this is because we do not know the polarity of populational evolution or the actual absolute growth rates. Whatever the case, it is interesting that the barnacle ecotype suggests both heterochronic and heterotopic mechanisms, depending on whether one considers shape or just whorl number.

The third lot, wave ecotypes from an exposed shore in Spain, shows a very different pattern of shape change. In this lot, shape variation along RW1 and RW2 is much smaller than in other lots, and this appears due to individual ontogenetic trajectories being much more similar to each other than in other groups (see Figs. 6.9B and 6.11). This reduction in variation leads to shells with less expansion of whorls in some parts of the shape space and less contraction in others. All of the shells from lot 3 appear to be more oval than the mix of rounded and oval shapes found in most other lots.

This stereotypy found in the third lot is interesting and may have a populational or ecotypic basis. Unfortunately, population and ecotype cannot be separated in this analysis; therefore, it is impossible to choose between these potential factors. Further work should help clarify this question and provide some important new insights into the causes and consequences of environmental and genetic factors on morphology. Animals might be responding to environments with higher stresses, such as the wave-exposed intertidal, with more stereotyped forms. Depending on how much outcrossing happens with other ecotypes, if this pattern of lower variation holds for shells from different exposed environ-

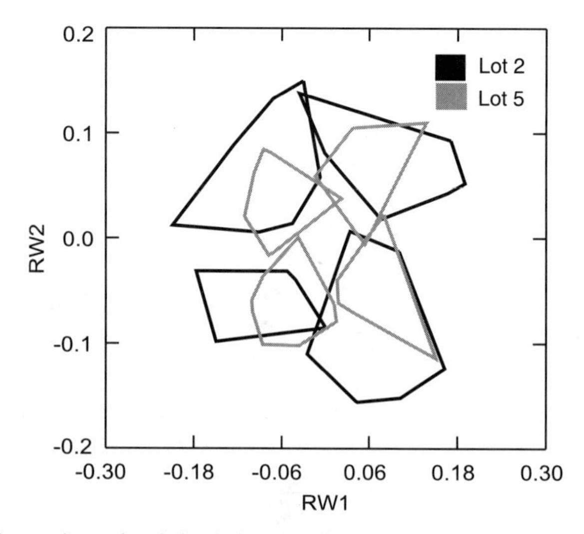

Figure 6.14. Comparison of variation in lots 2 and 5, with data points omitted and the hull representing the range of variation described by the data points. Note that the variation in the semi-exposed ecotypes, lot 5, is less than that for lot 2, a fairly sheltered ecotype.

ments, it might reflect either stabilizing selection or environmentally induced morphological conservation.

Importantly, lot 5, from a semi-exposed habitat (a moderate ecotype) and lot 7 (barnacle ecotype and assumed to be most closely associated with wave ecotypes) also show lower amounts of variation compared with the other lots (Fig. 6.9D for lot 7 and Fig. 6.14 for lot 5). In lot 7, most of the variation is in the 0- and 270-degree offsets, and examination of individual ontogenetic shape changes along RW1 and RW2 suggests that this variation is driven mostly by variation in the first 0-degree offset and last 270-degree offset (Fig. 6.13). When these outliers are removed, variation would be further reduced. Although the variation is greater for both lots 5 and 7 than for lot 3, all three lots suggest the pattern might be ecotypical and not just populational. The pattern is especially interesting because it suggests there might be a gradient in variation that covaries with a gradient in wave exposure, with variation decreasing with increasing exposure.

Growth Patterns

One of our main questions is whether *L. saxatilis* has the same stopping points during growth. This species had been assumed to be an indeterminate grower, although Reid (1996) noted that there is lip thickening and other features that suggest either a slowing down or stopping of growth once adulthood is reached.

Given the different localities and habitats sampled, we found five of the seven lots have whorls that are closely in phase to one another.

Another important piece of information for claiming that growth stops at the same point is number of whorls. If lots do stop at the same point, then both stopping point within the whorl *and* total number of whorls would be identical. From just visual inspection, the number of whorls varies in different lots. The variability might reflect real populational differences, but it might also be due to artifacts caused by wear. Lower variability in one or the other mode would argue for that assumption. Analysis of centroid size of the whorls supports that the five whorled lots (lots 3–6) are missing one whorl compared with lots 1 and 2.

Although the centroid size data suggest that perhaps lots 3–6 are missing the first visible whorl, another important piece of information related to apical wear is habitat itself. Apical wear is more likely in exposed habitats due to the higher environmental stresses like wave action. However, the number of whorls does not directly correlate with habitat/ecotype (Table 6.1). For example, lots 4 and 6 are sheltered/brackish ecotypes, yet they only have five whorls, not six like the first lot. The second lot has six whorls but comes from a more moderate environment.

Given that different inferences give different results about the number of whorls in the lots, it is difficult to make a claim one way or another about whether lots 3–6 are missing a whorl compared with lots 1 and 2. Further and more careful examination of larger samples may help solidify a claim. If it turns out that the whorl numbers in different lots are different, the obvious question is this: Why do the whorls in lots 1–5 stop at the same position?

Not all lots stop at the same position—there are two exceptions to the general pattern, lots 6 and 7. Lot 6, sheltered/brackish ecotype from Norfolk, England, has either an extra 270 degrees of growth or 90 degrees less compared with other lots. Because this lot shows a relative reduction in size compared with other lots, it is difficult to use centroid size to make arguments one way or another about whether it is a loss or gain of growth. The same amount of relative gain or loss of shell material shows up in every individual of the lot. The barnacle ecotype (lot 7) also has either an extra 270 degrees of growth or 90 degrees less compared with other lots, but it also appears to have fewer whorls than other specimens (3 compared with 5 or 6). This may be partly due to extreme apical wear, but the size of the presumed second whorl of the barnacle ecotype is not very different from the size of the second whorl in other lots.

This study documents unexpected growth patterns in *Littorina saxatilis* populations. It appears that, within lots, which in this case represent localities, growth is stereotyped to end—or at least to slow down to a snails pace—at similar spots after the deposition of six or so whorls. This stereotypical pattern also appears to be shared among many different populations from different habitats (sheltered and exposed) and locations (Maine, Spain, and Scotland). Despite these similarities, the barnacle ecotype, represented in our sample by lot

7, of *L. saxatilis* appears to have modified growth end points as discussed above. This suggests that, whereas *L. saxatilis as a whole* may have indeterminate growth, perhaps populations all have similar stopping points. Thus, indeterminate growth mode in *L. saxatilis* may be an emergent property based on different characteristics from the populations, which themselves stop growth at the same point.

Genetic and Environmental Interactions: A Perspective from Development

Littorina saxatilis is unique among *Littorina* because it broods its young up until the deposition of as many as two whorls, and we can use this information to address genetic vs. environmental interactions. The literature is rife with debates about the nature of morphological variation in *L. saxatilis* and the littorines in general. Some consider the morphological diversity to be true intraspecific plasticity, at least in some cases (e.g., Reid, 1993, on the barnacle ecotype). Other workers believe that some or all of that variation reflects selection (and in some isolated cases selection leading to incipient speciation) or a combination of plasticity and selection (Johannesson and Johannesson, 1996; Gosling et al., 1998). One of the ecotypes most often assigned its own species rank is the barnacle ecotype, referred to as *Littorina neglecta* in the literature (see Grahame et al., 1995, for one view, and Reid, 1996, for the other).

The question of environmental vs. genetic factors can be examined from a developmental perspective because brooding should buffer young snails from some of the vagaries of the environment. If patterns of shape change that lead to adult morphology show up early in development, before environmental exposure, this suggests that those shape differences are not caused by the direct effects of environment. However, it is perhaps conceivable that plasticity cues could still be transmitted to embryos through the parent, but this would be a more indirect effect.

The first question to address is this: Are the visible first whorls on the shells really the first deposited whorls? We know from comparisons of centroid size that at least one of the whorls has probably been worn off in some of the lots; however, is the first whorl, for example, in lots 1 and 2 the true beginning of shell deposition? In lieu of electron micrographs of our specimens, study of the brooded embryos can provide a fast and easy way to examine this question. We can measure the size of the whorls of embryos near hatching and compare these sizes with those of the first and second whorls in lots 1 and 2 and see whether they are similar or different. Similarity in size would suggest that the first whorl on the adult shell is the actual first whorl of the embryo. If the first whorls in lots 1 and 2 are larger, this would suggest that one or more of the first whorls are missing and that all shells from all lots have had some loss of material due to apical wear. As noted above, this method is approximate, and a more complete examination using an electron microscope would look for signs of the protoconch/teleoconch boundary.

Table 6.3. Measures of maximum diameter of whorls for the first whorl and for first and second whorl together for embryos (as photographed by Reid, 1996) and for shells used in this study from the first and second lot

Specimen	No. of Whorls	Total Diameter	Diameter of First Whorl
Embryo 1	1.6	436 μm	180 μm
Embryo 2	1.85	709 μm	345 μm
Embryo 3	1.85	600 μm	345 μm
Embryo 4	1.75	618 μm	327 μm
Lot 1, spec. 2	2	~680 μm	~320 μm
Lot 1, spec. 3	2	~600 μm	~330 μm
Lot 2, spec. 2	2	~1000 μm	~350 μm
Lot 2, spec. 4	2	~680 μm	~300 μm

spec., Specimen.

Table 6.3 shows the size of the first whorl for four embryos taken from different *L. saxatilis* individuals photographed by Reid (1996) and the size of the first whorls for lots 1 and 2. The sizes of the whorls compare very favorably, even given that lot 1 and 2 specimens tend to be rather large in overall size. These data suggest that, for at least lots 1 and 2, the first *visible* whorls are the *actual* first whorls. Therefore, it also seems likely that, in the other lots, the first visible whorl is actually the true second whorl. As mentioned earlier, embryos generally deposit approximately two whorls before being released from the brood pouch (Reid, 1996). Thus, part of the shell material in all lots was deposited when the animals were still protected from the external environment and therefore also from potential direct cues for plasticity.

If we examine the shape differences, for example, in the barnacle ecotype, the differences in shape manifest before the animals are exposed to the external environment. In fact, the first visible whorl of the barnacle ecotype starts out with the most extreme shape (Fig. 6.11A), and this is evidence that the trajectories in the barnacle ecotype are likely not due to direct environmental induction. However, at least one other explanation is possible and requires further testing. It is also possible that the maternal environment is very different in different habitats. In this scenario, the environment induces differences in the mother's body and shell shape and size. Body shape and size affect the size and shape of the ovaries and the brood pouch; therefore, both the number and size and shape of the embryos themselves are affected.

Our examination of ontogeny also produced one other pattern of relevance to the question of environmental and genetic causations. Plots of individual ontogenies generally show that individual ontogenies are relatively stereotyped—despite some allometries during growth. Once a shell begins with, for example, a more rounded whorl, that shape tends to continue through ontogeny (as compared with overall sample shape variation). There are few examples in *L. saxatilis* where a shell, during ontogeny, dramatically changes its shape based on both heuristic examination of the shells and on examination of the relative warp plots. More appropriate analyses (e.g., lot-based Procrustes distances) are needed to fully establish either observation because the relative warp plots are based on variation in all the samples and differences during one individual's growth may not be aligned with these axes. However, the preliminary data suggest that perhaps very early shell depositional factors, what Gould (1969) referred to as nucleating effects, have an important role in determining overall shape, which again suggests some genetic control of shell shape.

CONCLUSIONS

Our work resulted in the following six observations and conclusions.

First, a prerequisite for comparing the ontogenies of shells is to assure homologous units of comparisons. This prerequisite has two components: comparing the same whorls and comparing the same portions within whorls. For both determinate and indeterminate growers, making the right comparisons is difficult, especially if the earliest parts of ontogeny aren't preserved, which is common in intertidal gastropods. Morris and Allmon (1994) discussed morphometric techniques to solve part of this problem (missing whorls) in high-spired shells. However, no methods have ever been previously discussed that solve the problem of aligning whorls, and this problem is perhaps more serious than missing whorls. In some cases, it is not crucial to have an exact sequence of whorls because shells are known to grow isometrically and therefore shapes are similar throughout growth. However, as we have discussed, using different portions of the whorl in different specimens and assuming they are similar lead to incorrect interpretations of major shape differences, as shown in Figs. 6.2 and 6.4.

In the case of *Littorina saxatilis*, we used geometric morphometrics and the nature of spiral growth itself to test whether whorl segments were at the same or at different starting points—thus providing a method to answer whether growth in *L. saxatilis* has similar or different stopping points for different individuals. This method can be applied to any fossil or living group as a way to allow for valid comparisons of ontogenetic trajectories. Even in cases of indeterminate growth, using this method should allow researchers to align their shells to a common spot by first ordinating all the data, then choosing a common starting point, and finally removing the beginning landmarks from shells that ordinate in distinctly different regions of the shape space.

Second, most of the lots of *Littorina saxatilis* that we examined appear to have similar ending points in the shape space. Individuals in two lots have either deposited an extra 270 degrees or lost 90 degrees of growth. Thus, different populations of *L. saxatilis* can have different but fixed ending points in their growth trajectories.

Third, analyses of centroid sizes suggest that most lots have had the first whorl worn off. However, another important piece of indirect evidence for wear patterns—habitat—only partially supports the centroid size analysis. For example, lot 1 has six whorls but comes from a sheltered habitat, whereas lot 6, which is from the same habitat, has only five. Sheltered environments should generally have more of the early whorls visible because wear is reduced. A further examination of the shells, in a more controlled setting, may be necessary to give a more complete answer to the question of total number of whorls in each lot. However, it seems likely that most lots may have the same number of whorls. Given that most lots also have the same ending point, these two pieces of information suggest identical end points of ontogenetic trajectories. The barnacle ecotype is unique in unambiguously having fewer whorls than the others.

Fourth, examination of whorl shape within and between lots during ontogeny reveals that the barnacle ecotype has completely shifted its whorl shape toward greater expansion all along the whorl compared with the other lots. This shape difference should be considered heterotopic because it reflects a unilateral shift in direction along RW1. However, the barnacle ecotype's truncation of shell growth might be considered heterochronic, showing that different aspects of growth (shape change in whorls, number of whorls) might provide different means of achieving an adult shape.

Fifth, the exposed and semi-exposed lots have less variation than the others, and this variation is due to more stereotyped ontogenetic trajectories between individuals. This pattern requires further documentation but, if it holds, it would be an empirical demonstration that stressful environments like the wave-exposed intertidal directly affects variability through either stabilizing selection or another mechanism.

Sixth, genetic determination of shell shape or maternal environment are likely candidates for the ultimate cause of shell shape differences in *L. saxatilis*. For example, plasticity might be initiated with early cues, perhaps maternally mediated and resulting in a static morphology. Our data do not support direct influence of environment on *L. saxatilis* individuals as a mechanism of shell shape change.

ACKNOWLEDGMENTS

We thank David Reid, from the British Natural History Museum, for gracious attention when we visited and for providing material to study this interesting species. We also thank David Lindberg for seeing the interesting patterns and helping us understand the meaning of these patterns. All the photographs and data used in this analysis are available as JPEG files and text files in TPS format, respectively, at http://128.138.103.105/labdata.html.

REFERENCES

Ackerly SC (1989): Kinematics of accretionary shell growth, with examples from brachiopods and molluscs. Paleobiology 15: 147–164.

Gosling EM, Wilson IF, Andrews J (1998): A preliminary study on genetic differentiation in *Littorina saxatilis* from Galway Bay, Ireland. *Littorina tenebrosa* Montagu: a valid species or ecotype? Hydrobiologia 378: 21–25.

Gould SJ (1969): An evolutionary microcosm: Pleistocene and recent history of the land snail P (Poecilozonites) in Bermuda. Bull Mus Comp Zool 138: 407–532.

Grahame J, Mill PJ (1993): Shell shape variation in rough periwinkles: genotypic and phenotypic effects. In: Aldrich JC (ed), *Quantified Phenotypic Responses in Morphology and Physiology*. Ashford: JAPAGA, p. 25–30.

Grahame J, Mill PJ, Hull SL, Caley KJ (1995): Littornia neglecta Bean: ecotype or species? J Nat Hist. 1995. 29: 887–899.

Guralnick RP, Lindberg DR (1999): Integrating developmental evolutionary patterns and mechanisms: A case study using the gastropod radula. Evolution 53: 447–459.

Hayes B (1995): Computing science: space-time on a seashell. Am Sci 83: 214–218.

Hull SL, Grahame J, Mill PJ (1999): Reproduction in four populations of brooding periwinkle (*Littorina*) at Ravenscar, North Yorkshire: adaptation to the local environment? J Mar Biol Assoc UK 79: 891–898.

Johannesson B, Johannesson K (1996): Population differences in behaviour and morphology in the snail *Littorina saxatilis*: phenotypic plasticity or genetic differentiation? J Zool 240: 475–493.

Johannesson K, Rolan-Alvarez E, Erlandsson J (1997): Growth rate differences between upper and lower shore ecotypes of the marine snail *Littorina saxatilis* (Olivi) (Gastropoda). Biol J Linn Soc 61: 267–279.

Johnston MR, Tabachnick RE, Bookstein FL (1991): Landmark-based morphometrics of spiral accretionary growth. Paleobiology 17: 19–36.

Laurin B, Garcia-Joral F (1990): Miniaturization and heterochrony in *Homoeorhynchia meridionalis* and *Homoeorhynchia cynocephala* (Brachiopoda, Rhynchonellidae) from the Jurassic of the Iberian Range, Spain. Paleobiology 16: 62–76.

Lieberman BS, Brett CE, Eldredge N (1994): Patterns and processes of stasis in two species lineages of brachiopods from the Middle Devonian of New York State. Am Mus Novit 0: 1–23.

Mathworks (2000): MATLAB, version 6.0.0.88.

Morris PJ, Allmon WD (1994): Shell alignment for the morphometric analysis of high-spired gastropods. Nautilus 108: 15–22.

Reid DG (1993): Barnacle-dwelling ecotypes of three British Littorina species and the status of Littorina neglecta bean. J Molluscan Stud 59: 51–62.

Reid DG (1996): *Systematics and Evolution of Littorina*. London: The Ray Society.

Rohlf FJ (1998): TPSDIG version 1.20. Stoney Brook, NY: Department of Ecology and and Evolution, State University of New York.

Rohlf FJ (1998): TPSRELW version 1.18. Stoney Brook, NY: Department of Ecology and and Evolution, State University of New York.

Rolán-Alvarez E, Johannesson K (1996): Differentiation in radular and embryonic characters, and further comments on gene flow, between two sympatric morphs of *Littorina saxatilis* (Olivi). Ophelia 45: 1–15.

Stone JR (1998): Landmark-based thin-plate spline relative warp analysis of gastropod shells. Syst Biol 47: 254–263.

SYSTAT 8.0 for windows (1996): Evanston, IL: SPSS, Inc.

Tissot BN (1988): Geographic variation and heterochrony in two species of cowries genus Cypraea. Evolution 42: 103–117.

Zelditch ML, Fink WL, Swiderski DL (1995): Morphometrics, homology, and phylogenetics: quantified characters as synapomorphies. Syst Biol 44: 179–189.

Zelditch ML, Sheets HD, Fink WL (2000): Spatiotemporal reorganization of growth rates in the evolution of ontogeny. Evolution 54: 1363–1371.

7

PIGMENT PATTERNS OF ECTOTHERMIC VERTEBRATES: HETEROCHRONIC VS. NONHETEROCHRONIC MODELS FOR PIGMENT PATTERN EVOLUTION

David M. Parichy

Sections of Integrative Biology and Molecular, Cellular and Developmental Biology, Institute for Cellular and Molecular Biology, University of Texas at Austin, Austin, Texas

INTRODUCTION

A vast literature documents the prevalence of heterochronic changes during the evolution of morphology. Such heterochronies may be defined broadly as changes in the rate or timing of developmental events, compared with an ancestral ontogeny (Gould, 1977; Alberch et al., 1979; McKinney and McNamara, 1991). Typically, heterochronies have been identified at the whole organism level (so-called "global" heterochronies) or at the level of particular traits or organ systems ("local" heterochronies in the terminology of McKinney and McNamara, 1991). A classic example of a global heterochrony is the independently derived failure of metamorphosis in several species of salamander

Beyond Heterochrony: The Evolution of Development, Edited by Miriam Leah Zelditch
ISBN 0-471-37973-5

in the genus *Ambystoma*; this retention of a larva-like somatic morphology after sexual maturation has been considered a clear case of paedomorphosis (Shaffer and Voss, 1996; Voss and Shaffer, 1997; and see below). Another example, also from salamanders, is the extraordinary miniaturization observed in several lineages of plethodontid salamanders, which has resulted in radical changes in limb, skull, and central nervous system morphology, another example of paedomorphosis (Wake, 1991; Hanken and Wake, 1993).

The literature pertaining to heterochrony is vast (McKinney and McNamara, 1991), particularly in amphibians (Hanken, 1999). These studies have focused almost exclusively on patterns of morphological variation, as revealed by analyses of embryonic, larval, or adult morphology. Nevertheless, an explicitly hierarchical view of heterochrony has also been promoted by McKinney and McNamara (1991) in which organismal form and its evolutionary modification are seen as the product of underlying cellular events. In this framework, heterochrony can be considered an organizing principle at any biological level, as applicable to changes in the rate or timing of cellular events during development as to "global" changes in morphology. Pursuing this notion, McKinney and McNamara argued that heterochrony is both the cause of most developmental alterations and a source of major, morphological novelties (see p. 47). Despite these claims, relatively little attempt has been made to assess the frequency of heterochronic changes in cellular behaviors underlying the development of morphological traits or the morphological consequences such changes might produce. For example, we do not know whether a heterochrony identified at the whole organism level necessarily reflects a heterochrony at the underlying cellular or genetic levels. Likewise, there has been almost no effort to assess the extent to which seemingly nonheterochronic variation at a morphological level might nevertheless depend on heterochronic alterations to underlying cellular behaviors: Can continuous variation in a trait typically be explained by heterochronic cellular variation? Is discontinuous variation or morphological novelty often due to changes in the timing of cellular behaviors, as McKinney and McNamara postulated?

In this chapter, I seek to test the hypothesis that interspecific diversity in salamander pigment patterns is causally related to heterochronies at the cellular level. This hypothesis predicts that both continuous and discontinuous pigment pattern variation or novel pigment pattern variants can be related to changes in the rate or timing of underlying cellular mechanisms that are themselves conserved across taxa. Alternatively, if pigment pattern differences between species result from nonheterochronic changes in developmental mechanisms, for example, the appearance of a novel cellular behavior or a novel cue to which cells respond, this would call into question the utility of a heterochronic framework for understanding the mechanistic bases for pigment pattern variation.

To distinguish between these possibilities, I begin by reviewing some of the features of pigment patterns that make them a useful system for studying the evolution of form, as well as some of the cellular and developmental processes relevant to understanding pigment pattern evolution. I then briefly discuss some methodological issues relevant to studying the development of pigment patterns and other characters in an evolutionary context. Because heterochonic changes are only understandable with reference to an ancestral ontogeny, I then review our current understanding of phylogenetic relationships for several species in which pigment pattern development has been studied. These hypotheses of phylogeny then set the stage for examining how several features of pigment patterns develop and the extent to which phylogenetic transformations in these patterns may have resulted from heterochronic changes in developmental mechanisms. Finally, in the last part of this chapter, I ask whether a heterochronic framework at the cellular level is useful for understanding evolutionary changes in pigment patterns, and I suggest possible future directions for assessing heterochronies and other mechanistic changes relevant to pigment pattern evolution.

Pigment Patterns Are a Model System for Understanding Morphological Variation and Evolutionary Diversification in Vertebrates

Several features make pigment patterns a particularly useful system for studying the mechanistic bases for the evolution of form. Because pigment patterns are a classic and enduring system for studying development, there is a wealth of data on pigment cell morphogenesis and differentiation that helps guide studies of pigment pattern variation and evolution (DuShane, 1934; Bennett, 1993; Erickson, 1993; Reedy et al., 1998b). Some of these data are presented below.

Pigment patterns are also an attractive system because of their ecological and behavioral significance. For example, a host of studies have shown the importance of pigment patterns for thermoregulation, species recognition, locomotion, avoidance of predation, and mate choice (e.g., Endler, 1978; Houde, 1997). The selective factors that contribute to the evolution of specific pigment patterns are beyond the scope of this review. Nevertheless, the myriad functional roles served by pigment patterns offer the promise of truly integrative studies that span several levels of biological organization.

A final reason for using pigment patterns is the pigment cell lineage itself. Vertebrate pigment cells are derived from the neural crest, a transient population of embryonic precursor cells that arises during neurulation at the border between the neural plate and nonneural ectoderm and later are found dorsal to the neural tube (Le Douarin, 1982; Hall and Hörstadius, 1988; Erickson and Perris, 1993; Groves and Bronner-Fraser, 1999; Hall, 1999). In an evolutionary context, neural crest cells are of particular interest for at least two reasons: their differentiative potential and their epigenetic mode of development. Both of these features have implications for studying the potential roles heterochrony has played in phylogenetic transformations of morphology.

Neural crest cells give rise not only to pigment cells but also to a host of other cell types and tissues in the vertebrate embryo and adult (Hall and Hörstadius, 1988; Groves and Bronner-Fraser, 1999). For example, much of the peripheral nervous system, including both neurons and glia, originates from the neural crest. These cells also produce many of the dermal bones comprising the craniofacial skeleton and contribute to teeth, fin mesenchyme, and bony fin rays of teleosts. Finally, neural crest cells are the progenitors of endocardial cushion cells in the heart, adrenal chromaffin cells, and some smooth muscle. The many organ systems that arise in whole or in part from neural crest cells suggest that much of vertebrate evolution can be understood in terms of changes in the patterning of these cells and their derivatives (Gans and Northcutt, 1983; Gerhart and Kirschner, 1997; Hall, 1999). In vitro clonal analyses show that some neural crest cells are pluripotent (i.e., can differentiate into more than one derivative), whereas other neural crest cells differentiate into just one or two cell types (Sieber-Blum and Cohen, 1980; Baroffio et al., 1988, 1991). Likewise, in vivo lineage tracing has demonstrated marked heterogeneity in the fates acquired by neural crest cells (Bronner-Fraser and Fraser, 1988; Frank and Sanes, 1991; Fraser and Bronner-Fraser, 1991). In the zebrafish *Danio rerio*, for example, Raible and Eisen (1994) showed that some neural crest cells produce progeny that ultimately differentiate into pigment cells, sensory and sympathetic neurons, as well as glia, whereas other neural crest cells produce type-restricted progeny differentiating as only one or another cell type. The developmental potential of neural crest cells becomes more restricted over time. Thus, early-appearing neural crest cells produce a more diverse array of fates than later-appearing neural crest cells. The factors responsible for this progressive restriction of cell fate remain a matter of investigation (Groves and Bronner-Fraser, 1999). Nevertheless, the existence of such a pattern and the observation that neural crest cells at different states of specification have different morphogenetic requirements and abilities (Erickson and Goins, 1995; Erickson and Reedy, 1998a) raise the possibility that heterochronic changes in mechanisms of neural crest specification could have downstream consequences for the allocation of these cells to particular fates. For example, the zebrafish *colourless* mutation specifically eliminates pigment cell, neuronal, and glial fates but does not affect fin mesenchyme or craniofacial derivatives, suggesting an early segregation of these cell lineages (Kelsh and Eisen, 1999). Thus, a naturally occurring temporal change in the activity of a gene required for some general aspect of neural crest development (e.g., proliferation) could conceivably have different effects across neural crest-derived lineages, depending on precisely when this change occurs. The notion of whether changes in timing affecting fate specification have roles in generating naturally occurring morphological variation remains almost completely unexplored, but a potential example in which changes in neural crest differentiation may have mediated the evolutionary loss of stripes will be presented below (also see Smith, 2000).

Besides the extraordinary range of derivatives to which neural crest cells contribute, this population of embryonic precursor cells is particularly intriguing because of the epigenetic nature of its development. Although cell movements characterize the morphogenesis of numerous cell types and organ systems in both vertebrate and invertebrate embryos (Bard, 1990), neural crest cells are truly exceptional in this regard. As noted above, neural crest cells arise at the border between the neural plate and epidermis and then are found along the dorsal neural tube (or neural keel in teleosts; Moury and Jacobson, 1990; LaBonne and Bronner-Fraser, 1999; Nguyen et al., 2000). These cells then disperse from this location and migrate along stereotypical pathways throughout the embryo (Fig. 7.1; Löfberg et al., 1980; Loring and Erickson, 1987; Raible et al., 1992; Erickson and Perris, 1993). Most neural crest cells that will contribute to externally visible pigment patterns migrate along a dorsolateral migratory pathway, between the ectoderm and the somites (or, at later stages, the myotomes). In contrast, trunk neural crest cells that contribute to the peripheral nervous system take a ventromedial migratory pathway, between the somites and the neural tube or notochord. During this migration, the cells interact with one another and adjacent cell types, and these interactions can influence both the morphogenetic behavior of the cells and their state of differentiation. A large body of literature addresses the mechanisms by which migration may commence at the level of the neural tube, factors that contribute to the rate and pattern of cell movement as migration is underway, and the controls that determine when and where cells should cease their migration and take up residence (Erickson and Perris, 1993; Perris and Perissinotto, 2000; Sela-Donenfeld and Kalcheim, 2000). There is thus ample opportunity for changes in the patterning of neural crest derivatives as a result of changes in the timing or rate of migration or the interactions that occur during this migration. Several examples of such interactions will be presented in detail below. The epigenetic nature of neural crest development also suggests that phylogenetic changes could be either autonomous to the neural crest cells themselves or could arise as nonautonomous changes in the extracellular environment that the cells encounter. As discussed below, these different possibilities can have different evolutionary implications.

One intraspecific example of a nonautonomous, heterochronic change in neural crest patterning comes from the axolotl *Ambystoma mexicanum* (itself a famous example of a global paedomorphosis). Nearly a century ago, the spontaneous white (*d*) mutant of *A. mexicanum* was first described (Häcker, 1907). Whereas neural crest-derived pigment cells give wild-type axolotls a dark color as both larvae and adults, white mutants are brilliant white (Fig. 7.2). A series of studies have since identified some of the factors that contribute to this striking phenotype and have shown that it first arises as a defect in neural crest migration. In contrast to wild-type axolotls, white mutants completely lack differentiated pigment cells in the skin as adults, and, in larvae, these cells are largely confined to the region immediately dorsal to the neural tube. Scanning electron microscopy shows that most neural crest cells fail to enter the dorsolateral migratory pathway, and a series of embryological grafting experiments revealed that the defect lies within the epidermis, rather than the neural crest cells themselves (DuShane, 1935, 1939; Keller et al., 1982; Spieth and Keller, 1984; Keller and Spieth, 1984). Subsequent histological, ultrastructural, and biochemical studies, as well as heterochronic grafting experiments, showed that white mutant embryos exhibit a retardation in the development of the subepidermal extracellular matrix (Löfberg et al., 1989; Perris et al., 1990), in-

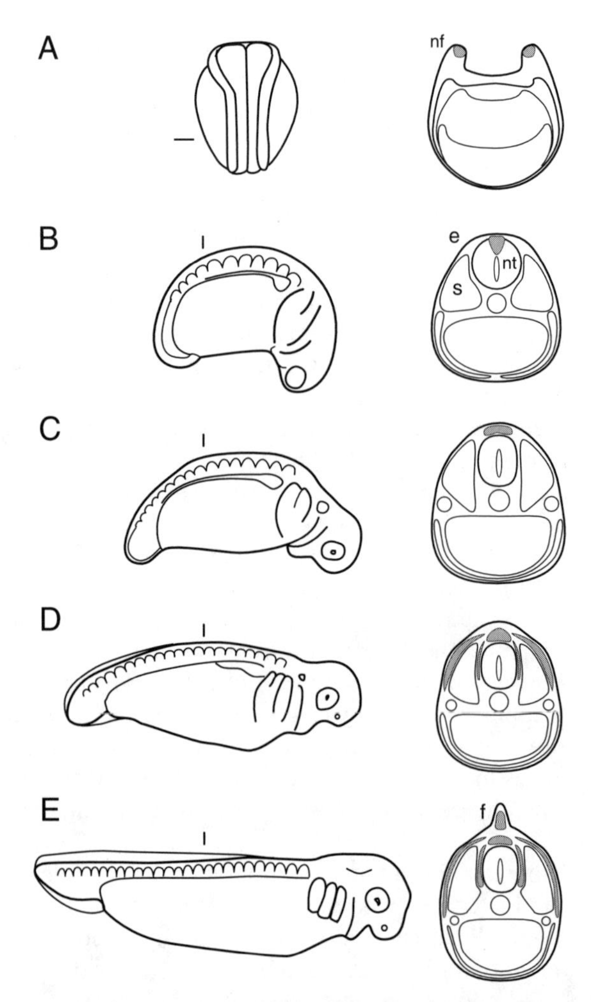

Figure 7.1. Development of neural crest cells in salamanders. Shown is a schematic of a salamander embryo at several stages of development. On the left are dorsal (A) and lateral (B–E) views; on the right are cross sections at the level indicated by the bar on the left. Positions of neural crest cells, their precursors, or derivatives are indicated in grey. A: at neurula stages, prospective neural crest cells are found within the rising neural folds (nf). B: subsequently, after the neural folds fuse, neural crest cells form a wedge within the dorsal neural tube (nt). Also shown is the position of bilateral somites (s) and the epidermis (e). The notochord is found immediately ventral to the neural tube. C: by early tailbud stages, neural crest cells have segregated from the neural tube, and, in salamanders, these cells form a distinctive cord running anterior to posterior along the dorsal midline immediately above the neural tube. Differentiation of neural crest cells into pigment cells can begin in this premigratory position. D: with further development, neural crest cells (or their derivatives) disperse from the neural tube and follow stereotypical pathways to destinations throughout the embryo. Cells that contribute to externally visible pigment patterns typically migrate between the somites and overlying epidermis. E: by hatching stages, neural crest cell migration is essentially complete, and derivatives of these cells are found throughout the body of the embryo as well as in the median fin fold (f).

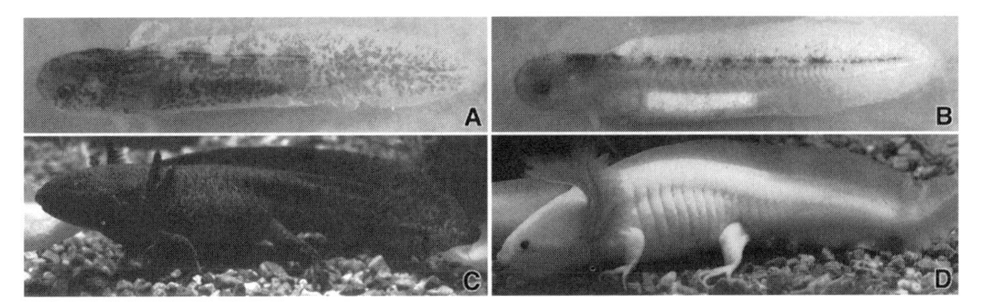

Figure 7.2. A mutant of the laboratory axolotl, *A. mexicanum*, provides a model for an intraspecific heterochrony affecting neural crest patterning. Wild-type axolotls exhibit a dark coloration as larvae (A) and adults (C). White mutant axolotls have a pigment cell deficiency as larvae (B) and an essentially complete absence of pigment cells as adults (D), owing to a heterochronic defect in extracellular matrix development that inhibits the normal dispersal of pigment cells from the neural crest. [From Parichy et al. (1999b).]

cluding reduced levels of a specific proteoglycan during the stages when neural crest cells normally disperse from the neural crest (Stigson et al., 1997a, 1997b). This delay in extracellular matrix development is postulated to result in an environment that is nonpermissive for neural crest cell migration when these cells are competent to disperse from their position above the neural tube; at a later stage, the environment becomes permissive, but neural crest cells are no longer competent to disperse. Thus, a heterochrony in the development of the environment through which neural crest cells migrate has been implicated in a change in neural crest patterning, in this instance, an essentially complete failure of pigment cells to colonize the flank. This interpretation can be tested further once the gene corresponding to the white mutation is identified (Parichy et al., 1999b). Although the phenotype of the white mutant is extreme, and probably not relevant to most naturally occurring variation, the underlying defect illustrates how even transient developmental changes that affect neural crest migration may have large phenotypic effects. A possible instance in which interspecific differences in the timing of cell migration determine whether a particular pigment pattern arises is presented below.

Approaches to Studying Heterochronic and Nonheterochronic Mechanisms in Pigment Pattern Development and Evolution

One advantage to pigment patterns for studying the developmental bases for evolutionary changes in form is that pigment cells are externally visible, and, in oviparous vertebrates like many amphibians and fishes, these cells may be followed essentially throughout the life of the organism. This has allowed for time lapse and imaging series in which the behaviors of individual cells are recorded and interspecific differences in cell behavior are identified by simply watching as the phenotypes develop (see below). In some instances, phylogenetic changes in morphology can thus be immediately interpreted as differences in the morphogenetic behaviors of cells, a major advantage over other traits in which differences in final form may provide little insight into the morphogenetic events by which the form was constructed.

Although the pigment in pigment cells serves as an autonomous marker of cell lineage, studies of neural crest cells and pigment cell precursors (i.e., before visible pigmentation has developed) typically require methods to distinguish these cells from surrounding cell types. A variety of approaches has been employed, including molecular markers for neural crest cells or their descendent cell types (Sadaghiani and Vielkind, 1990; Erickson and Goins, 1995; Lister et al., 1999; Parichy et al., 1999a, 2000a, 2000b), ultrastructural analyses to identify nascent pigment-containing subcellular organelles (Frost et al., 1984), neural crest grafts between embryos using species- or strain-specific characteristics (e.g., nuclear morphology; Sadaghiani and Thiébaud, 1987; Couly et al., 1993), and a variety of lineage tracers including radioactive isotopes (Weston, 1963; Chibon, 1967), retroviruses (Frank and Sanes, 1991), and fluorescent vital dyes (Bronner-Fraser and Fraser, 1988; Collazo et al., 1993; Raible and Eisen, 1994; Erickson and Goins, 1995; Olsson and Hanken, 1996). In addition, some pigment cell precursors are identifiable by their endogenous autofluorescence shortly before the development of externally visible pigmentation (Epperlein et al., 1988; Raible and Eisen, 1994). Thus, a number of different strategies may now be used to assess the development of pigment cell precursors even before they acquire visible pigment. By and large, however, these reagents have yet to be exploited for assessing the mechanistic bases for interspecific variation in the development of pigment patterns or other neural crest derivatives. Although a histological technique has been widely used for many years in which initially unmelanized melanoblasts darken after administration of the melanin synthesis intermediate DOPA (Epperlein et al., 1988), this technique is relatively nonspecific and also can stain a variety of other cell types that are not derived from the neural crest (Tucker and Erickson, 1986a). Moreover, because this method only reveals melanophore precursors that are already competent to produce melanin and the earliest time at which these cells are visible differs considerably among species (Epperlein and Löfberg, 1984; Olsson, 1993), an increased use of other, more specific markers will be required for further analyzing the distribution of pigment cell precursors during development and evolution.

Rigorous inferences concerning developmental mechanisms require not only careful descriptive analyses but manipulative experiments as well. This is particularly true in an interspecific context in which causality is even more difficult to demonstrate than during the ontogeny of a single species and more factors can contribute to confounding interpretations of developmental process (e.g., differences in overall development rate and morphology, failure of molecular probes to cross-react between species, unexpected differences in developmental genetic pathways). For instance, numerous studies of nonmodel organisms have documented the expression of genes thought to pattern various tissues, organs, or body axes (reviewed in Carroll et al., 2001). Nevertheless, these studies often have not demonstrated the relevance of the expression patterns in question for the final form of the character. Indeed, there are many instances in which tissues express a gene yet do not require it for their development (e.g., Parichy et al., 2000a). Only through manipulative experiments or mutational analyses can this possibility be addressed. Likewise, descriptive analyses of cellular distributions during development may suggest cellular and tissue level controls, but perturbational approaches must be used to challenge these mechanistic hypotheses. Fortunately, both amphibians and fishes offer many opportunities for experimental approaches to dissecting the evolution of form, via transgenesis, mutational analyses, various kinds of microsurgical manipulations, and other approaches (e.g., Solnica-Krezel et al., 1994; Parichy, 1996a; Huang et al., 1999). In the long term, the use of such experimental methods, employed within the context of well-resolved hypotheses of phylogenetic relationships, will be essential for rigorous evolutionary inferences regarding changes in developmental mechanisms.

CASE STUDIES: SALAMANDER PIGMENT PATTERN DEVELOPMENT

Salamanders and Their Pigment Patterns

Salamanders exhibit a diverse array of pigment patterns. Moreover, many species of salamanders undergo a metamorphosis in which an aquatic larval morphology is transformed into a terrestrial adult morphology, and this often entails radical changes in externally visible pigment patterns. Figure 7.3 illustrates some of the diversity in pigment patterns at early larval stages, shortly after salamanders hatch and when they are especially vulnerable to predators (examples of adult pigment patterns are shown below). Recent work suggests that at least some of these patterns serve a role in crypsis (Storfer et al., 1999).

An important step in analyzing the development or evolution of any character is to identify its component parts. At early larval stages, salamander pigment patterns comprise principally two classes of neural crest-derived pigment cell: black melanophores and yellow xanthophores, the latter of which can be visualized most easily by their autofluorescence under ultraviolet light of an appropriate wavelength (Epperlein et al., 1988; Fig. 7.4). As described in detail below, much of the variation in early larval pigment patterns across species can be understood in terms of the presence or absence of two pattern elements that reflect the spatial arrangements of melanophores and xanthophores. These elements are (1) alternating patches of melanophores and xanthophores extending ventrally from the dorsal flank so as to form a series of vertical bars and (2) a region running horizontally along the middle of the flank in which xanthophores are present but melanophores are not found, thereby forming a melanophore-free horizontal stripe. Both vertical bars and a melanophore-free region are shown in Figure 7.4. Although vertical bars and melanophore-free regions have sometimes been considered alternative states of a single character (i.e., a larval pigment pattern; Epperlein et al., 1996), these different elements

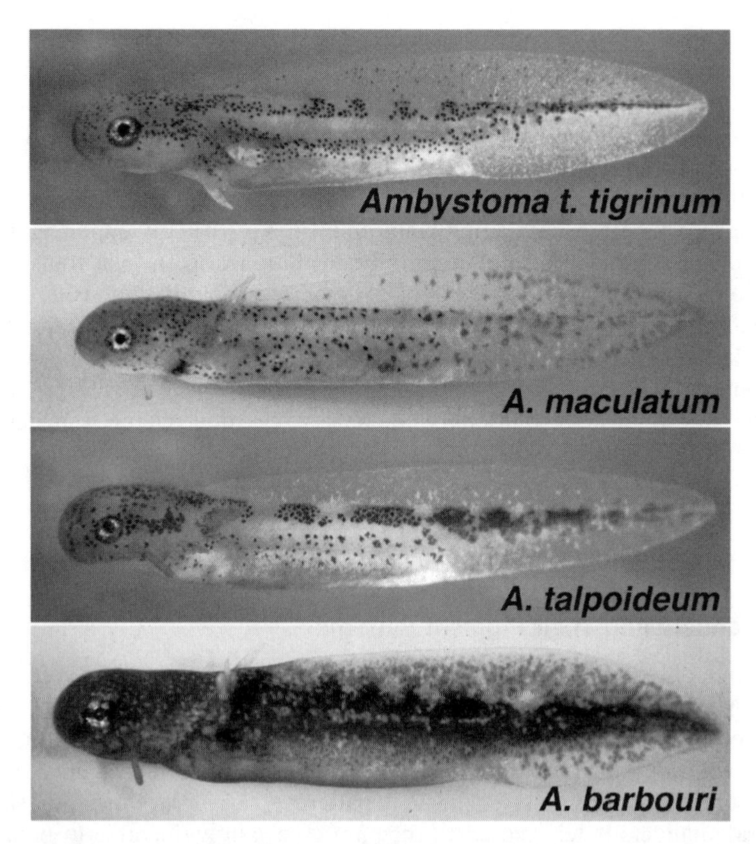

Figure 7.3. Salamanders exhibit diverse early larval pigment patterns. Shown are four species of *Ambystoma* that exhibit various combinations of vertical barring and a horizontal melanophore-free region in the middle of the flank.

depend at least partly on distinct developmental mechanisms, as described below (although the extent of such independence is an empirical question that can be addressed only using experimental approaches). Thus, I treat these pigment pattern elements as discrete characters for the purpose of assessing evolutionary changes in developmental mechanisms but also briefly discuss mechanisms that may be shared by these elements during their formation. How prevalent are these different pattern elements in salamanders, and what is the evolutionary history of the mechanisms that produce them? A complete answer to these questions requires a hypothesis of phylogenetic relationships.

Phylogenetic Relationships of Salamander Taxa

The evolutionary relationships of salamanders have been studied extensively. For the purpose of understanding the evolution of developmental mechanisms underlying pigment pattern formation, several different taxonomic levels must

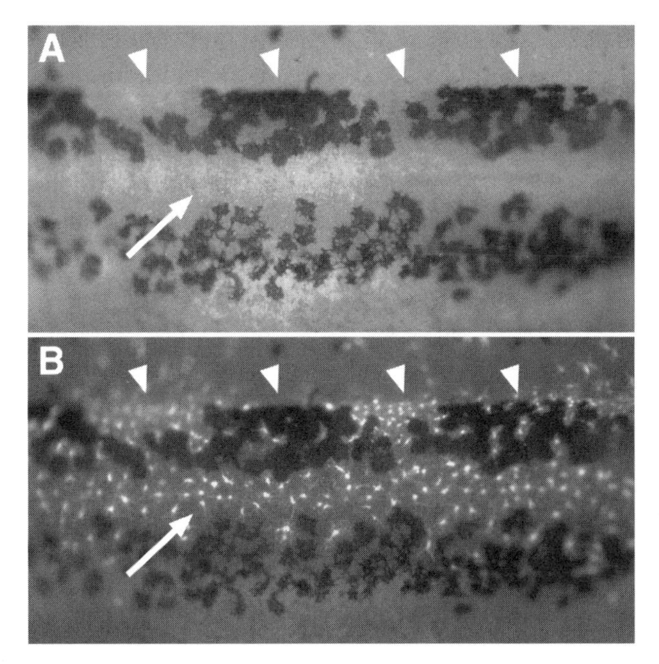

Figure 7.4. Pigment pattern elements in hatchling salamander larvae. Shown is the flank of a hatching stage *A. t. tigrinum* larva. Anterior is to the right. A: brightfield view shows the distribution of black melanophores. B: ultraviolet illumination of the same field reveals yellow xanthophores by their autofluorescence. Two distinctive pigment pattern elements are present in this species: a series of alternating vertical bars of melanophores and xanthophores (arrowheads) and a melanophore-free region in the middle of the flank (arrow) that defines a horizontal stripe pattern.

be considered. Here, I briefly review some of the current hypotheses for phylogenetic relationships among salamander families, as well as relationships within two families (Ambystomatidae, Salamandridae) that have figured prominently in studies of pigment pattern development.

Salamanders, Order Caudata, comprise 10 families of approximately 350 total species (Duellman and Trueb, 1986). Early studies of family-level relationships were based on morphological characters, which in salamanders are often plagued by homoplasies (Wake, 1991). More recently, molecular analyses by Larson (1991), which used large and small subunit rRNA sequence data, as well as by Larson and Dimmick (1993), which used these molecular data and additional morphological data, have suggested the hypothesis of phylogenetic relationships shown in Figure 7.5. The overall features of this hypothesis are well supported and are based on 209 phylogenetically informative characters comprising 177 rRNA nucleotide positions and 32 anatomical characters (derived from head, trunk, and cloacal morphology) from 20 species. Analyses of these characters as a combined data set support the monophyly of all salamander families, and examination of character state transformations revealed

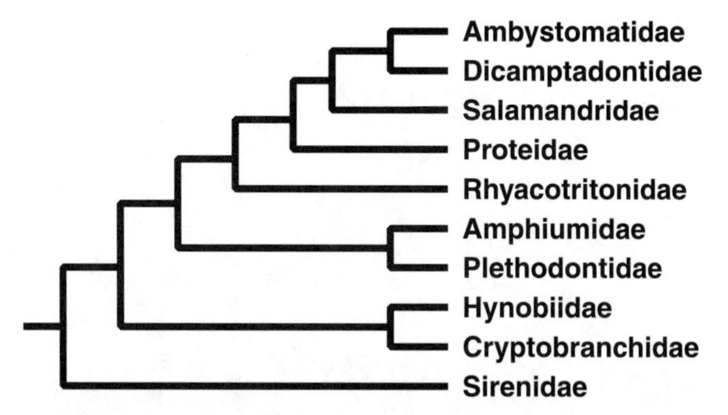

Figure 7.5. Phylogenetic relationships of the living salamander families. Shown is a recent hypothesis of relationships based on molecular and morphological data (Larson and Dimmick, 1993). Most studies of pigment pattern development have focused on species within Ambystomatidae and Salamandridae.

relatively little incongruence between molecular and morphological subsets of the data. With respect to pigment pattern analysis, this hypothesis differs in one important placement, compared with older hypotheses based strictly on morphological data. Specifically, phylogenies based on morphological criteria alone placed the speciose family Plethodontidae (~250 species; Wake and Hanken, 1996) in a relatively derived position as a sister taxon to Ambystomatidae (33 recognized species; Shaffer et al., 1991). In contrast, the phylogeny of Larson and Dimmick (1993) places Plethodontidae in a more basal position, with a terminal clade comprising Ambystomatidae, the three species of Dicamptadontidae, and Salamandridae. Thus, with the exception of three dicamptadontids that are close relatives of Ambystomatidae (and occupy a similar, unique larval habitat, discussed below), the two families in which pigment pattern development has been studied most extensively are sister taxa.

The phylogenetic relationships of taxa within Salamandridae and Ambystomatidae have been matters of contention. An abbreviated hypothesis of relationships is shown in Figure 7.6. Salamandridae comprise 15 genera and 53 recognized species in North America, Europe, and Asia (Titus and Larson, 1995). Studies of salamandrid phylogeny have employed a variety of characters, including anatomical traits, particularly of the hyobranchial apparatus (Wake and Özeti, 1969), courtship behavior (Salthe, 1967; Macgregor et al., 1990), and 12S and 16S rRNA sequences (Titus and Larson, 1995). A phylogenetic issue of debate has concerned the relationship between the genera of newts (e.g., *Pleurodeles*, *Taricha*, *Triturus*, and *Notophthalmus*) and the "true" salamanders (*Salamandra*, *Mertensiella*, *Chioglossa*) and specifically whether newts are monophyletic. A recent analysis (Titus and Larson, 1995) combining both molecular data (431 informative nucleotides) and morphological data (44 informative characters) indicates that newts are monophyletic and are a sister taxon to true salamanders. To date, studies of pigment pattern development and evolution in salamandrids have been concerned solely with newts; thus it will ultimately be necessary to examine the "true" salamanders to rigorously infer mechanisms within the family as a whole. Within the newts, most studies of pigment pattern evolution have focused on species in the genera *Taricha* and *Triturus*. Relationships within the North American genus *Taricha* (formerly classified at *Triturus*) are well resolved (Riemer, 1958). Relationships within *Triturus* remain somewhat unclear, although currently pigment pattern development in only a single species, *Tr. alpestris*, has received extensive study (see below).

All species within Ambystomatidae are now generally considered to comprise a single genus, *Ambystoma* [reviewed by Shaffer (1993); "*Amblystoma*" in some older literature] and are found exclusively in North America. Disagreements over relationships within *Ambystoma* have centered on the appropriateness of combining morphological and molecular data sets vs. analyzing data sets separately, the extent and meaning of character correlations within morphological and molecular data sets, as well as the methods used for coding and analyzing these various data. Much of the controversy has revolved around the subgenus *Linguaelapsus* identified initially by Tihen (1958) and supported by Kraus (1988) as monophyletic based on morphological criteria and comprising *A. annulatum*, *A. cingulatum*, *A. barbouri*, *A. texanum*, and *A. mabeei*. In contrast, Shaffer et al. (1991) found only equivocal support for this clade, as allozyme-based phylogenies (data from 26 loci) suggested a polyphyletic grouping of these species, whereas a combined molecular and morphological data set (the allozyme loci plus 32 informative anatomical characters) supported monophyly. However, a more recent reanalysis of these combined molecular and morphological data supports the monophyly of *Linguaelapsus* (Jones et al., 1993; Fig. 7.6). In the present context, these alternative hypotheses suggest different reconstructions for the evolution of vertical barring patterns (see below). A similar conflict between morphological and molecular data sets concerns *A. maculatum* (formerly *A. punctatum*) and *A. gracile*, which are only distant relatives in morphologically based hypotheses but sister taxa in hypotheses from allozymes alone as well as from combined allozyme and morphological data sets (Kraus, 1988; Shaffer et al., 1991; Jones et al., 1993). Once again, these alternative hypotheses have different implications for the evolution of vertical barring patterns. The relatively large genetic distances among ambystomatids (excluding the *A. tigrinum* complex, see below) suggest an ancient divergence, and, given the disparities among current data sets, additional analyses using sequence data from sufficiently slowly evolving loci may be needed to fully resolve these lineages (Shaffer et al., 1991).

Finally, I briefly consider relationships within the *A. tigrinum* complex of ambystomatid salamanders. This group comprises 5–7 subspecies of *A. tigrinum* distributed throughout the continental United States and southern Canada, as well as 15 species extending into central Mexico. Seven of the Mexican

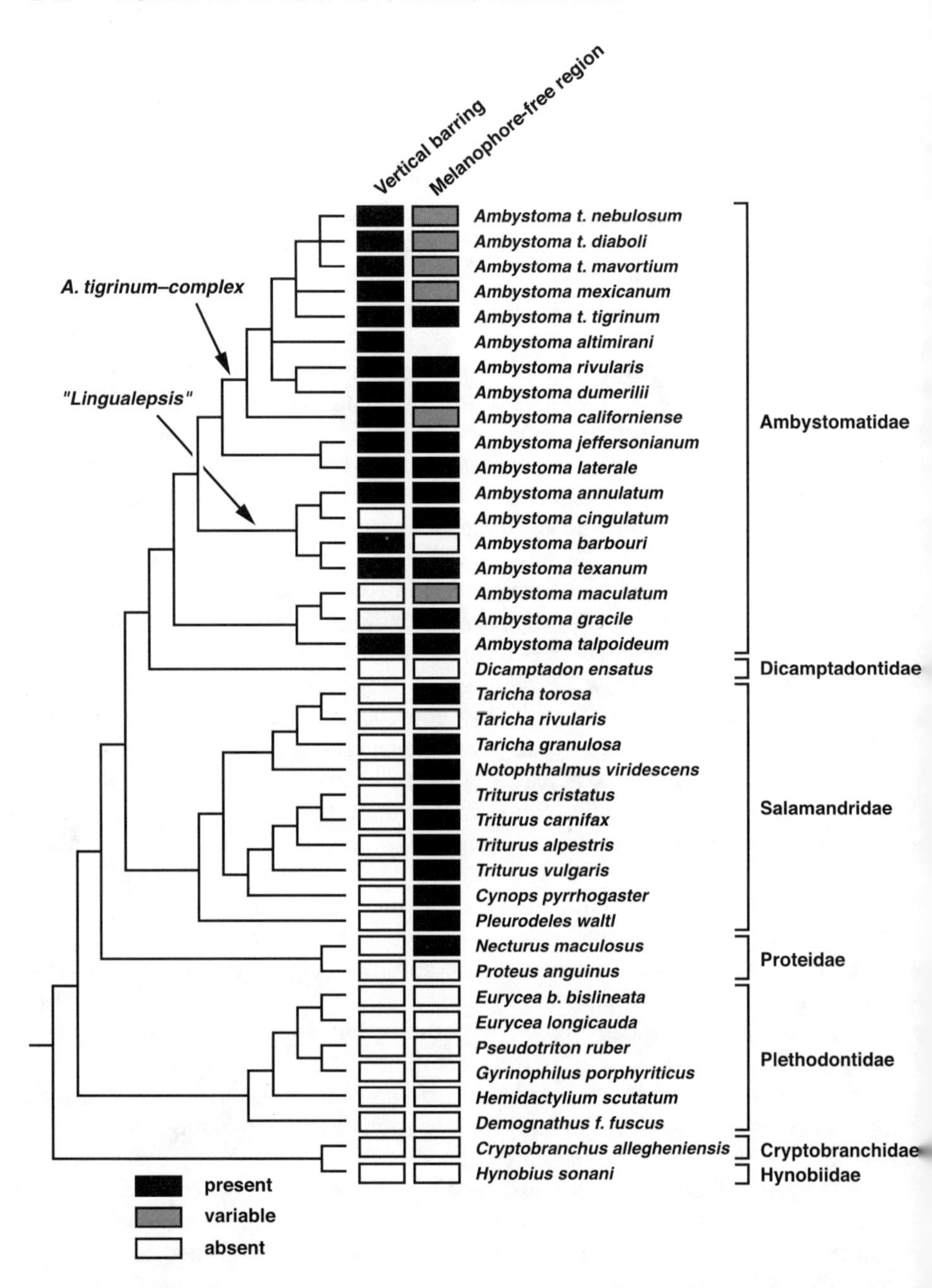

Figure 7.6. Phylogenetic distribution of salamanders in which early larval pigment patterns have been described. The phylogeny indicates the presence, absence, or variability of vertical bars and melanophore-free regions and is a composite of several hypotheses from different studies that used molecular or morphological characters other than pigment patterns. For interpretations of pigment pattern evolution under different hypotheses of phylogenetic relationships, see text. Vertical bars in salamanders are limited to Ambystomatidae and are most likely to have arisen in

taxa exhibit an obligately paedomorphic (nonmetamorphosing) life history, including the Mexican axolotl *A. mexicanum* (Shaffer, 1993). An additional six Mexican taxa exhibit facultative paedomorphosis (i.e., some individuals metamorphose, whereas others do not). Although recognized as closely related to one another, the precise phylogenetic relationships of these taxa have remained obscure. In an attempt to resolve relationships within the *A. tigrinum* complex, Shaffer (1984) used allozyme electrophoresis and Shaffer and McKnight (1996) analyzed 840 nucleotides of the mitochondrial D loop as well as a mitochondrial intron from 83 individuals representing 77 localities. Despite analyzing rapidly evolving sequences of mitochondrial DNA, these authors found little divergence among taxa (0–8.5%). Although eight reasonably well-defined clades were identified, they were unable to further resolve relationships among these lineages. It is likely that most of these taxa split from one another in the relatively recent past (0.02–5 million years; Shaffer and McKnight, 1996; Shaffer, personal communication), which is striking from the perspective of pigment pattern evolution, as many of these taxa have markedly different larval and adult pigment patterns and these differences seem to have evolved very rapidly.

Larval Pigment Patterns. I. Evolutionary Novelty and the Development of Vertical Bars

Vertical barring is a distinctive pigment pattern in several salamanders and is especially apparent when larvae are viewed from above. This pattern may serve a cryptic function by allowing larvae to blend in with a mottled substrate, although this notion has not been tested. A survey of several species based on direct observations and data from the literature is summarized in Figure 7.6, which maps the presence or absence of vertical barring onto a phylogeny that includes representatives from 7 of the 10 recognized salamander families. This survey reveals that vertical barring within salamanders is limited to Ambysto-

◄

the common ancestor of *Ambystoma*, with a subsequent loss of this pattern element in the lineages leading to *A. maculatum* and *A. gracile*, as well as *A. cingulatum*. Melanophore-free regions are found in both Ambystomatidae and Salamandridae but have been lost independently in taxa within these clades. A lateral line-dependent stripe-forming mechanism is inferred to be ancestral for Ambystomatidae and Salamandridae, whereas evolutionarily derived stripe-forming mechanisms are present in *T. torosa* (see text). Data for pigment patterns are from Twitty, 1936; Henry and Twitty, 1940; Bishop, 1941; Orton, 1942; Fox, 1955; Riemer, 1958; Brandon, 1961; Durand and Vandel, 1968; Brandon, 1972; Brandon and Altig, 1973; Kakegawa et al., 1989; Pfingsten and Downs, 1989; Epperlein and Löfberg, 1990; Liozner and Dettlaff, 1991; Olsson, 1993; Olsson, 1994; Collazo and Marks, 1994; Parichy, 1996a, 1996b. Phylogenies are from Wake and Özeti, 1969; Shaffer, 1984; Macgregor et al., 1990; Larson, 1991; Shaffer et al., 1991; Jones et al., 1993; Larson and Dimmick, 1993; Titus and Larson, 1995; Shaffer and McKnight, 1996.

matidae, although at least three species, *A. maculatum*, *A. gracile*, and *A. cingulatum*, lack vertical bars. This distribution of vertical barring suggests several equally parsimonious scenarios for the evolution of this pattern element. Vertical bars could have arisen independently in both *A. talpoideum* and the common ancestor of all other *Ambystoma*, excluding *A. maculatum* and *A. gracile*, and then have been lost in *A. cingulatum*. Consistent with the notion that barring might be independently derived multiple times within *Ambystoma* are the observations made in several species of frogs, which exhibit apparently similar barring patterns during the tadpole stage (Gosner and Black, 1957; Gaudin, 1965; Altig and Johnston, 1989). An alternative and probably more likely scenario is that vertical barring first arose in the common ancestor of all *Ambystoma* but was subsequently lost in the ancestor of *A. maculatum* and *A. gracile* and independently lost within *A. cingulatum*. In contrast, the morphologically based phylogeny of Kraus (1988) places *A. gracile* as the basal most ambystomatid and both *A. maculatum* and *A. cingulatum* deeply nested in different clades within *Ambystoma*, thereby raising the possibility that an absence of vertical barring in *A. gracile* (as in other salamander families) is plesiomorphic, with bars first arising in the ancestor to all other *Ambystoma* and then being lost independently in both *A. maculatum* and *A. cingulatum*. Under any of these phylogenetic hypotheses, however, vertical barring is a novelty for *Ambystoma* relative to other salamanders and has been either lost or gained in multiple lineages.

The developmental events of bar development were first described by Lehman (1954, 1957) for *A. mexicanum*. The photographic series in Figure 7.7 illustrates the development of vertical barring in *A. t. tigrinum*, which is similar to that seen in *A. mexicanum* and other species. At early stages, dispersing neural crest cells differentiate as melanophores that become relatively uniformly scattered over the flank. Meanwhile, xanthophores begin to differentiate in clusters within a premigratory cord of neural crest cells, immediately dorsal to

Figure 7.7. Image series reveal cellular events of pigment pattern development in *A. t. tigrinum*. Shown is a single *A. t. tigrinum* embryo during 110 h of pigment pattern development. Anterior is to the right. Images on the left are under brightfield illumination and reveal melanophores. Corresponding images on the right are under epifluorescent illumination and reveal the positions of xanthophores and the developing lateral line sensory system (see text). Vertical bar development is associated with the formation of aggregates of cells containing xanthophores in the premigratory neural crest (arrowheads, Fig. 7.7G), at stages when melanophores already have dispersed over the flank. Subsequent emigration of xanthophores from these aggregates and interactions between xanthophores and melanophores already on the flank generate the vertical barring pattern seen at hatching stages (bottom). Formation of a melanophore-free region occurs as initially uniformly dispersed melanophores retreat from the advancing lateral line primordium (arrow, Fig. 7.7H). Subsequently a distinctive stripe forms (large arrow, Fig. 7.7F), probably in response to persistent effects of the lateral line, interactions between melanophores and xanthophores, and passive movements of melanophores due to growth of the flank. Anterior is to the right. [From Parichy (1996a).]

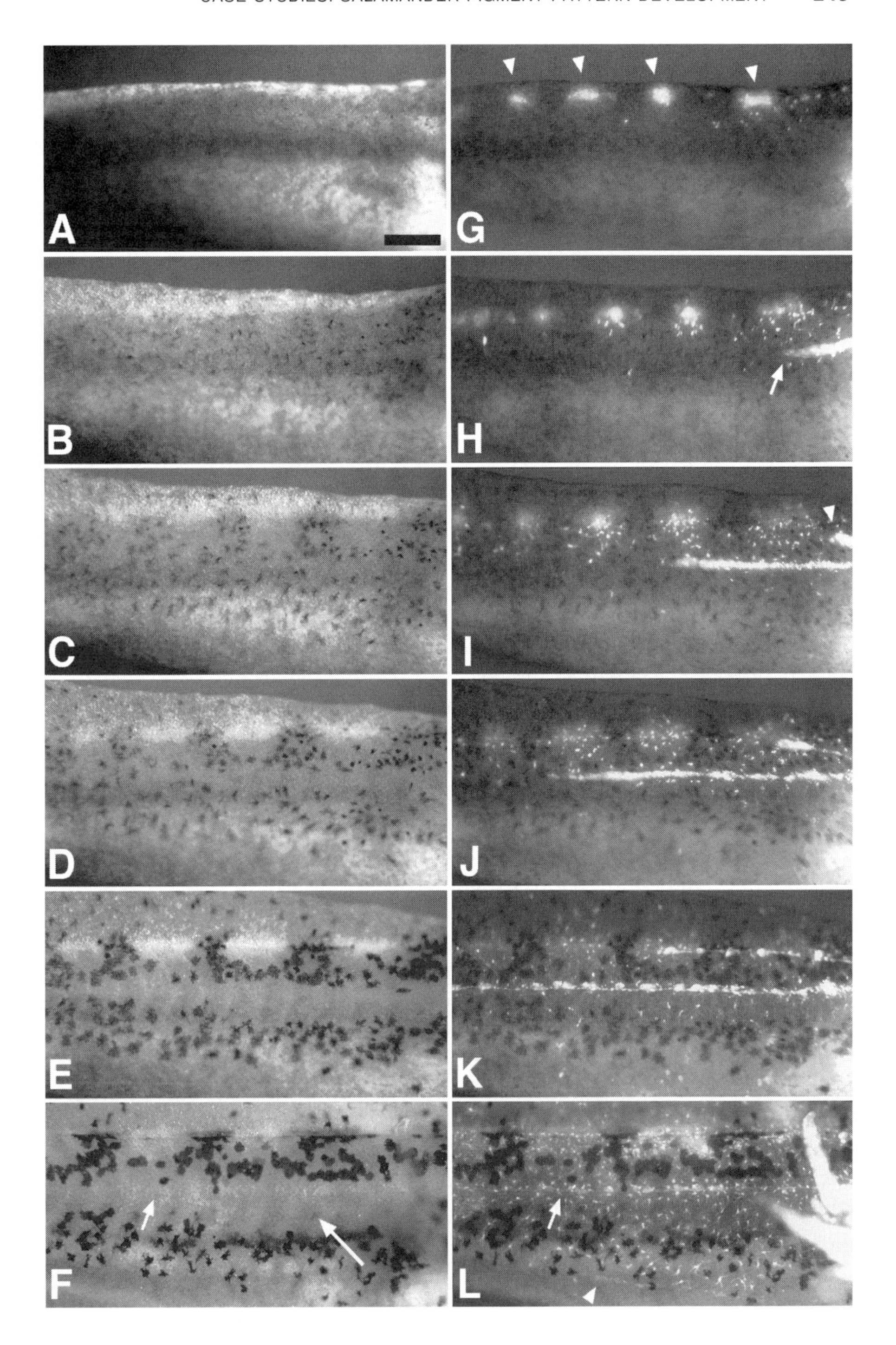
A
B
C
D
E
F
G
H
I
J
K
L

the neural tube. These clusters are not externally visible; however, with techniques that were not available to Lehman, xanthophores within the clusters can be readily identified using epifluorescent illumination even in living embryos. These xanthophore groups represent a prepattern that is subsequently translated into the definitive vertical barring pattern as xanthophores disperse from these groups and melanophores in their path recede short distances.

The mechanisms underlying vertical bar development are poorly understood. Early scanning electron microscopy revealed that xanthophores occur in morphologically distinctive groups of cells within the premigratory neural crest cord of *A. mexicanum*, and these "aggregates" of cells were further shown to contain not only autofluorescing xanthophore precursors but also cells containing organelles typical of both melanophores and xanthophores, the fates of which remain unresolved (Epperlein and Löfberg, 1984). Morphologically similar aggregates have since been identified in *A. t. tigrinum*, *A. barbouri*, *A. talpoideum*, and *A. annulatum* (Olsson and Löfberg, 1992; Olsson, 1993). Other taxa with vertical barring patterns presumably also develop aggregates at early stages, although this inference must be made cautiously because even superficially similar patterns sometimes depend on different underlying mechanisms (see below). In contrast, the two species lacking bars that have been examined, *Tr. alpestris* and *A. maculatum*, also lack cellular aggregates within the premigratory neural crest. Thus, an understanding of how cellular aggregates form should provide insights into the mechanisms underlying the phylogenetic appearance of vertical barring and what mechanisms could have changed to eliminate this evolutionary novelty in the lineages leading to *A. maculatum* and *A. gracile*, or *A. cingulatum*, or both.

Might an underlying heterochrony in cellular behaviors determine whether vertical bars develop? Perhaps. Lehman (1957) proposed a model in which the presence or absence of vertical barring depends on the relative timing of melanophore and xanthophore emigration from the neural crest. Specifically, Lehman's (1954, 1957) observations of *A. mexicanum* as well as subsequent studies of barred species (Epperlein and Löfberg, 1984; Olsson and Löfberg, 1992; Olsson, 1993; Parichy, 1996a) indicate that melanophores are visible on the flank before xanthophores, which remain initially at the level of the premigratory neural crest. In contrast, Lehman suggested that melanophores and xanthophores disperse simultaneously from the neural crest in species that fail to form vertical bars. Lehman was unaware of cellular aggregates in the premigratory neural crest of barred species, and his mechanism for translating delayed xanthophore migration into a barring pattern probably is incorrect. Nevertheless, with the finding of xanthophore aggregates in the neural crest, one may postulate either that delayed migration allows xanthophore precursors to form aggregates (e.g., by directed cell motility), or that xanthophores differentiate initially in the position of future aggregates, then proliferate until visible aggregates are formed. Consistent with a relationship between the timing of xanthophore dispersal and bar formation, Lehman (1957) found that outgrowths from explanted trunk neural crest of *A. mexicanum* embryos differentiate principally as melanophores after brief periods in culture, whereas increasing proportions of xanthophores were found after a longer time in culture. In contrast, outgrowths from *A. maculatum* neural crest explants produced relatively large proportions of xanthophores from the earliest times in culture. If the in vitro data reflect events in the embryo, these results could indicate a staggered emigration of melanophores and xanthophores in *A. mexicanum* (supported by direct observation) and simultaneous emigration of these cell types in *A. maculatum*.

An identical interpretation of the evolutionary change in cellular behavior was advanced independently in a more recent comparison of pigment pattern development between *A. maculatum* and barred species (Olsson, 1993). In this study, melanophores and xanthophores were observed already dispersed over the flank when each cell type was first visible, and this was inferred to indicate a simultaneous dispersal of these cells from the neural crest. It was further suggested (Olsson, 1993, 1994) that melanophores and xanthophores disperse simultaneously from the neural crest in *Tr. alpestris*, in which these cells are first visible already dispersed together over the flank (Epperlein and Claviez, 1982; Epperlein, 1982). Despite these inferences, no direct evidence has been gathered to indicate precisely when melanophores and xanthophores (or their precursors) actually leave the neural crest in species that lack bars. Thus, the relationship between timing of xanthophore emigration and subsequent patterning remains unresolved; Lehman's (1957) in vitro data remain the most compelling evidence for an underlying heterochrony in cellular behavior. Future studies using molecular markers for melanophore and xanthophore precursors (e.g., Wehrle-Haller and Weston, 1995; Reedy et al., 1998a; Parichy et al., 1999a, 2000a, 2000b) or other methods of lineage analysis (e.g., Raible and Eisen, 1994) may be able to resolve these issues. Clearly, multiple taxa with and without bars need to be examined to statistically support or refute the hypothesized phylogenetic relationship between differential timing of xanthophore emigration and vertical bar development.

An assessment of heterochronies in the dispersal of xanthophores and melanophores will provide insights into the evolutionary acquisition and loss of vertical barring patterns. So would other data. Grafting of neural folds containing prospective neural crest cells between species that either form bars or fail to form bars indicates that the ability to generate a barring pattern lies within either the neural crest or the epidermis, which also is found within the neural folds (DuShane, 1943; Lehman, 1957; Hirano and Shirai, 1984; Epperlein and Löfberg, 1990; Parichy, unpublished data). What this property reflects is unclear. As suggested above, an intrinsic difference in the timing of xanthophore dispersal may have a causal relationship in determining whether aggregates form. At the level of molecular mechanisms, several different heterochronic and nonheterochronic hypotheses could account for this behavior. For example, it has been suggested repeatedly (Epperlein and Löfberg, 1990, 1996; Epperlein et al., 1996) that the formation of cellular aggregates reflects a sorting out of differentially adhesive cell types (sensu Steinberg, 1970). No direct evidence has been gathered to assess the strength of adhesive interactions between xanthophores compared with interactions between melanophores or between melanophores and xanthophores. Nevertheless, such a differential adhesion hypothesis is plausible based on studies of other systems and the finding that, in *A. mexicanum* embryos denuded of epidermis (which prevents all neural crest dispersal), xanthophores occur in the middle of aggregates, adjacent to the neural tube, whereas melanophores occur on the periphery (Epperlein and Löfberg, 1990). Thus, it is conceivable that similar cell-cell adhesion molecules might be expressed by melanophores and xanthophores, with a delayed down-regulation of these molecules in xanthophores that prevents their early dispersal into the periphery. Such a finding would constitute a molecular heterochrony resulting in a corresponding heterochrony at the cellular level. Alternatively, different classes of pigment cells may express different adhesion molecules at the cell surface, thereby conferring different cellular behaviors. Consistent with this possibility, melanophores and xanthophores of an anuran and a teleost express different cell-cell adhesion molecules (Fukuzawa and Obika, 1995), and different neural crest derivatives in other species exhibit a broad spectrum of adhesion receptors (Erickson and Perris, 1993; Parichy, 1996b). An additional possibility is that very generalized differences in the dynamics of neural crest cell dispersal or the number of cells within the premigratory neural crest cord (i.e., not limited to melanophore or xanthophore lineages) results in aggregate formation and a microenvironment that is favorable for xanthophore differentiation. Indeed, morphologically similar aggregates form in the avian premigratory neural crest (which does not produce xanthophores) when neural tubes are explanted in culture, and prolonged proximity to the neural tube can affect the specification of these cells (Vogel and Weston, 1988; Rogers et al., 1990). Examination of the few salamander taxa that have been studied thus far suggests that aggregate formation correlates with the total number of neural crest cells in the premigratory cord (Detwiler, 1937; Epperlein and Löfberg, 1990; Olsson, 1993, 1994). Thus, several different nonheterochronic mechanisms could account for aggregate and vertical bar formation. It will be interesting to see how many mechanisms have changed phylogenetically to enable aggregation to occur and whether some or all of these mechanisms have reverted to a pleisiomorphic state during the evolutionary loss of a vertical barring pattern. In a broader context, identification of molecular and cellular bases for bar development in anurans should provide insights into whether this superficially similar phenotype has evolved via the same (parallel) or different (convergent) underlying mechanisms.

Larval Pigment Patterns. II. Evolutionary Elaboration of Developmental Mechanisms and the Ontogeny of Horizontal Stripes

The second distinctive pattern element in several species of salamanders is a horizontal melanophore-free region in the middle of the flank. Because this region is bordered by melanophores dorsally, and usually ventrally as well, the overall pattern consists of a horizontal stripe. Figure 7.6 maps the presence, absence, or variability of melanophore-free regions and reveals that this pattern element often is present in both ambystomatids and salamandrids. However, some taxa exhibit variable or indistinctive melanophore-free regions (e.g., *A. maculatum* and several taxa within the *A. tigrinum* complex), and other taxa within the clade [Ambystomatidae, Dicamptadontidae, Salamandridae, Proteidae (ADSP)] essentially lack melanophore-free regions (i.e., *A. barbouri*, *D. ensatus*, *T. rivularis*, *Proteus anguinus*). A strict parsimony reconstruction of the evolution of melanophore-free regions suggests that this pattern element may have arisen independently across families or just once in the common ancestor of ADSP. As detailed below, this latter hypothesis is reasonable, based on an understanding of how this pattern element develops. In turn, if melanophore-free regions first arose in the ancestor of ADSP, the phylogeny in Figure 7.6 further suggests that a melanophore-free region has been lost independently in *A. barbouri*, *D. ensatus*, *T. rivularis*, and *P. anguinus*, and variation in melanophore-free regions has arisen independently in *A. maculatum* and within the *A. tigrinum* complex. Thus, melanophore-free regions appear to be evolutionary labile pattern elements. Superficially similar melanophore-free regions are also found in tadpoles of the anuran *Xenopus laevis*, as well as in many fish larvae, although the mechanisms underlying the development of melanophore-free regions in these taxa appear to differ from those in salamanders (see below).

The developmental mechanisms underlying horizontal stripe development in salamander larvae have been studied for nearly three-quarters of a century. Early studies by Twitty and colleagues focused on stripe formation (and loss) in newts of the genus *Taricha* (Twitty and Bodenstein, 1939; Twitty, 1945; Twitty and Niu, 1948). Because more recent analyses indicate that some mechanisms in *Taricha* are likely to be derived evolutionarily, I first describe what appears to be a more general, plesiomorphic, mechanism for stripe development in several ambystomatids and other salamandrids. I use *A. t. tigrinum* as an example of this mode of stripe development. I then describe two types of evolutionary transition involving melanophore-free regions: an elaboration of developmental mechanisms in *T. torosa* and the loss of stripes in *T. rivularis* and *A. barbouri*. For both sorts of transition, I assess the explanatory and heuristic power of heterochronic vs. nonheterochronic models of developmental evolution. Finally, I describe briefly what appear to be convergent mechanisms of stripe development in larval *X. laevis* and teleosts.

Recent studies reveal that melanophore-free regions in several ambystomatids and salamandrids depend on interactions between pigment cells and the developing lateral lines (Parichy, 1996a, 1996c). In fishes and aquatic amphibian larvae, the lateral lines comprise a bilateral sensory system for detecting mechanical stimuli and function in orientation, feeding, and predator avoidance (Stone, 1933; Blaxter and Fuiman, 1990; Northcutt et al., 1994; Montgomery et al., 1997; Higgs and Fuiman, 1998). A potential role for the lateral lines in stripe development was suggested by the position of the midbody lateral line, which lies approximately in the middle of the melanophore-free region. As a first step in testing this possibility, the developing lateral lines were labeled with a fluorescent vital dye to distinguish them from surrounding tissues. Lateral lines arise from ectodermal placodes in the head of the embryo. Each placode that contributes to a lateral line on the trunk produces a missile-shaped primordium of cells that subsequently migrates along the inner epidermis toward the tail. Vital dye labeling in *A. t. tigrinum* showed that the midbody lateral line primordium enters onto the trunk after melanophores have already dispersed uniformly over the flank, and, as the primordium advances, melanophores in its path retreat dorsally and ventrally (Fig. 7.7). These movements inferred from static image series were subsequently confirmed by time-lapse analyses of lateral line and melanophore behaviors. To test whether the correlation between lateral line development and stripe development reflects a causal, epigenetic interaction between cell types, lateral line development was prevented by microsurgically ablating the lateral line placodes on one side of *A. t. tigrinum* embryos. Analyses of resulting melanophore distributions showed that melanophores more completely colonized sides without lateral lines and that most of this effect was attributable to changes in melanophore positions rather than overall changes in melanophore number (Fig. 7.8). Thus, lateral line development is essential for stripe development in *A. t. tigrinum* (a fuller treatment of the mechanisms underlying these interactions, as well as contributions of melanophore-xanthophore interactions and passive movements due to growth, can be found in Parichy, 1996c).

To further test the generality of lateral line-dependent mechanisms, lateral

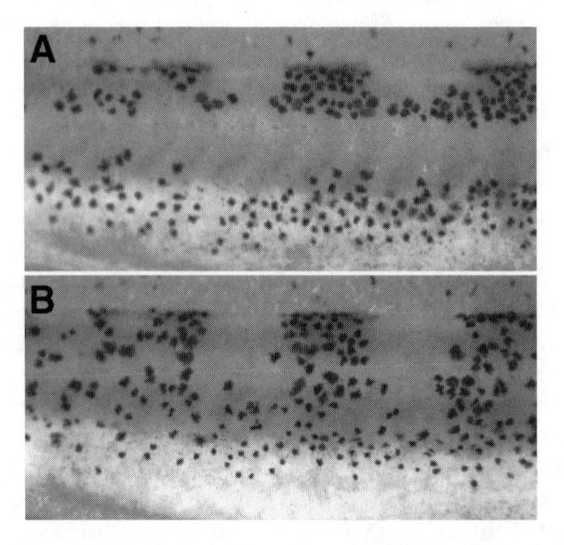

Figure 7.8. Lateral line ablation eliminates the melanophore-free region in *A. t. tigrinum*. Shown are opposite sides of a single larva. One image is reversed so that anterior is to the right in both A and B. A: a distinctive melanophore-free region is present on the lateral line-intact side. B: on the side without a lateral line, melanophores readily colonize the middle of the flank, implicating lateral line-dependent mechanisms in stripe development.

line ablations were repeated in several taxa chosen to represent a wide range of stripe morphologies, as well as distinct phylogenetic lineages within Ambystomatidae and Salamandridae: *A. mexicanum*, *A. barbouri*, *A. maculatum*, *A. talpoideum*, *T. torosa*, *T. rivularis*, *T. granulosa*, *Nothophthalmus viridescens*, and *Pleurodeles waltl* (Parichy, 1996a, and unpublished data). With the exception of *T. torosa* (see below), similar lateral line effects on melanophore distributions were found in each species, even though some lack distinctive melanophore-free regions. These findings strongly suggested that lateral line effects on melanophores are an ancestral way to make a stripe, and presumably arose in the common ancestor of Ambystomatidae and Salamandridae (and perhaps the ancestor to ADSP). In an ecological context, the presence of melanophore-lateral line interactions raises the possibility that horizontal stripes in many of these species might even be regarded as side effects of lateral line development. As such, they might or might not contribute to some overall function of the pigment pattern (like predator avoidance). This possibility highlights the importance of understanding the mechanistic bases for character development, as different interdependencies among traits can sometimes suggest very different hypotheses for the selective factors responsible for evolutionary changes in morphology. Here, stripes could have arisen as a direct response to selection on the pigment pattern or as a correlated response to selection on attributes of the lateral line sensory system (e.g., selection for more mechanoreceptive cells would result in a larger lateral line primordium with a bigger effect on the pigment pattern). These hypotheses await empirical challenge but would not have been apparent without an experimental approach to understanding trait development. Nor do these developmental-evolutionary insights depend on a heterochronic framework of trait evolution.

Stripes do not depend on lateral lines in all species, however. In *T. torosa* (Fig. 7.9), melanophores initially scatter uniformly over the flank and then segregate to form stripes, like *A. t. tigrinum*. However, in contrast to other ambystomatids and salamandrids tested, lateral line ablations do not perturb the horizontal stripe pattern in *T. torosa*. This in turn suggested two alternative hypotheses: (1) ancestral lateral line-dependent mechanisms could have been lost in *T. torosa* as different, lateral line-independent mechanisms evolved or (2) the ancestral mechanisms could have been retained in *T. torosa* as novel and redundant lateral line-independent mechanisms were layered over them. Interspecific chimeras were used to distinguish between these hypotheses (Parichy, 1996a). Specifically, if lateral line-dependent mechanisms have been retained in *T. torosa*, then its lateral lines should be competent to make a stripe, and its melanophores should be competent to respond to cues provided by lateral lines. If either or both of these conditions are not met, this would be consistent with the loss of the ancestral lateral line-dependent mechanisms. Transplanting *T. torosa* lateral line placodes in place of *A. t. tigrinum* lateral line placodes demonstrated that *T. torosa* lateral lines can generate a melanophore-free region. Moreover, transplanting *T. torosa* neural folds (containing prospective neural crest cells) in place of *A. t. tigrinum* neural folds, then ablating the lateral lines on one side of each chimera, showed that *T. torosa* melanophores more com-

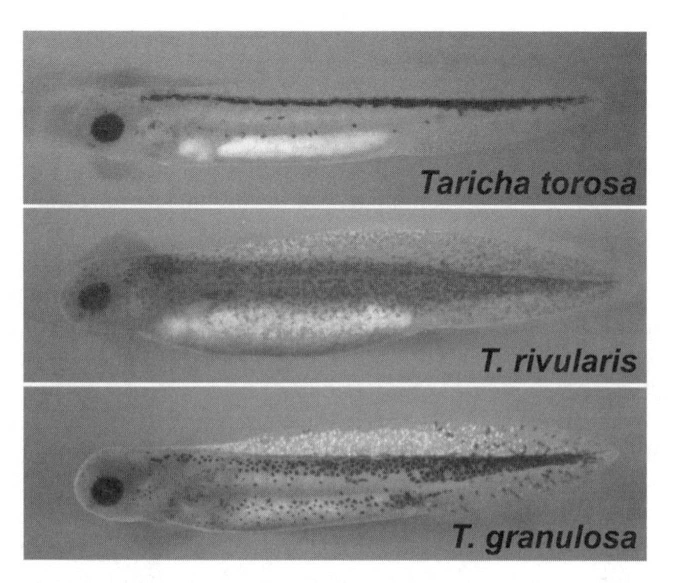

Figure 7.9. Pigment patterns of larvae in the genus *Taricha*. Top: *T. torosa*. Middle: *T. rivularis*. Bottom: *T. granulosa*.

pletely colonized sides without lateral lines. Thus, *T. torosa* melanophores are competent to respond to cues associated with the lateral lines. Together, these experiments suggest that *T. torosa* has retained ancestral lateral line-dependent mechanisms as novel and redundant lateral line-independent mechanisms for stripe formation evolved. Consistent with this idea, studies of taxa in which the lateral lines are required for stripe development show that the anti-adhesive extracellular matrix molecule tenascin is transiently deposited in a lateral line-dependent manner in the middle of the flank (i.e., in the developing melanophore-free region). These and other data suggest a model in which the future melanophore-free region arises in part because the local extracellular matrix is made unfavorable for melanophore localization (also see Epperlein and Löfberg, 1990; Parichy, 1996b, 1996c). Microsurgical manipulations of *T. torosa* show that transient lateral line-dependent deposition of tenascin still occurs, although it is not required for stripe development in this species (Parichy, 2001).

Can a heterochronic framework shed light on the phylogenetic transformation in patterning mechanisms in *T. torosa* relative to ambystomatids and other salamandrids? Image series and time-lapse videos of pigment pattern formation in *T. torosa* revealed, unlike in *A. t. tigrinum*, a melanophore-free region that begins to form several hours in advance of the migrating lateral line primordium. Conceivably, this timing difference could reflect a novel mechanism for stripe development. For example, in a series of elegant embryological manipulations, Twitty (1936, 1945) showed that, at stages before neural crest migration, the somites are able to specify the future site of stripe formation. This potential later shifts to the epidermis (Tucker and Erickson, 1986a). Together, these findings suggest the hypothesis that a cue for melanophore localization is produced by somitic mesoderm and associates with the epidermis or that mesoderm induces the epidermis to produce such a cue. Although several candidates for such a cue can be suggested (Tucker and Erickson, 1986b; Epperlein and Löfberg, 1990; Wehrle-Haller and Weston, 1995), its identity remains unknown. It also remains unclear whether this phylogenetic transformation reflects the novel production of a cue by the extracellular environment or a novel ability of melanophores to respond to a preexisting cue that they encounter during their migration. The ability of *T. torosa* melanophores (and those of other species) to generate stripes in species that normally lack distinctive melanophore-free regions has been interpreted to mean that an autonomous change in the melanophore lineage has occurred and stripe-forming cues are present in other species (Twitty, 1936, 1945; Epperlein and Löfberg, 1990). These conclusions are premature for two reasons. First, these experiments sometimes have relied on neural fold transplantation and thus cannot distinguish effects that are autonomous to the neural crest cells from effects that are associated with epidermis also contained within the neural folds. Second, the nature of the stripes that form in heterospecific hosts remains unresolved. These stripes have been inferred to be homologous to endogenous stripes in *T. torosa*. Nevertheless, our current understanding of stripe-forming mechanisms, the morphology of the stripes themselves (Twitty, 1945), and results of *T. torosa*-*A. t. tigrinum* chimeras (Parichy, 1996a) suggests that these stripes may actually reflect an ability of *T. torosa* cells to respond to the lateral lines of a heterospecific host, not a latent cue to which only *T. torosa* cells are competent to respond. Likewise, inferences regarding the role of somitic mesoderm in producing stripe-forming cues in *Tr. alpestris* (Epperlein and Löfberg, 1990) may need to be reexamined in light of lateral line-dependent mechanisms, since manipulations of epidermis and somites can perturb lateral line development (Smith et al., 1990), resulting in a potential confounding of effects due to different tissues.

A second heterochrony is evident in *T. torosa* during a terminal phase of stripe development, and this may also reflect a novel mechanism for melanophore patterning. Specifically, time-lapse videos and static image series indicate that *T. torosa* melanophores remain motile at a later stage than melanophores of other taxa (Parichy, 1996a). During this time, *T. torosa* melanophores migrate in a ventral-to-dorsal direction and eventually form persistent contacts with melanophores already at the site of dorsal stripe formation (Fig. 7.10). Similar contacts have been observed in *T. torosa* melanophores in culture, and a melanophore-autonomous reaggregation was suggested by Twitty (1945) to be responsible for stripe development in *T. torosa*. Nevertheless, the failure of *T. torosa* melanophores to form normal stripes in lateral line-ablated *A. t. tigrinum* argues that, although melanophore-melanophore interactions probably contribute to stripe development, they are not sufficient to ensure that stripes will form. The nature of this heterochrony in the persistence of melanophore motility remains unresolved. For example, differences in the expression

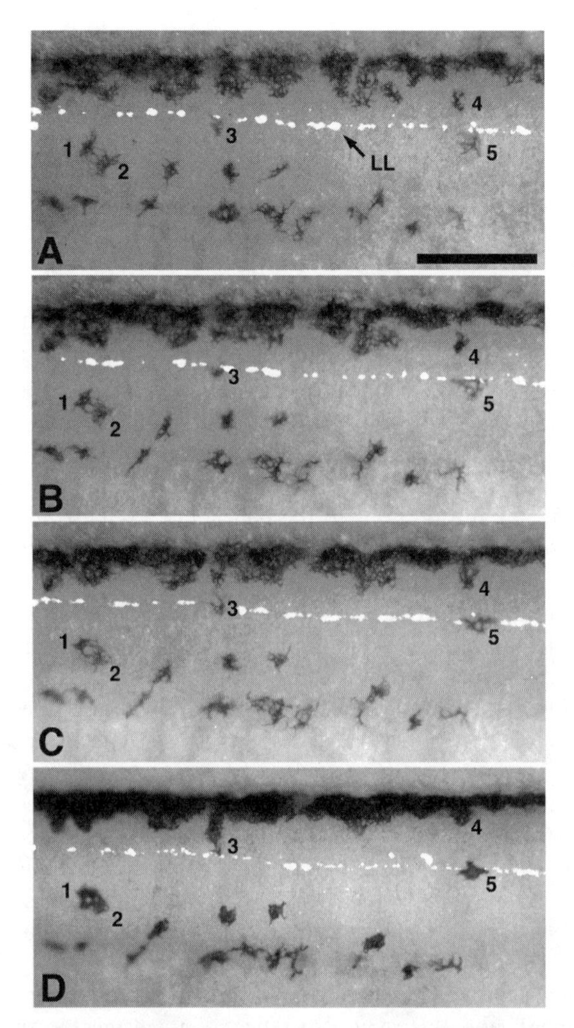

Figure 7.10. Melanophore behaviors differ in *T. torosa* compared with other taxa. Shown are representative images recorded over 48 h during the terminal stages of larval pigment pattern development. The position of the vital dye-labeled midbody lateral line is indicated in white (LL). Unlike other species, *T. torosa* melanophores remain motile during relatively late stages of development. For example, cell 3 travels in a ventral-to-dorsal direction, contacts melanophores further dorsally, and ultimately joins the dorsal stripe Anterior is to the right. [From Parichy (1996a).]

of cell adhesion molecules or other receptors (e.g., Parichy et al., 1999a) may enable these cells to form melanophore-melanophore contacts prohibited in other species because melanophores already have settled. Or, novel cues in the extracellular environment, or an absence of arresting extracellular matrix molecules (Parichy, 2001) may allow melanophores to wander and interact at those late stages. Additional experimental manipulations will be needed to distinguish between these possibilities. Finally, the relationships between heterochronic changes, an elaboration of developmental mechanisms, and possible selective factors contributing to the evolution of redundancy remain wholly unexplored but should be exciting to uncover.

The other interesting evolutionary transformation in melanophore-free regions is the independent loss of this pattern element in *T. rivularis* (Fig. 7.9), *A. barbouri*, and possibly *Dicamptadon* and *P. anguinus*. Intriguingly, larvae of *T. rivularis*, *A. barbouri*, and *Dicamptadon* all occur in rapidly moving streams (an autapomorphic life style at least for *T. rivularis* and *A. barbouri*), and each shares several other correlated traits in addition to the absence of a stripe: all lack or have reduced balancer organs, all have reduced dorsal fin folds, and all develop slowly from relatively large eggs. This convergent stream-type morphology suggests that selection may have occurred similarly across lineages to modify these various characters (Duellman and Trueb, 1986). Can a change in the rate or timing of cellular behaviors explain the loss of stripes in each of these species? Virtually nothing is known of pigment pattern development in dicamptadontids (Henry and Twitty, 1940). In *T. rivularis* and *A. barbouri*, however, the most noticeable feature of these pigment patterns is the apparent increase in total melanophore number relative to close relatives with distinctive melanophore-free regions. Compared with *T. torosa*, melanophores of *T. rivularis* differentiate more slowly and may stop their differentiation prematurely, at an early stage (Twitty, 1945). *Taricha rivularis* melanophores also have a higher overall rate of proliferation than *T. torosa* melanophores (Youngs, 1957). Conceivably, both of these factors could contribute to the development of a relatively uniform distribution of these cells (e.g., if less differentiated cells proliferate more extensively and essentially "fill" even unfavorable regions of the flank or if these cells never reach a state of differentiation at which they are competent to recognize cues for stripe formation). Although the mechanisms of pigment pattern development in *A. barbouri* are not known, the relatively uniform pattern in this species may have arisen via convergent developmental mechanisms: in *T. rivularis*, the increased number of melanophores is matched by a similarly increased number of xanthophores interspersed among them; in *A. barbouri*, the proportion of melanophores appears to have increased at the expense of xanthophores (Parichy, 1996b, unpublished data). The change in the proportions of different classes of pigment cells presumably also explains a less distinctive vertical barring pattern in *A. barbouri* and suggests a partial coupling between stripe development and vertical bar formation that may be present more generally (MacMillan, 1976; Parichy, 1996a). Thus, a variety of mechanisms may have contributed to the evolutionary loss of a horizontal stripe pattern. A heterochronic framework provides a context for understanding some of the relevant changes, but their mechanistic underpinnings remain unknown.

A final example of convergent stripe-forming mechanisms may be found in other species. For example, larvae of the anuran *X. laevis* exhibit a melanophore-free region that is superficially similar to that of salamanders. It arises by very different mechanisms. Specifically, neural crest cells in *X. laevis* exhibit an evolutionarily derived pattern of migration (Andres, 1963; Tucker, 1986; Collazo et al., 1993). In contrast to other species that have been studied, these cells typically do not travel between the somites and epidermis. Instead, melanophores that comprise a dorsal stripe represent cells that failed to disperse over the flank, whereas melanophores comprising a ventrolateral stripe arrive there only after emerging from more medial regions of the embryo. Thus, stripe formation in *X. laevis* appears to result from a heterotopy (McKinney and McNamara, 1991) in the pattern of neural crest cell migration. Likewise, stripes in larvae of several teleosts resemble stripes in salamanders, particularly *T. torosa*, and melanophores exhibit somewhat similar behaviors during stripe development (Orton, 1953; Raible and Eisen, 1994; Kimmel et al., 1995; Parichy et al., 1999a). The lateral lines probably are not required for stripe formation in teleost larvae (Parichy, unpublished data), but a role for somitic mesoderm is suggested by mutants in which somite patterning is perturbed, and melanophores localize ectopically in regions where stripes do not form normally (Kelsh et al., 1996). Lastly, analyses of zebrafish mutants are also beginning to reveal mechanisms of stripe development in adults of this species (Kawakani et al., 2000; Parichy et al., 2000b), though it is still too soon to know the extent to which these stripe-forming mechanisms are shared with amphibians.

Adult Pigment Patterns: Morphological Reorganization Across Metamorphosis

Salamanders exhibit a wide variety of adult pigment patterns (Fig. 7.11). At a cellular level, these patterns typically include melanophores and xanthophores, as well as a third class of neural crest-derived pigment cell, the silvery iridophore. To date, very little is known about how these patterns develop, and concerted efforts have not been made to either identify the component elements of adult pigment patterns or reconstruct the evolution of these elements phylogenetically. Early studies revealed roles for epidermis and somitic mesoderm in specifying where spots will form in some species (Lehman, 1953; Lehman and Youngs, 1959), and a few studies have documented the cellular changes that occur during pigment pattern metamorphosis (Stearner, 1946; Niu and Twitty, 1950). A more recent study showed that, in *A. t. tigrinum*, the adult pigment pattern arises largely independently of both the distribution of pigment cells comprising the early larval pattern and the larval lateral line sensory system (Parichy, 1998). Thus, larval and adult pigment patterns, at least in this species, appear to be relatively uncoupled from one another. Despite this paucity of information, the diversity of adult pigment patterns, particularly within the context of life history variation across species, suggests a fruitful area of inquiry.

An intriguing question related to the role of heterochrony in adult pigment pattern evolution is whether the mechanisms of adult pattern development are

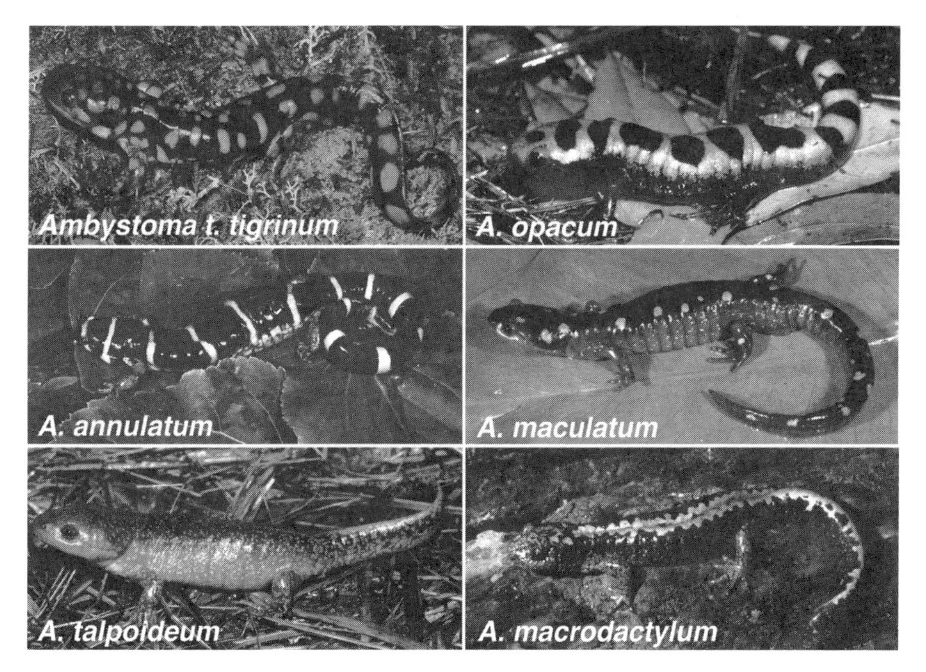

Figure 7.11. Adult pigment pattern diversity in salamanders. Shown are several species of adult *Ambystoma* that exhibit various spots, stripes, and mottled patterns.

integrated with the mechanisms controlling the metamorphosis of other characters (e.g., resorption of the gills and fin fold, as well as changes in the craniofacial skeleton and integument). More specifically, metamorphosis in these animals depends on tissue-level responses to circulating thyroid hormone (Gilbert et al., 1996; Shaffer and Voss, 1996; Brown, 1997). Although thyroid hormone can influence pigment cell behaviors (Reedy et al., 1998b), an in vivo role in pigment pattern metamorphosis remains unresolved. In support of a role for thyroid hormone in determining whether an adult pattern develops, when *A. mexicanum* are induced to metamorphose by exogenously administering thyroid hormone (or pituitary extract), these salamanders develop a pigment pattern of bright spots on a dark background, somewhat similar to that seen in other ambystomatids (Woronzowa, 1932; Brown, 1997). Whether this pigment pattern represents a true atavism remains unclear. In contrast, an independence of adult pigment pattern development from other aspects of somatic metamorphosis is suggested by several facultatively or obligately paedomorphic taxa within the *A. tigrinum* complex (e.g., *A. t. diaboli*, *A. ordinarium*) in which an apparently adult pattern is expressed despite the retention of an otherwise larval-like morphology (S. R. Voss, personal communication; Parichy, unpublished data). Different thresholds of sensitivity to thyroid hormone across tissue types might account for this apparent heterochrony in pigment pattern development, but this idea has yet to be tested.

IMPLICATIONS

Utility of a Heterochronic Framework in Understanding the Evolution of Pigment Pattern Development

The studies reviewed above suggest that evolutionary changes in pattern-forming mechanisms may be understandable, at least in part, as changes in the rate or timing of developmental events and cellular behaviors. Nevertheless, in no instance is there conclusive evidence that heterochrony provides the definitive change resulting in a phylogenetic transformation. In the vertical barring example, a heterochrony in cell migration associated with the presence or absence of this pattern element is suspected but not yet proven. In the elaboration of developmental mechanisms in *T. torosa*, at least two heterochronies are identifiable in melanophore behavior, but their causal significance for stripe development remains unclear. Thus, despite many years of study, we still do not have a definitive answer as to whether heterochronies at the cellular level are decisively involved in salamander pigment pattern evolution.

Does a heterochronic framework then provide a useful device for understanding the evolution of patterning mechanisms? Given the many ways in which cell behaviors and molecular mechanisms can change, it is unlikely that broad patterns of heterochrony will be identified as causally related to pigment pattern evolution. Even where heterochronies are suspected, they are not universally paedomorphic or peramorphic in nature. Indeed, the literature reviewed above made almost no use of the many and varied terms that have been introduced to describe different sorts of heterochronic change (Alberch et al., 1979; McKinney and McNamara, 1991; Reilly et al., 1997). These studies did not explicitly search for heterochronies either: where they are suspected, they were stumbled on in more general investigations of developmental and cellular mechanisms. These observations suggest that a heterochonic framework is not essential for understand evolutionary changes in developmental mechanisms. In some instances, it can be positively misleading (e.g., the early evacuation of *T. torosa* melanophores from the middle of the flank does not indicate an inability of the lateral lines to influence their behavior). Clearly, a heterochronic framework can be a useful heuristic device as it ensures consideration of various possibilities for rate and timing changes that otherwise might be overlooked. Nevertheless, investigations directed solely toward testing for heterochronies may provide relatively little insight on their own: heterochronic patterns are easy to identify, but establishing their significance for trait development or evolution requires a rigorous experimental approach to understanding the mechanisms themselves.

Future Directions for Studying Pigment Pattern Evolution and Development

Pigment patterns offer considerable opportunities to identify the mechanistic bases for the evolution of form in vertebrates. The studies reviewed above have

Figure 7.12. The zebrafish, *D. rerio*, undergoes a pigment pattern metamorphosis. An embryonic/early larval pigment pattern (top, 72 h, 3 mm) is transformed into an adult pigment pattern (bottom, >28 days, 4 cm).

focused principally on events at the cellular level, but an area for future studies is the identification of the molecular mechanisms that control these cellular behaviors. Although salamanders and other amphibians pose some difficulties for molecular and genetic analyses (but see Parichy et al., 1999b; Voss et al., 2001), investigations in teleosts may help to shed light on these matters. Specifically, the zebrafish, *D. rerio*, has become an increasingly important model organism for developmental and genetic studies on a variety of vertebrate traits, including pigment patterns. Similar to salamanders, pigment patterns in *D. rerio* comprise principally three classes of neural crest-derived pigment cell, and these fish undergo a pigment pattern metamorphosis in which a relatively simple early larval pattern is transformed into a more complex adult pigment pattern (Fig. 7.12; Haffter et al., 1996; Parichy et al., 2000a, 2000b). Also, as in salamanders, fishes in the genus *Danio* exhibit considerable adult pigment pattern variation (Fig. 7.13; Fang, 1997a, 1997b, 1998; McClure, 1999). Although the existing hypothesis of phylogenetic relationships (Meyer et al., 1995) is taxonomically depauperate, more comprehensive phylogenies will ultimately allow rigorous inferences regarding the frequency and polarity of changes in developmental mechanisms. Moreover, because one of these danios is a developmental genetic model organism, a host of resources are available that can be applied to understanding pigment pattern evolution within this genus. For example, a large number of mutants affecting pigment pattern development have been isolated in *D. rerio*, and several of these have pigment patterns that resemble the naturally occurring patterns of other danios, making them candidate genes for pigment pattern diversification (Fig. 7.13). Indeed, recent genetic analyses identify the *panther/fms* gene as a good candidate for contributing to an evolutionary loss of stripes in *D. albolineatus* (Parichy et al., 2000b; Parichy and Johnson, 2001). Some of these changes may also be interpretable within a heterochronic framework. For example, *endothelin receptor b1/rose* mutant *D. rerio* lack a

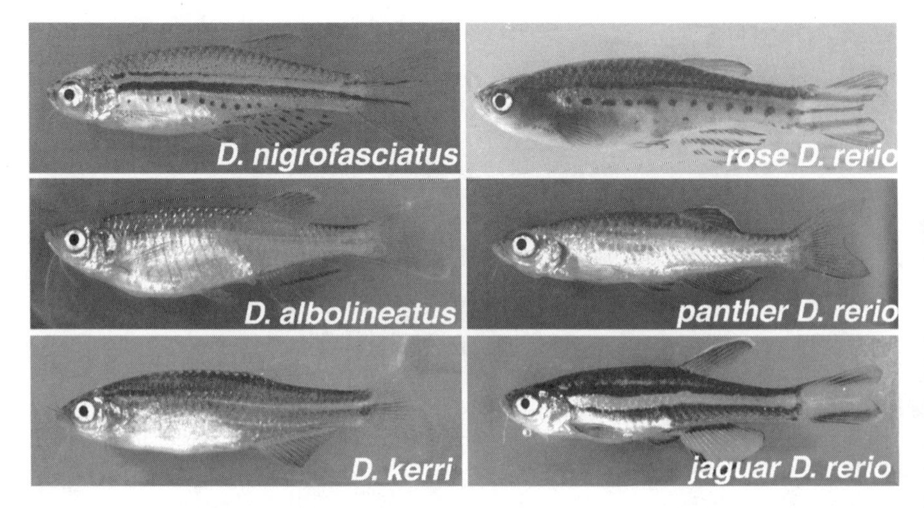

Figure 7.13. 1Fishes in the genus *Danio* offer an opportunity to investigate the genetic bases for pigment pattern evolution. Top left: zebrafish *D. rerio* has a relatively simple early larval pigment pattern comprised principally of melanophores and xanthophores. Top right: wild-type adult *D. rerio* exhibits a series of alternating light and dark horizontal stripes. Bottom left: adult pigment patterns of danios differ from *D. rerio* (although larval patterns are indistinguishable). Bottom right: pigment pattern mutants of *D. rerio* resemble other danios and identify candidate pigment pattern diversification genes.

late-developing subpopulation of adult melanophores (Johnson et al., 1995; Parichy et al., 2000a, 2000b). Although *endothelin receptor b1* probably is not responsible for the superficially similar phenotype in *D. nigrofasciatus* (Parichy and Johnson, 2001; Parichy, unpublished data), a similar cellular change may have occurred (essentially a local paedomorphosis, limited to the pigment pattern). Analyses of these mechanisms in danios should ultimately provide insights that can be applied to understanding pigment pattern development and evolution in salamanders, as well as other nonmodel organisms. Some of these insights may involve heterochrony.

REFERENCES

Alberch P, Gould SJ, Oster GF, Wake DB (1979): Size and shape in ontogeny and phylogeny. Paleobiology 5: 296–317.

Altig R, Johnston GF (1989): Guilds of anuran larvae: relationships among developmental modes, morphologies and habitats. Herpetol Monogr 3: 81–109.

Andres GM (1963): Eine experimentelle analyse der Entwicklung der larvalen Pigmentmuster von fünf Anurenarten. Zoologica (Stuttg) 40: 1–112.

Bard J (1990): *Morphogenesis: The Cellular and Molecular Processes of Developmental Anatomy.* Cambridge, UK: Cambridge University Press.

Baroffio A, Dupin E, Le Douarin NM (1988): Clone-forming ability and differentiation potential of migratory neural crest cells. Proc Natl Acad Sci USA 85: 5325–5329.

Baroffio A, Dupin E, Le Douarin NM (1991): Common precursors for neural and mesectodermal derivatives in the cephalic neural crest. Development 112: 301–305.

Bennett DC (1993): Genetics, development, and malignancy of melanocytes. Int Rev Cytol 146: 191–260.

Bishop SC (1941): The salamanders of New York. NY State Mus Bull 324: 1–365.

Blaxter JHS, Fuiman LA (1990): The role of the sensory systems of herring larvae in evading predatory fishes. J Mar Biol Assoc UK 70: 413–427.

Brandon RA (1961): A comparison of the larvae of five northeastern species of Ambystoma (Amphibia, Caudata). Copeia 961: 377–383.

Brandon RA (1972): Hybridization between the Mexican salamanders *Ambystoma dumerilii* and *Ambystoma mexicanum* under laboratory conditions. Herpetologica 28: 199–207.

Brandon RA, Altig RG (1973): Eggs and small larvae of two species of *Rhyacosiredon*. Herpetologica 29: 329–351.

Bronner-Fraser M, Fraser SE (1988): Cell lineage analysis reveals multipotency of some avian neural crest cells. Nature 335: 161–164.

Brown DD (1997): The role of thyroid hormone in zebrafish and axolotl development. Proc Natl Acad Sci USA 94: 13011–13016.

Carroll SB, Grenier JK, Weatherbee SD (2001): From DNA to diversity: molecular genetics and the evolution of animal design. Blackwell Science, Malden MA.

Chibon P (1967): Marquage nucléaire par la thymidine tritiée des dérivés de la crête neurale chez l'amphibien urodèle *Pleurodeles waltlii* Michah. J Embryol Exp Morphol 18: 343–358.

Collazo A, Bronner-Fraser M, Fraser SE (1993): Vital dye labelling of *Xenopus laevis* trunk neural crest reveals multipotency and novel pathways of migration. Development 118: 363–376.

Collazo A, Marks SB (1994): Development of *Gyrinophilus porphyriticus*: identification of the ancestral developmental pattern in the salamander family Plethodontidae. J Exp Zool 268: 239–258.

Couly GF, Coltey PM, Le Douarin NM (1993): The triple origin of skull in higher vertebrates: a study in quail-chick chimeras. Development 117: 409–429.

Detwiler SR (1937): Observations upon the migration of neural crest cells, and upon the development of the spinal ganglia and vertebral arches in *Amblystoma*. Am J Anat 61: 63–94.

Duellman WE, Trueb L (1986): *Biology of Amphibians*. New York: McGraw-Hill.

Durand JP, Vandel A (1968): Proteus: an evolutionary relic. Sci J Feb 1968: 44–49.

DuShane GP (1934): The origin of pigment cells in Amphibia. Science 80: 620–621.

DuShane GP (1935): An experimental study of the origin of pigment cells in Amphibia. J Exp Zool 72: 1–31.

DuShane GP (1939): The role of embryonic ectoderm and mesoderm in pigment production in Amphibia. J Exp Zool 82: 193–217.

DuShane GP (1943): The embryology of vertebrate pigment cells. Part I: Amphibia. Q Rev Biol 18: 109–127.

Endler JA (1978): A predator's view of animal color patterns. Evol Biol 11: 319–364.

Epperlein HH (1982): Different distribution of melanophores and xanthophores in early tailbud and larval stages of *Triturus alpestris*. Wilhelm Roux's Arch 191: 19–27.

Epperlein HH, Claviez M (1982): Changes in the distribution of melanophores and xanthophores in *Triturus alpestris* embryos during their transition from the uniform to banded pattern. Wilhelm Roux's Arch 1982: 5–18.

Epperlein HH, Löfberg J (1984): Xanthophores in chromatophore groups of the premigratory neural crest initiate the pigment pattern of the axolotl larva. Wilhelm Roux's Arch Dev Biol 193: 357–369.

Epperlein H-H, Löfberg J (1990): The development of the larval pigment patterns in *Triturus alpestris* and *Ambystoma mexicanum*. Adv Anat Embryol Cell Biol 118.

Epperlein H-H, Löfberg J (1996): What insights into the phenomena of cell fate determination and cell migration has the study of the urodele neural crest provided? Int J Dev Biol 40: 695–707.

Epperlein HH, Löfberg J, Olsson L (1996): Neural crest cell migration and pigment pattern formation in urodele amphibians. Int J Dev Biol 40: 229–238.

Epperlein H-H, Ziegler I, Perris R (1988): Identification of pigment cells during early amphibian development (*Triturus alpestris*, *Ambystoma mexicanum*). Cell Tissue Res 253: 493–505.

Erickson CA (1993): From the crest to the periphery: control of pigment cell migration and lineage segregation. Pigm Cell Res 6: 336–347.

Erickson CA, Goins TL (1995): Avian neural crest cells can migrate in the dorsolateral path only if they are specified as melanocytes. Development 121: 915–924.

Erickson CA, Perris R (1993): The role of cell-cell and cell-matrix interactions in the morphogenesis of the neural crest. Dev Biol 159: 60–74.

Erickson CA, Reedy MV (1999): Neural crest development: the interplay between morphogenesis and cell differentiation. Curr Top Dev Biol 40: 177–209.

Fang F (1997a): *Danio maetaengensis*, a new species of cyprinid fish from northern Thailand. Ichthyol Explor Freshwaters 8: 41–48.

Fang F (1997b): Redescription of *Danio kakhienensis*, a poorly known cyprinid fish from the Irrawaddy basin. Ichthyol Explor Freshwaters 7: 289–298.

Fang F (1998): *Danio kyathit*, a new species of cyprinid fish from Myitkyina, northern Myanmar. Ichthyol Explor Freshwaters 8: 273–280.

Fox H (1955): Early development of two subspecies of the salamander *Triturus cristatus*. Copeia 1955: 131–133.

Frank E, Sanes JR (1991): Lineage of neurons and glia in chick dorsal root ganglia: analysis in vivo with a recombinant retrovirus. Development 111: 895–908.

Fraser SE, Bronner-Fraser M (1991): Migrating neural crest cells in the trunk of the avian embryo are multipotent. Development 112: 913–920.

Frost SK, Epp LG, Robinson SJ (1984): The pigmentary system of developing axolotls. I. A biochemical and structural analysis of chromatophores in wild-type axolotls. J Embryol Exp Morphol 81: 105–125.

Fukuzawa T, Obika M (1995): N-CAM and N-cadherin are specifically expressed in xanthophores, but not in the other types of pigment cells, melanophores, and iridophores. Pigm Cell Res 8: 1–9.

Gans C, Northcutt RG (1983): Neural crest and the origin of vertebrates: a new head. Science 220: 268–274.

Gaudin AJ (1965): Larval development of the tree frogs *Hyla regilla* and *Hyla calirniae.* Herpetologica 21: 117–130.

Gerhart J, Kirschner M (1997): *Cells, Embryos, and Evolution: Toward a Cellular and Developmental Understanding of Phenotypic Variation and Evolutionary Adaptability.* Malden, MA: Blackwell Science.

Gilbert LI, Tata JR, Atkinson BG (eds.) (1996): Metamorphosis: Postembryonic reprogramming of gene expression in amphibian and insect cells. New York: Academic Press.

Gosner KL, Black IH (1957): Larval development in New Jersey Hylidae. Copeia 1957: 31–36.

Gould SJ (1977): *Ontogeny and Phylogeny.* Cambridge, MA: Harvard University Press.

Groves AK, Bronner-Fraser M (1999): Neural crest diversification. Curr Top Dev Biol 43: 221–258.

Häcker V (1907): Uber Mendelschen Vererbung bei Axolotln. Zool Anz 31: 99–102.

Haffter P, Odenthal J, Mullins MC, Lin S, Farrell MJ, Vogelsang E, Haas F, Brand M, van Eeden FJM, Furutani-Seiki M, Granato M, Hammerschmidt M, Heisenberg C-P, Jiang Y-J, Kane DA, Kelsh RN, Hopkins N, Nüsslein-Volhard C (1996): Mutations affecting pigmentation and shape of the adult zebrafish. Dev Genes Evol 206: 260–276.

Hall BK (1999): *The Neural Crest in Development and Evolution.* New York: Springer-Verlag.

Hall BK, Hörstadius S (1988): *The Neural Crest.* New York: Oxford University Press.

Hanken J (1999): Larvae in amphibian development and evolution. In: Hall BK and Wake MH (eds), *The Origin and Evolution of Larval Forms.* New York: Academic Press, p. 61–108.

Hanken JH, Wake DB (1993): Miniaturization of body size: organismal conssequences and evolutionary significance. Annu Rev Ecol Syst 24: 501–519.

Henry WV, Twitty VC (1940): Contributions to the life histories of *Dicamptadon ensatus* and *Ambystoma gracile*. Copeia 1940: 247–250.

Higgs DM, Fuiman LA (1998): Associations between behavioural ontogeny and habitat change in clupeoid larvae. J Mar Biol Assoc UK 78: 1281–1294.

Hirano S, Shirai T (1984): Morphogenetic studies on the neural crest of *Hynobius* larvae using vital staining and India ink labelling methods. Arch Histol Japan 47: 57–70.

Horigome N, Myojin M, Ueki T, Hirano S, Aizawa S, Kuratani S (1999): Development of cephalic neural crest cells in embryos of *Lampetra japonica*, with special reference to the evolution of the jaw. Dev Biol 207: 287–308.

Houde AE (1997): *Sex, Color, and Mate Choice in Guppies.* Princeton, NJ: Princeton University Press.

Huang H, Marsh-Armstrong N, Brown DD (1999): Metamorphosis is inhibited in trangenic *Xenopus laevis* tadpoles that overexpress type III deiodinase. Proc Natl Acad Sci USA 96: 962–967.

Johnson SL, Africa D, Walker C, Weston JA (1995): Genetic control of adult pigment stripe development in zebrafish. Dev Biol 167: 27–33.

Jones TR, Kluge AG, Wolf AJ (1993): When theories and methodologies clash: a phylogenetic reanalysis of the North American ambystomatid salamanders (Caudata: Ambystomatidae). Syst Biol 42: 92–102.

Kakegawa M, Iizuka K, Kuzumi S (1989): Morphology of egg sacs and larvae just after hatching in *Hynobius sonani* and *H. formosanus* from Taiwan, with an analysis of skeletal muscle protein compositions. Curr Herpetol East Asia: 147–155.

Keller RE, Löfberg J, Spieth J (1982): Neural crest cell behavior in white and dark embryos of *Ambystoma mexicanum*: epidermal inhibition of pigment cell migration in the white axolotl. Dev Biol 89: 179–195.

Keller RE, Spieth J (1984): Neural crest cell behavior in white and dark larvae of *Ambystoma mexicanum*: time-lapse cinemicrographic analysis of pigment cell movement in vivo and in culture. J Exp Zool 229: 109–126.

Kelsh RN, Brand M, Jiang Y-J, Heisenberg C-P, Lin S, Haffter P, Odenthal J, Mullins MC, van Eeden FJM, Furutani-Seiki M, Granato M, Hammerschmidt M, Kane DA, Warga RM, Beuchle D, Vogelsang L, Nusslein-Volhard C (1996): Zebrafish pigmentation mutations and the processes of neural crest development. Development 123: 369–389.

Kelsh RN, Eisen JS (1999): The zebrafish *colourless* gene regulates development of ectomesenchymal neural crest derivatives. Development 127: 515–525.

Kimmel CB, Ballard WW, Kimmel SR, Ullmann B, Schilling TF (1995): Stages of embryonic development of the zebrafish. Dev Dyn 203: 253–310.

Kraus F (1988): An empirical evaluation of the use of the ontogeny polarization criterion in phylogenetic inference. Syst Zool 37: 106–141.

LaBonne C, Bronner-Fraser M (1999): Neural crest induction in *Xenopus*: evidence for a two signal model. Development 125: 2403–2414.

Larson A (1991): A molecular perspective on the evolutionary relationships of the salamander families. Evol Biol 25: 211–278.

Larson A, Dimmick WM (1993): Phylogenetic relationships of the salamander families: an analysis of congruence among morphological and molecular characters. Herpetol Monogr 6: 77–93.

Le Douarin N (1982): *The Neural Crest.* Cambridge, UK: Cambridge University Press.

Lehman HE (1953): Analysis of the development of pigment patterns in larval salamanders, with special reference to the influence of epidermis and mesoderm. J Exp Zool 124: 571–617.

Lehman HE (1954): An experimental study of the "barred" pigment pattern in Mexican axolotl larvae. J Elisha Mitchell Sci Soc 70: 218–221.

Lehman HE (1957): The developmental mechanics of pigment pattern formation in the black axolotl, *Amblystoma mexicanum.* J Exp Zool 135: 355–386.

Lehman HE, Youngs LM (1959): Extrinsic and intrinsic factors influencing amphibian pigment pattern formation. In: Gordon M (ed), *Pigment Cell Biology.* New York: Academic Press, p. 1–36.

Liozner LD, Dettlaff TA (1991): The newts *Triturus vulgaris* and *Triturus cristatus.* In: Dettlaff TA, Vassetzky SG (eds), *Animal Species for Developmental Studies.* New York: Plenum Publishing Company.

Lister JA, Robertson CP, Lepage T, Johnson SL, Raible DW (1999): *nacre* encodes a zebrafish microphthamlia-related protein that regulates neural crest-derived pigment cell fate. Development 126: 3757–3767.

Löfberg J, Ahlfors K, Fällström C (1980): Neural crest cell migration in relation to extracellular matrix organization in the embryonic axolotl trunk. Dev Biol 75: 148–167.

Löfberg J, Perris R, Epperlein HH (1989): Timing in the regulation of neural crest cell migration: retarded "maturation" of regional extracellular matrix inhibits pigment cell migration in embryos of the white axolotl mutant. Dev Biol 131: 168–181.

Loring JF, Erickson CA (1987): Neural crest cell migratory pathways in the trunk of the chick embryo. Dev Biol 121: 220–236.

Macgregor HC, Sessions SK, Arntzen JW (1990): An integrative analysis of phylogenetic relationships among newts of the genus *Triturus* (gamily Salamandridae), using comparative biochemistry, cytogenetics and reproductive interactions. J Evol Biol 3: 329–373.

MacMillan GJ (1976): Melanoblast-tissue interactions and the development of pigment pattern in *Xenopus* larvae. J Embryol Exp Morphol 35: 463–484.

McClure M (1999): Development and evolution of melanophore patterns in fishes of the genus *Danio* (Teleostei: Cyprindae). J Morphol 241: 83–105.

McKinney ML, McNamara KJ (1991): *Heterochrony: the Evolution of Ontogeny.* New York: Plenum Press.

Meyer A, Ritchie PA, Witte K-E (1995): Predicting developmental processes from evolutionary patterns: a molecular phylogeny of the zebrafish (*Danio rerio*) and its relatives. Phil Trans R Soc Lond B 349: 103–111.

Montgomery JC, Baker CF, Carton AG (1997): The lateral line can mediate rheotaxis in fish. Nature 389: 960–963.

Moury JD, Jacobson AG (1990): The origins of neural crest cells in the axolotl. Dev Biol 141: 243–253.

Nguyen VH, Trout J, Conner SA, Andermann P, Weinberg E, Mullins MC (2000): Dorsal and intermediate neuronal cells types of the spinal cord are established by a BMP signaling pathway. Development 127: 1209–1220.

Niu MC, Twitty VC (1950): The origin of epidermal melanophores during metamorphosis in *Triturus torosus.* J Exp Zool 113: 633–648.

Northcutt RG, Catania KC, Criley BB (1994): Development of lateral line organs in the axolotl. J Comp Neurol 340: 480–514.

Olsson L (1993): Pigment pattern formation in the larval salamander *Ambystoma maculatum.* J Morphol 215: 151–163.

Olsson L (1994): Pigment pattern formation in larval ambystomatid salamanders: *Ambystoma talpoideum*, *Ambystoma barbouri*, and *Ambystoma annulatum.* J Morphol 220: 123–138.

Olsson L, Hanken J (1996): Cranial neural-crest migration and chondrogenic fate in the oriental fire-bellied toad *Bombina orientalis*: defining the ancestral pattern of head development in anuran amphibians. J Morphol 229: 105–120.

Olsson L, Löfberg J (1992): Pigment pattern formation in larval ambystomatid salamanders: *Ambystoma tigrinum tigrinum.* J Morphol 211: 73–85.

Orton G (1942): Notes on the larvae of certain species of *Ambystoma.* Copeia 1942: 170–172.

Orton GL (1953): Development and migration of pigment cells in some teleost fishes. J Morphol 93: 69–99.

Parichy DM (1996a): Pigment patterns of larval salamanders (Ambystomatidae, Salamandridae): the role of the lateral line sensory system and the evolution of pattern-forming mechanisms. Dev Biol 175: 265–282.

Parichy DM (1996b): Salamander pigment patterns: how can they be used to study developmental mechanisms and their evolutionary transformation? Int J Dev Biol 40: 871–884.

Parichy DM (1996c): When neural crest and placodes collide: interactions between melanophores and the lateral lines that generate stripes in the salamander *Ambystoma tigrinum tigrinum* (Ambystomatidae). Dev Biol 175: 282–300.

Parichy DM (1998): Experimental analysis of character coupling across a complex life cycle: pigment pattern metamorphosis in the tiger salamander, *Ambystoma tigrinum.* J Morphol 237: 53–67.

Parichy DM (2001): Homology and evolutionary novelty in the deployment of extracellular matrix molecules during pigment pattern formation in salamanders (*Taricha, Ambystoma*). Mol Dev Evol (J Exp Zool) 291: 13–24.

Parichy DM, Johnson SL (2001): Zebrafish hybrids suggest genetic mechanisms for pigment pattern diversification in *Danio*. Dev Genes Evol, in press.

Parichy DM, Mellgren EM, Rawls JF, Lopes SS, Kelsh RN, Johnson SL (2000a): Mutational analysis of *endothelin receptor b1* (*rose*) during neural crest and pigment pattern development in the zebrafish *Danio rerio*. Dev Biol 227: 294–306.

Parichy DM, Ransom DG, Paw B, Zon LI, Johnson SL (2000b): An orthologue of the *kit*-related gene *fms* is required for development of neural crest-derived xanthophores and a subpopulation of adult melanocytes in the zebrafish, *Danio rerio*. Development 127: 3031–3044.

Parichy DM, Rawls JF, Pratt SJ, Whitfield TT, Johnson SL (1999a): Zebrafish *sparse* corresponds to an orthologue of c-*kit* and is required for the morphogenesis of a subpopulation of melanocytes, but is not essential for hematopoiesis or primordial germ cell development. Development 126: 3425–3436.

Parichy DM, Stigson S, Voss SR (1999b): Genetic analysis of *Steel* and the PG-M/versican-encoding gene *AxPG* as candidate genes for the *white* (*d*) pigmentation mutant in the salamander *Ambystoma mexicanum.* Dev Genes Evol 209: 349–356.

Perris R, Löfberg J, Fällström C, von Boxberg Y, Olsson L, Newgreen DF (1990): Structural and compositional divergencies in the extracellular matrix encountered by neural crest cells in the white mutant axolotl embryo. Development 109: 533–551.

Perris R, Perissinotto D (2000): Role of the extracellular matrix during neural crest cell migration. Mech Dev 95: 3–21.

Pfingsten RA, Downs FL (1989): Salamanders of Ohio. Bull Ohio Biol Surv 7: 1–315.

Raff RA (1996): *The Shape of Life.* Chicago, IL: University of Chicago Press.

Raible DW, Eisen JS (1994): Restriction of neural crest cell fate in the trunk of the embryonic zebrafish. Development 120: 495–503.

Raible DW, Wood A, Hodsdon W, Henion PD, Weston JA, Eisen JS (1992): Segregation and early dispersal of neural crest cells in the embryonic zebrafish. Dev Dyn 195: 29–42.

Reedy MV, Faraco CD, Erickson CA (1998a): The delayed entry of thoracic neural crest cells into the dorsolateral path is a consequence of the late emigration of melanogenic neural crest cells from the neural tube. Dev Biol 200: 234–246.

Reedy MV, Parichy DM, Erickson CA, Mason KA, Frost-Mason SK (1998b): The regulation of melanoblast migration and differentiation. In: Nordland JJ, Boissy RE, Hearing VJ, King RA, Ortonne JP (eds), *The Pigmentary System: Physiology and Pathophysiology*. New York: Oxford University Press, chapt. 5, p. 75–95.

Reilly SM, Wiley EO, Meinhardt DJ (1997): An integrative approach to heterochrony: the distinction between interspecific and intraspecific phenomena. Biol J Linn Soc 60: 119–143.

Riemer WJ (1958): Variation and systematic relationships within the salamander genus *Taricha*. Univ Calif Publ Zool 56: 301–390.

Rogers SL, Bernard L, Weston JA (1990): Substratum effects on cell dispersal, morphology, and differentiation in cultures of avian neural crest cells. Dev Biol 141: 173–182.

Roopnarine, PD (2001): A history of diversification, extinction, and invasion in tropical America as derived from species-level phylogenies of chionine genera (Family Veneridae). Journal of Paleontology 75: 644–658.

Sadaghiani B, Thiébaud CH (1987): Neural crest development in the *Xenopus laevis* embryo, studied by interspecific transplantation and scanning electron microscopy. Dev Biol 124: 91–110.

Sadaghiani B, Vielkind JR (1990): Distribution and migration pathways of HNK-1-immunoreactive neural crest cells in teleost fish embryos. Development 110: 197–209.

Salthe SN (1967): Courtship patterns and phylogeny of the urodeles. Copeia 1967: 100–117.

Sela-Donenfeld D, Kalcheim C (2000): Inhibition of noggin expression in the dorsal neural tube: a mechanism for coordinating the timing of neural crest migration. Development 127: 4845–4854.

Shaffer HB (1984): Evolution in a paedomorphic lineage. I. An electrophoretic analysis of the Mexican ambystomatid salamanders. Evolution 38: 1194–1206.

Shaffer HB (1993): Systematics of model organisms: the laboratory axolotl, *Ambystoma mexicanum*. Syst Biol 42: 508–522.

Shaffer HB, Clark JM, Kraus F (1991): When molecules and morphology clash: a phylogenetic analysis of the North American ambystomatid salamanders (Caudata: Ambystomatidae). Syst Zool 40: 284–303.

Shaffer HB, McKnight ML (1996): The polytypic species revisited: genetic differentiation and molecular phylogenetics of the tiger salamander *Ambystoma tigrinum* (Amphibia: Caudata) complex. Evolution 50: 417–433.

Shaffer HB, Voss SR (1996): Phylogenetic and mechanistic analysis of a developmentally integrated character complex: alternate life history modes in ambystomatid salamanders. Am Zool 36: 24–35.

Sieber-Blum M, Cohen AM (1980): Clonal analysis of quail neural crest cells: they are pluripotent and differentiate in vitro in the absence of noncrest cells. Dev Biol 80: 96–106.

Smith KK (2000): Early cranial development in marsupial mammals: the origins of heterochrony. Am Zool 39: 13A.

Smith SC, Lannoo MJ, Armstrong JB (1990): Development of the mechanoreceptive lateral-line system in the axolotl: placode specification, guidance of migration, and the origin of neuromast polarity. Anat Embryol 182: 171–180.

Solnica-Krezel L, Schier AF, Driever W (1994): Efficient recovery of ENU-induced mutations from the zebrafish germline. Genetics 136: 1401–1420.

Spieth J, Keller RE (1984): Neural crest cell behavior in white and dark larvae of *Ambystoma mexicanum*: differences in cell morphology, arrangement and extracellular matrix as related to migration. *J Exp Zool* 229: 91–107.

Stearner SP (1946): Pigmentation studies in salamanders, with especial reference to the changes at metamorphosis. *Physiol Zool* 19: 375–404.

Steinberg MS (1970): Does differential adhesion govern self-assembly processes in histogenesis? Equilibrium configurations and the emergence of a hierarchy among populations of embryonic cells. *J Exp Zool* 173: 395–434.

Stigson M, Löfberg J, Kjellén L (1997a): Reduced epidermal expression of a PG-M/versican-like proteoglycan in embryos of the white mutant axolotl. Exp Cell Res 236: 57–65.

Stigson M, Löfberg J, Kjellén L (1997b): PG-M/versican-like proteoglycans are components of large disulfide-stabilized complexes in the axolotl embryo. J Biol Chem 272: 3246–3253.

Stone LS (1933): The development of lateral-line sense organs in amphibians observed in living and vital-stained preparations. J Comp Neurol 57: 507–540.

Storfer A, Cross J, Rush V, Caruso J (1999): Adaptive coloration and gene flow as a constraint to local adaptation in the streamside salamander, *Ambystoma barbouri.* Evolution 53: 889–898.

Thompson, D'AW (1942): On Growth and Form, abridged edition. JT Bonner (ed.). Cambridge University Press. 346p.

Tihen JA (1958): Comments on the osteology and phylogeny of the ambystomatid salamanders. Fla State Mus Biol Sci 3: 1–50.

Titus T, Larson A (1995): A molecular phylogenetic perspective on the evolutionary radiation of the salamander family Salamandridae. Syst Biol 44: 125–151.

Tucker RP (1986): The role of glycosaminoglycans in anuran pigment cell migration. J Embryol Exp Morphol 92: 145–164.

Tucker RP, Erickson CA (1986a): Pigment cell pattern formation in amphibian embryos: a reexamination of the DOPA technique. J Exp Zool 240: 173–182.

Tucker RP, Erickson CA (1986b): Pigment cell pattern formation in *Taricha torosa:* the role of the extracellular matrix in controlling pigment cell migration and differentiation. Dev Biol 118: 268–285.

Tucker RP, Erickson CA (1986c): The control of pigment cell pattern formation in the California newt, *Taricha torosa*. J Embryol Exp Morphol 97: 141–168.

Twitty VC (1936): Correlated genetic and embryological experiments on *Triturus*. I. Hybridization: development of three species of *Triturus* and their hybrid combinations. II. Transplantation: the embryological basis of species differences in pigment pattern. J Exp Zool 74: 239–302.

Twitty VC (1945): The developmental analysis of specific pigment patterns. J Exp Zool 100: 141–178.

Twitty VC, Bodenstein D (1939): Correlated genetic and embryological experiments on *Triturus*. III: further transplantation experiments on pigment development. IV: the study of pigment cell behavior in vitro. J Exp Zool 81: 357–398.

Twitty VC, Niu MC (1948): Causal analysis of chromatophore migration. J Exp Zool 108: 405–437.

Vogel KS, Weston JA (1988): A subpopulation of cultured avian neural crest cells has transient neurogenic potential. Neuron 1: 569–577.

Voss SR, Shaffer HB (1997): Adaptive evolution via a major gene effect: paedomorphosis in the Mexican axolotl. Proc Natl Acad Sci USA 94: 14185–14189.

Voss SR, Smith JJ, Gardiner DM, Parichy DM (2001): Conserved vertebrate chromosome segments in the large salamander genome. Genetics, in press.

Wake DB (1991): Homoplasy: the result of natural selection, or evidence of design limitations. Am Nat 138: 543–567.

Wake DB, Harken J (1996): Direct development in the lurgless salamanders: what are the consequences for developmental biology, evolution and phylogenesis. Int. J. Dev Biol 40: 858–869.

Wake DB, Özeti N (1969): Evolutionary relationships in the family Salamandridae. Copeia 1969: 124–137.

Wehrle-Haller B, Weston JA (1995): Soluble and cell-bound forms of steel factor activity play distinct roles in melanocyte precursor dispersal and survival on the lateral neural crest migration pathway. Development 121: 731–742.

Weston JA (1963): A radioautographic analysis of the migration and localization of trunk neural crest cells in the chick. Dev Biol 6: 279–310.

Woronzowa MA (1932): Analyse der wiessen Flecken bei Amblystomen. Biol Zentb 52: 676–684.

Youngs LM (1957): Experimental analysis of quantitative determination of primary melanophores in *Triturus torosus*. J Exp Zool 134: 1–31.

Zelditch ML, Swiderski DL, Fink WL (2000): Discovery of phylogenetic characters in morphometric data. p. 37–83. *in* Phylogenetic Analysis of Morphological Data. JJ Wiens (ed.). Smithsonian Institution Press.

8

TESTING THE HYPOTHESIS OF HETEROCHRONY IN MORPHOMETRIC DATA: LESSONS FROM A BIVALVED MOLLUSK

Peter D. Roopnarine
Department of Invertebrate Zoology and Geology, California Academy of Sciences, San Francisco, California

INTRODUCTION

Heterochrony can be defined broadly as a change in the timing or rate of development of a descendant relative to its ancestor (de Beer, 1958). This is frequently interpreted in terms of individual traits or characters of organisms, in recognition of the dissociative (Gould, 1977) or mosaic nature of evolution. The widely accepted and theoretically satisfying framework developed by Alberch et al. (1979) specifies (quantitative morphological) heterochrony as a three-dimensional phenomenon; the developmental process moves through a space defined by time, size, and shape, tracing out an ontogenetic trajectory. The perception, both historical and recent, that temporal alterations of these trajectories can lead to evolutionary change both mundane and profound, has led many to accept heterochrony as one of or perhaps the major mechanism for effecting morphological evolution (Gould, 1977; Gould, 1988; McKinney and McNamara, 1991).

Beyond Heterochrony: The Evolution of Development, Edited by Miriam Leah Zelditch
ISBN 0-471-37973-5

However, the framework is problematic in several practical ways. The first and most fundamental problem is a frequent failure to formulate a critical test of the heterochrony hypothesis. Neither the Alberch et al. (1979) model nor any of its modified derivatives are tests of the validity of heterochrony as an explanatory hypothesis of the data at hand, unless the three dimensions of time, size, and shape are measured and independent. A proper examination of heterochrony would consist of a test of the hypothesis of heterochrony, and rejection of the null would then permit categorization of the heterochronic phenomenon using the Alberch et al. (1979) framework. Many derived models instead present frameworks for interpreting phenomena that are already accepted as heterochronic. Failure to recognize this fundamental aspect of the heterochrony models often results in the ability to see heterochrony everywhere (incorrect failure to reject the null hypothesis), the fragmentation of any uniform or operational concept of heterochrony, and the failure to consider alternative hypotheses such as heterotopy (Raff, 1996; Zelditch and Fink, 1996; also see Preface).

The second problem stems from the fact that, regardless of the specific operational definition of heterochrony used, it is generally accepted that some measure of time is essential to studies of heterochrony. The Alberch et al. (1979) framework relies explicitly on knowledge of organismal age or developmental time and correctly so (Blackstone and Yund, 1989; Rice, 1997); however, obtaining these data often presents an insurmountable barrier to the interpretation of size and shape in a heterochronic framework. It does not preclude tests of the hypothesis of heterochrony, simply categorization of the type of heterochrony being observed. The failure to understand the ramifications of this problem has lead to the long and continuing fallacy of "allometric heterochrony" (McKinney, 1988; see also Klingenberg, 1996, and Zelditch and Fink, 1996), in which any phylogenetically oriented change in the relationship between size and shape or growth and differentiation can be interpreted as heterochrony. One can indeed compare allometries and use size as a proxy for time in particular situations, but the notion that "one can appropriate the categories of heterochronic perturbations for changes in allometry" is "invalid" (Zelditch, personal communication). Finally, every study of heterochrony requires a phylogenetic hypothesis within which to operate (Fink, 1982), and the conclusions of those studies are only as robust as the phylogenetic hypotheses on which they are based.

In this paper, I will argue that, to gain a quantitative understanding of the role of heterochrony in morphological evolution, we must proceed by testing the biological null hypothesis of an absence of heterochrony and seek alternative hypotheses when necessary (I discriminate here between heterochrony as a biological hypothesis of evolutionary change and heterochrony as a statistical null hypothesis; that is of no difference in the geometry of ontogeny). The appropriate way to do this is to deconstruct the three-dimensional framework of time, size, and shape and recognize that, when testing for the presence of heterochrony, time is the least important of the three dimensions. The relevance of chronological time to evolutionary change in this morphometric context is limited to an interpretation of the bases for developmental similarities; time alone provides no demonstration of developmental homology. This logical independence dictates that one discard the temporal dimension, at least temporarily, and examine the spatial dimensions of scale and geometry. Furthermore, if these latter dimensions can be quantified individually, and indeed they can be as size and shape, respectively, then we should restrict ourselves initially to examinations of the homologous geometries of clade members. Rejection of the null hypothesis requires at least the existence of shared morphologies during the development of ancestor and descendant or putative sister taxa. Subsequent reintroduction of time then permits a refinement of the type of heterochrony detected by ordinating the temporal relationships of ancestral and descendant developmental morphologies.

The first and most important step in demonstrating quantitative morphological heterochrony is to determine whether ancestral and descendant morphologies overlap in the shape dimension (that is, the dimension of developmental morphological similarity). This is a quantitative implication of the sharing of developmental morphologies (see also Chapter 1). If they do not overlap, then heterochrony is not a logical explanation for the observed evolutionary changes because there is no similarity of morphology at any given size or at any time during development, no developmental morphologies are shared, and alternative explanations must be sought. Even if similar stages exist, one must ask whether the descendant's development adheres to the ancestral pattern and is merely an extension, truncation, or change of rate. Or does the descendant development represent a divergence between the two taxa? The latter is the result of different ontogenetic trajectories between ancestor and descendant, resulting in descendant spatial relationships among characters that cannot be predicted from ancestral development (heterotopy). One moves from this preliminary comparison of shape distributions or developmental stages to the construction of growth trajectories by ordinating the stages ontogenetically, using either size or time. It is only in the case of demonstrable heterochrony (altering of the timing or duration of common stages or overlap in spatial developmental dimensions and covariance with time) that time is finally required to determine which mode of heterochrony best explains the observed preliminary phenomenon.

In the following section, I will outline a geometric morphometric approach for such preliminary tests of heterochrony and heterotopy and demonstrate the approach with a morphometric analysis of *Chione*, a tropical American genus of venerid bivalves. Use of a geometric morphometric approach as developed by Bookstein (1991) is essential because the above arguments rely on this vital assumption: the developmental dimensions of time, size, and shape can be measured separately and independently. Correct application of the null hypothesis of nonheterochrony requires comparison of the sets of ancestor and descendant morphologies or, more technically, of shape (morphological) distributions that bear no temporal components. Any measure of shape derived from a size component, such as the comparison of bivariate allometric trajectories or the first principal component under appropriate circumstances, is disqualified by the positive covariance of size and time. Therefore, we require a system capable of quantifying shape and size as potentially independent factors.

Geometric Morphometrics and Development

The past 10 years has witnessed a revolution in morphometrics, as the geometric framework due largely to Bookstein (1991) has reshaped the ways in which we describe biological form. The basic operational premise of geometric morphometrics is that biological form can be archived and quantified using the ordinary Cartesian coordinates of geometrically homologous landmarks. Various manipulations beyond archiving the landmarks (for example, Procrustes, partial warp, and relative warp analyses) are very powerful and descriptive tools for morphological analysis. However, one of the most potent steps underlying all these analyses is the separate recording of geometric shape and scale. Whether one converts landmark coordinates to shape coordinates by translation, rotation, and rescaling to a baseline (Bookstein, 1991; Rohlf, 1996) or rescaling to unit centroid size (Rohlf, 1996) (a measure of size that does not covary with shape measures in the absence of allometry), scale information is effectively removed from the data. Allometric information is not removed however because, in the presence of allometry (covariation of shape and size), shape coordinates will covary with centroid size.

Bookstein (1991) demonstrates this by noting that the expected distribution of a landmark, in the absence of external cofactors and relative to a standard baseline, is circular. Standardization to a baseline removes differences of scale among specimens, and variations of individual measures about the mean are therefore essentially the result of random error only (developmental noise, measurement error, and so forth). Allometry, in this case the covariation of shape and size, causes deviation from circularity and a positive correlation between size and the x and/or y coordinates of the landmark. Although this formulation of allometry is methodologically straightforward, it represents a bit more; it also describes the ontogenetic sequence of shape change.

The separate recording of shape, and its subsequent ordination by size, is exactly what we seek for constructing a test of heterochrony. The only necessary assumption is that size and ontogenetic age are positively correlated (that is, the organisms or their individual characters do not decrease in size as they develop). Notice that this relationship holds regardless of the exact quantitative relationship between size and time (growth rate). The covariation of shape and size is a result of morphological differentiation, with shape change being a function of both changes in spatial orientation as well as size (growth); the latter two phenomena are themselves functions of the passage of time. A trajectory of ontogenetic shape change is therefore a series of ontogenetic shapes ordinated by the sequence of development, whether that sequence is measured as a size increase or elapsing time.

Given this conclusion, we can make several precise statements about heterochrony and heterotopy in a geometric framework derived from a baseline (Fig. 8.1):

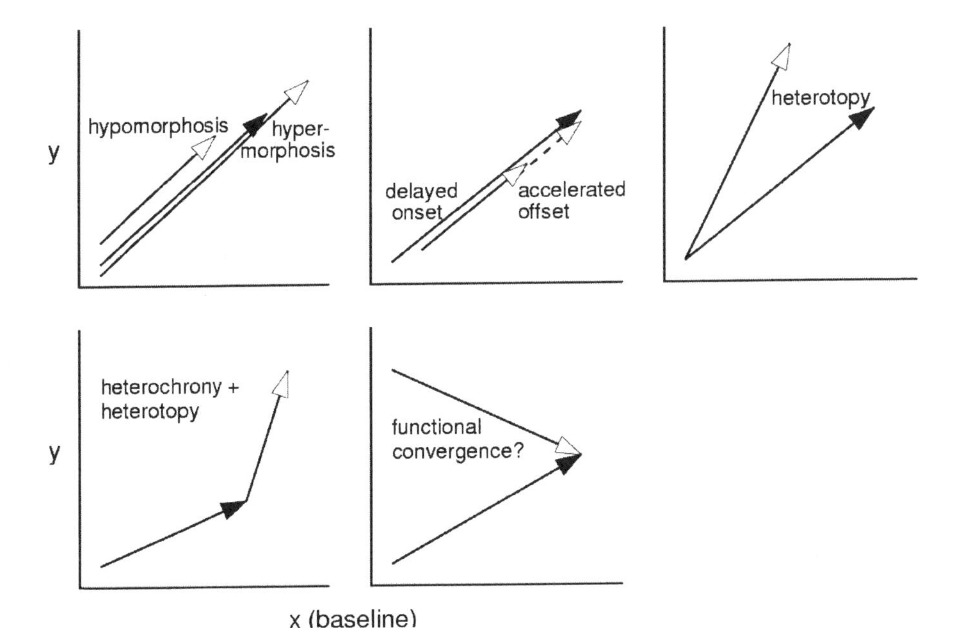

Figure 8.1. Relationships possible between ontogenetic vectors, as represented in a baseline shape coordinate space. The vectors represent movement through this shape space as developmental time elapses. Ancestral vectors are represented by solid arrows. It is not possible to distinguish all models of heterochronic change as outlined by Gould (1977) or Alberch et al. (1979), but heterochronic patterns can be distinguished from nonheterochronic patterns. Furthermore, it is important to note that this space is not equivalent to an ontogenetic space, where one dimension is temporal, and not equivalent to an "allometric" plot of one morphological dimension against another.

1. Ontogenetic trajectories can be represented in the *xy* plane by vectors that describe allometry, that is, vectors that describe shape changes between small and large size (Zelditch and Fink, 1996).
2. Ancestral and descendant vectors must overlap, or be continuous, to support a claim of heterochrony as the mechanism of shape change. For example,
   a. Ancestral and descendant vectors may be coincident at initiation (youngest age, smallest size measured). A shorter descendant vector (that is, a subset of the ancestral vector) corresponds to hypomorphosis, or arrested development. A descendant vector that extends the ancestral vector is representative of hypermorphosis.
   b. Noncoincident origins of overlapping vectors may be representative of delayed onset or early onset. These can in turn be associated with early or delayed offset.
3. Heterotopic vectors can be identified as those diverging from an ancestral vector. The vectors themselves are not coincident. Rather, these divergent vectors describe changes in the shape of homologous structures, caused

by the same developmental processes, but differing in the spatial distribution of growth rates (rates of mineralization in the case of a molluscan shell).

4. Descendant vectors that are not even partially coincident with ancestral or sister-taxon vectors may still have been derived heterochronically or heterotopically, but the connecting vectors are lost because of taxon extinction or losses of developmental stages. This conjecture is untestable without additional ontogenetic or phylogenetic information.
5. Vectors that begin apart but converge as size increases most likely reflect similar functional or architectural constraints or adaptations to the ecological consequences of increasing size. Conversely, they may represent a release from constraints or divergent adaptation at small size, with subsequent convergence because of developmental constraints or canalization. These conjectures are untestable in the present framework.

Ontogenetic trajectories can therefore be visualized as vectors in *xy* shape coordinate space, ordinated by covariation of *x* and *y* shape coordinates with centroid size. Furthermore, the hypotheses of heterochrony and heterotopy can be tested within this framework. The procedures proposed here are outlined below (see **MATERIALS AND METHODS**) but is preceded by an introduction to the organisms with which they are demonstrated.

Case Study: Genus *Chione*

The venerid bivalve genus *Chione* s.s. is strictly tropical American in distribution (both Atlantic and Pacific) and has been since its first appearance in the Early Miocene of the tropical western Atlantic (Roopnarine, 1996). The genus spread to the eastern Pacific by at least the Early Pliocene (Roopnarine, 1996; Roopnarine and Vermeij, 2000) and is represented today in the Atlantic by three species and in the Pacific by six species (Roopnarine, 1996). At least two branches of the clade invaded the eastern Pacific by middle Pliocene time (Roopnarine, 2001). A recent phylogenetic analysis of the extant species, plus the oldest fossil species, reveals several interesting patterns (Roopnarine and Vermeij, 2000; Roopnarine, 2001) (Fig. 8.2). First, the stratigraphically oldest species, *C. chipolana* (Early Miocene, the Chipola Formation of northwest Florida), is the most basal taxon in the ingroup and is a sister taxon to the rest of the genus. Second, there is a distinct subclade comprising the extant shallow water taxa (*C. californiensis*, *C. cancellata*, *C. compta*, *C. elevata*, and *C. undatella*), with the exception of the morphologically aberrant trio of species represented in the analysis by *C. tumens* (the other two species are *C. subimbricata* and the Pliocene *C. vaca*). *C. mazyckii* and *C. guatulcoensis* are smaller species and inhabit deeper waters. Third, there has been a definite trend toward the evolution of larger species within the genus during the Neogene, although, as mentioned, the extant offshore species are quite small (the "increased variance" pattern; see Jablonski, 1997). Finally, Roopnarine and Vermeij (2000) identified

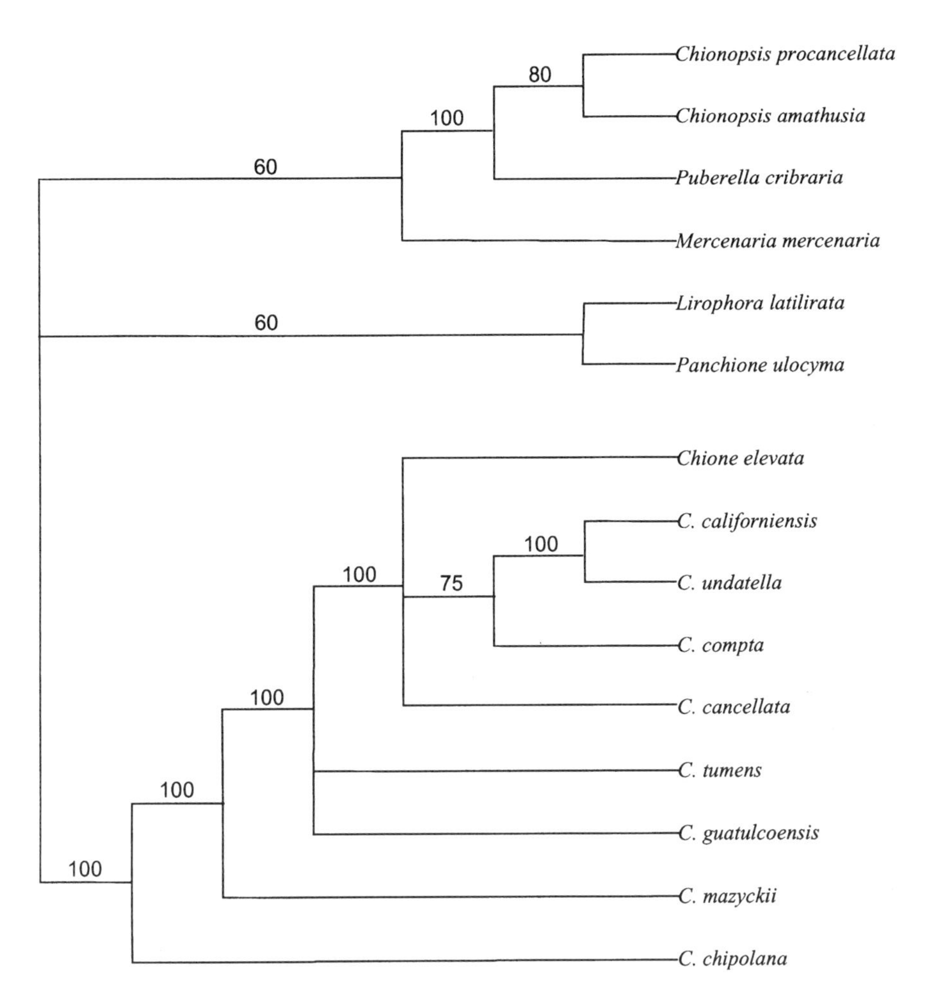

Figure 8.2. 50% majority-rule consensus tree resulting from an analysis of extant *Chione* species. The data set consists of 23 conchological characters (selected to accommodate the addition of fossil taxa; most fossil taxa were removed from current analysis), which were analyzed as unordered and unweighted with a branch and bound algorithm (20 equally parsimonious cladograms, consistency index = 0.5778, length = 49). Numbers on branches indicate the percentage of cladograms supporting a particular branch. Details of the analysis are presented in Roopnarine and Vermeij (2000) and Roopnarine (2001).

several characters of external radial sculpture that are suggestive of a hypermorphic trend within the genus. Radial sculpture in *C. chipolana* consists of raised, rounded cords that do not generally bifurcate during growth. This plesiomorphic condition is retained by *C. guatulcoensis* and *C. mazyckii*. In *C. elevata*, the cords tend to bifurcate (and therefore become more numerous) later in shell growth, whereas in the other shallow water taxa bifurcation begins almost immediately.

Chione is an ubiquitous and abundant genus throughout tropical America, but surprisingly little is known about its development, soft anatomy, life history, or ecology (Jackson, 1973; Jones, 1976). Morphometric analysis of the shell has proven to be a useful tool for delineating species (Roopnarine, 1995) and understanding patterns of evolution (Roopnarine, 1995; Roopnarine and Vermeij, 2000) and paleoecology (Roopnarine and Beussink, 1999). Studies of postembryonic shell growth rate within the subfamily Chioninae have relied primarily on trace element and stable oxygen isotopic analyses (Dettman et al., 1998), as well as sclerochronological analysis (Berry and Barker, 1975; Jones, 1991; Goodwin, 2000; Schoene, 2000). These studies indicate that, whereas there is significant size variation within tropical and subtropical members of the subfamily, most growth takes place during the first 3–5 yr of life, declining significantly beyond that (although never halting fully).

Postembryonic development (juvenile and adult growth) of the *Chione* shell has proven useful in morphometric studies, and it is this stage that will be focused on in this chapter. Shell morphometric development and geometric morphometrics will be used to demonstrate a test for the detection of heterochrony and heterotopy at individual landmarks. The results of these tests will be interpreted in the framework of the existing phylogeny but will also serve to inform and enhance the phylogenetic analysis. The procedure requires extensive morphometric sampling of the species, application of the morphometric results to the construction of ontogenetic trajectories, and an examination of the relationship between those trajectories and the cladistically implied evolutionary history of the genus.

MATERIALS AND METHODS

Several criteria must be met for a proper analysis of the sort proposed here. Species should be represented by sample sizes adequate to the types of statistical analyses performed on the morphometric data collected, and samples should provide geographic coverage of a species' range. The latter criterion ensures the recognition of geographically based morphometric and developmental variation, enhances the discrimination of species (Roopnarine and Vermeij, 2000), and distinguishes intra- and interspecific variation. Finally, one should use the broadest size range available for a species. Details of the phylogenetic analysis used in this paper are given in Roopnarine and Vermeij (2000).

Materials

Sample sizes exceeding 50 specimens were available for the following species: *Chione chipolana*, *C. elevata*, *C. cancellata*, *C. californiensis*, *C. compta*, and *C. undatella*. Samples provided broad geographic coverage of the range of each species, with the exception of *C. undatella* (Table 8.1). No samples of this species collected in southern California were used because specimens there differ

Table 8.1. Samples and localities used in morphometric analysis

Species	Locality	Sample Number	*n*
Chione californiensis	Bahia Concepcion, Baja California	A	42
	Bahia de Los Angeles, Baja California	CASIZ 92331, 092278	23, 21
	San Pedro Bay, Los Angeles, California	CASIZ 092036	18
C. cancellata	Aruba	GJV	10
	St. Thomas, USVI	ANSP 53593	19
	Tobago	ANSP 209819	30
	Venezuela	ANSP 232143, GJV	38, 20
C. chipolana	Chipola Formation, Florida	A	43
C. compta	Bahia Concepcion	CASIZ	5
	Carmen Island, Baja California	CASIZ (Stanford 49384)	13
	Guaymas, Sonora, Mexico	CASIZ 8425	7
	Port Parker, Costa Rica	CASIZ 17926	13
C. elevata	Aransas, Texas	ANSP 149494, 189437, 232616, CASIZ 33348	15
	Bahamas	ANSP 367271, 299345, 326356, 325648	10
	Myrtle Beach, South Carolina	ANSP 145701, 179497	13
	Sanibel, Florida	FLMNH 12135	23
	Wilmington, North Carolina	A	8
C. undatella	Islas Res Maria, Baja California	CASIZ 089499	20
	San Ignacio Lagoon	CASIZ 4969, 4970	25

ANSP, Academy of Natural Sciences, Philadelphia; CASIZ, California Academy of Sciences, Department of Invertebrate Zoology and Geology; FLMNH, Florida Museum of Natural History; A, author's collection; GJV, collection of Geerat Vermeij. n = sample sizes.

obviously and consistently from specimens located further to the south off both the Pacific and Gulf of California coasts of Baja California and may represent a separate taxon. *Puberella cribraria* (subfamily Chioninae), representing a closely related and also strictly tropical American Tertiary genus, was selected as a representative outgroup taxon. Absent from the analysis are *C. guatulcoensis*, *C. mazyckii*, and *C. tumens*, the former two species because of inadequate available sample sizes and the latter because of a lack of broad geographic coverage in available samples.

Data Collection and Analysis

Homologous landmarks digitized for each specimen include umbo, lunule, escutcheon, hinge tooth, and muscle scar features and are described in Roopnarine and Vermeij (2000) (Fig. 8.3). The landmarks were digitized by first imaging each specimen on a flatbed scanner at 300–400 dpi (Hp ScanJetIICx) and subsequent *xy* coordinate capture using image analysis software (Sigma-

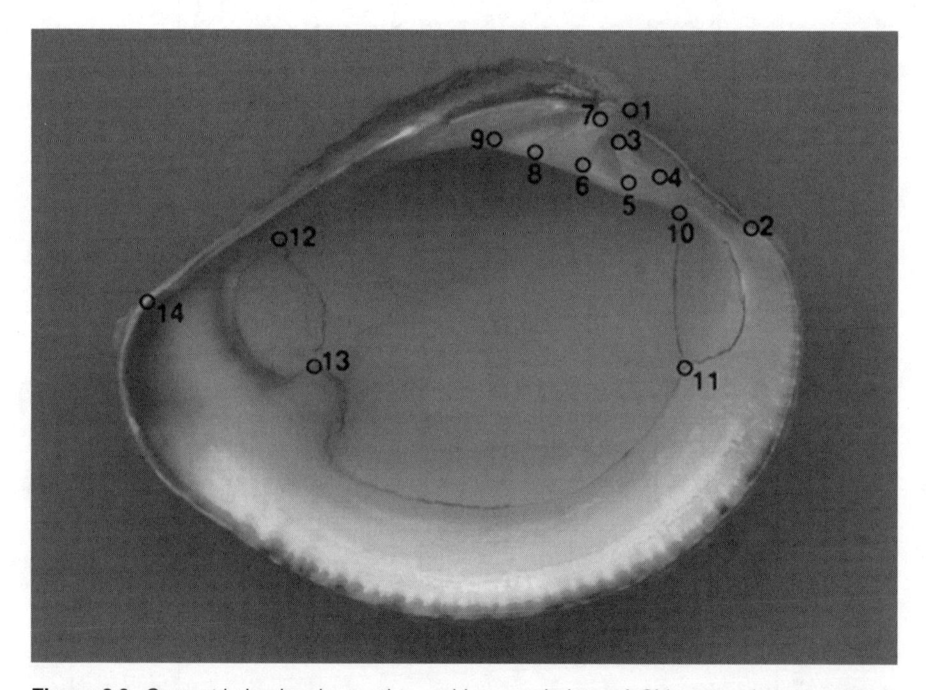

Figure 8.3. Geometric landmarks used to archive morphology of *Chione* species. 1, umbo; 2, ventral terminus of lunule margin; 3, dorsal tip of anterior cardinal tooth; 4, anteroventral tip of anterior cardinal tooth; 5, posteroventral tip of anterior cardinal tooth; 6, anteroventral tip of middle cardinal tooth; 7, dorsal tip of middle cardinal tooth; 8, posteroventral tip of middle cardinal tooth; 9, ventral tip of posterior-most hinge tooth socket; 10, dorsal tip of anterior retractor/adductor muscle scar; 11, intersection of pallial line and anterior adductor muscle scar; 12, dorsal tip of posterior retractor/adductor muscle scar; 13, intersection of pallial line and posterior adductor muscle scar; 14, ventral terminus of escutcheon.

Scan, SPSS) (see Roopnarine and Beussink, 1999, and Roopnarine and Vermeij, 2000, for details).

The methodology utilizes the concepts of geometric morphometrics, whereby landmark coordinates are converted to shape coordinates, designating location in a Kendall shape space (Bookstein, 1991; Rohlf, 1996). The comparison of shapes or description of shape change is accomplished by examining global, affine, or uniform transformations and nonuniform transformations. Uniform transformations leave parallel lines parallel after transformation, therefore describing global and overall differences of shape, such as stretching and shearing (for example, the transformation of a square to a nonsquare rectangle). Nonuniform transformations are more localizable and occur on scales ranging from single landmarks to more global nonlinear transformations. Nonuniform transformations may be understood as changes in the relative positions of landmarks, in whichever directions necessary to transform one shape to another. They lend a certain quantitative reality to the Cartesian grid transformations postulated by D'Arcy Thompson (1942). These transformations can be calculated relative to a reference form and are represented in a simplified Euclidean space, tangential to the curved Kendall space, as principal warp axes (the dimensions of the tangential space) and partial warp scores (the locations of specimens in this new space) (Rohlf, 1996) (see Roopnarine and Vermeij, 2000, for an example applicable to bivalves). Both types of transformations can be visualized quite effectively as the now familiar thin-plate splines of landmark coordinate configurations.

Within-sample and within-species sample homogeneity and allometry were tested with partial warp and subsequent relative warp analyses of the raw landmark data. All specimens of a species were first aligned maximally using the Procrustes alignment procedure in the tpsRelw program, version 1.17 (Rohlf, 1999); partial warps were extracted (by comparison to a generalized least-squares consensus form), and the resulting partial warp scores were subjected to relative warp analysis. Technically, because relative warp analysis is a modified principal component extraction procedure, analysis of heterogeneous multigroup assemblages will not always result in independent or necessarily meaningful relative warps (principal components) (Bookstein, 1996; Klingenberg, 1996). Like general principal component analysis, however, it is a useful first step in ordinating dissimilar samples within the data. If a species is represented by dissimilar samples, then proper ordination may be accomplished with multivariate analysis of variance (MANOVA) and canonical variate analysis of the partial warp scores (Bookstein, 1996; Roopnarine and Beussink, 1999), or, if potential causal factors can be identified, their effects may be quantified by regression of partial warp scores on those factors. Allometry is identified as the covariance of any relative warp factor (summary shape factors) with centroid size and can be visualized with thin-plate splines (Bookstein, 1991).

Ontogenetic trajectories were described using a simpler procedure of relating baseline-derived shape coordinates to centroid size. Specimens were first aligned and rescaled to a common baseline consisting of the ventral terminuses of the escutcheon and lunule, by a process of translation, rotation, and rescaling. The resulting shape coordinates are scale free, and size is documented separately as centroid size. The ontogenetic trajectories of individual homologous landmarks of single homogeneous groups, identified by the warp analyses described above, were then described by regression of x and y shape coordinates on centroid size. The resulting functions describe the allometry of a single landmark (Bookstein, 1991) (although the regression coefficients are not interpretable as conventional coefficients of allometry). Landmark allometry was visualized in the baseline shape space by calculating the growth vector of the landmark with the regression function(s) and the minimum and maximum centroid sizes recorded for that group. Use of minimum and maximum centroid sizes serves simply to restrict vector visualization to the ranges of sizes actually observed, with no extrapolation to smaller or larger size. Similarity of vectors was then assessed as the coincidence of beginning or end points, as suggested in Figure 8.1, using the expected variances and 90% confidence interval of each end of a vector, which in turn is derived from the Bonferroni confidence intervals of the vector. Baseline shape coordinates and centroid size were calculated with programs written for Linux in C/C++, regression analyses were performed using Stata 6.0 for Linux, and warp analyses (partial and relative) were performed with the tpsRelw program of Rohlf (1999).

There are two caveats to this type of baseline analysis. First, the derivation of ontogenetic trajectories is somewhat deficient because it does not account for interlandmark covariance in the way that a partial warp analysis does. As stated by Zelditch and Fink (1996) (see also Zelditch et al., 2000), we are interested in "changes in rates of growth here relative to there." Therefore, we can understand the growth of an integrated organism only by synthesis of the individual measurement. Although the vector maps of the shape coordinates presented here are intended as a visual synthesis and are easily interpreted and comparable among species, quantitative tests can be applied. The vector comparison method of Zelditch (see Preface) was used to construct multivariate ontogenetic trajectories of each species and perform pairwise comparisons of trajectories among species. This method measures the angle (correlation) between species trajectories, and significance of the correlation is assessed relative to an estimate of within-species trajectory variance. Regression of the shape coordinates on size, as explained above, yields landmark ontogenetic trajectories, and consideration of all the landmarks together yields a sum ontogenetic trajectory for the species in morphospace. Zelditch's comparative method seeks to measure the relationship between species vectors as the angle between them. The method operates by comparing the angle between species trajectories to within-species trajectory variation (within-species variation is measured with a resampling approach). Statistically similar trajectories will have a zero angle.

Second, one should be mindful of the relative sizes of the species being dealt with. Although many of the species differ significantly in maximum adult sizes (Roopnarine, 1996), their size ranges do overlap. Maximum size disparities are not critical in the procedure (as these serve simply to extend vectors), although

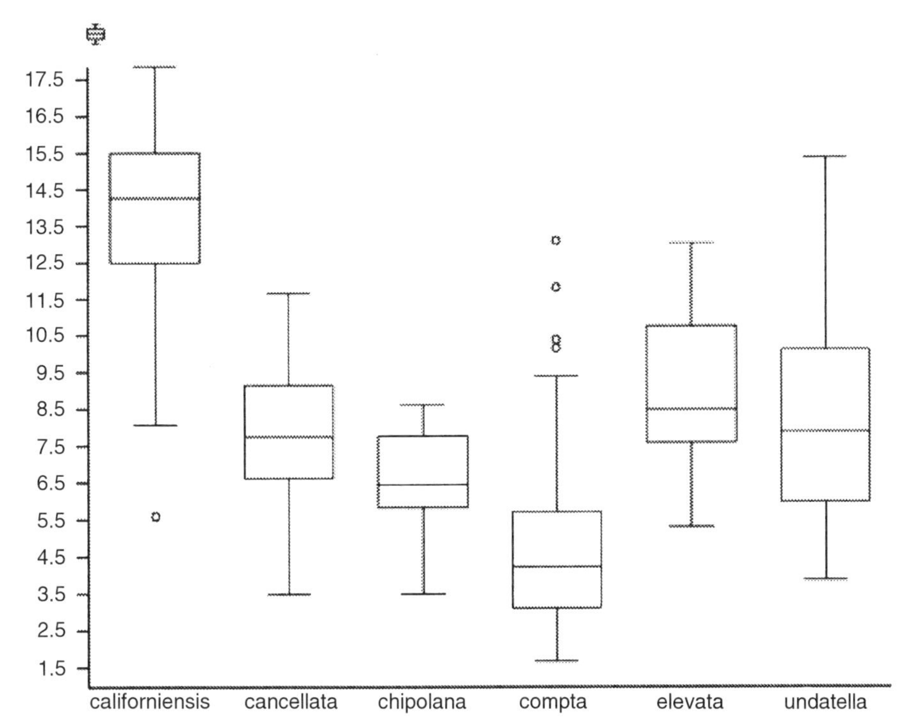

Figure 8.4. Box plots of species centroid size distributions. All species have overlapping ranges, although *C. californiensis*' distribution is biased toward larger specimens.

minimum size disparities could lead to a failure to recognize coincident vector origins.

RESULTS

Size

Figure 8.4 illustrates the centroid size distributions for all the species. The distributions are obviously heterogeneous, but the important factor here is understanding the ranges. Minimum size is comparable among all species except *Chione compta*, which has significantly smaller individuals, and *C. aliforniensis*, whose smallest individuals are close to or above the median size of the other species. Both these exceptions are sample driven; smaller individuals of *C. compta* were available than were for the other species, and the number of small specimens available for *C. californiensis* was limited.

The upper ends of the distributions are more reflective of real differences of maximum size among the species. Eastern Pacific chionine taxa are on average larger than their western Atlantic congeneric counterparts (Roopnarine, 1996).

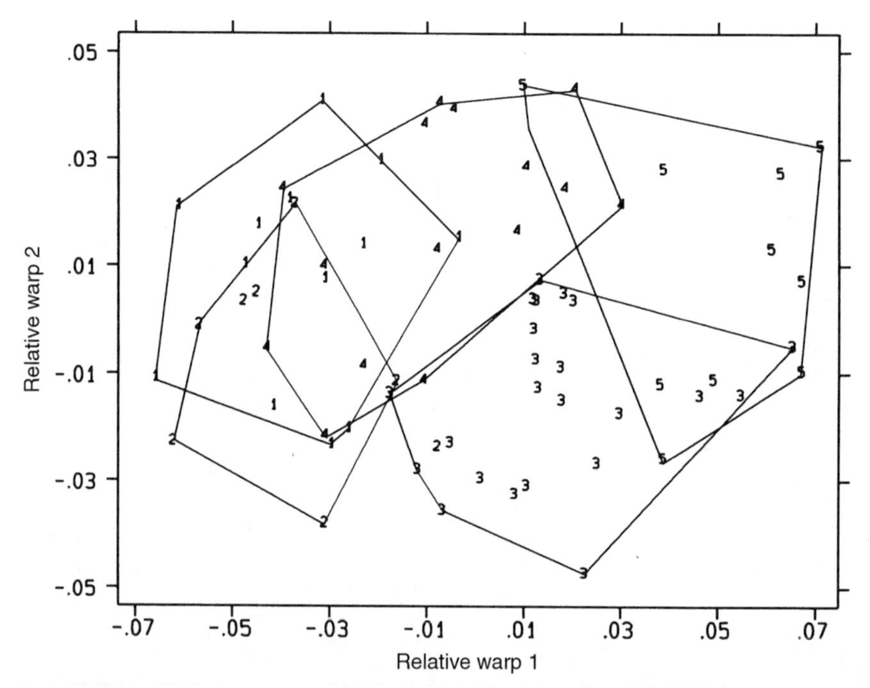

Figure 8.5. Relative warp analysis of *Chione elevata*. Relative warps 1 and 2 (33.39% and 13.56% of variation explained, respectively) are illustrated, with sample differentiation oriented along relative warp 1. Symbols represent localities: 1, North Carolina; 2, South Carolina; 3, Sanibel, Florida (Gulf Coast); 4, Bahamas Islands; 5, Galveston, Texas.

The extinct *C. chipolana* is the smallest documented shallow water member of the genus, whereas *C. californiensis* and *C. undatella* attain the largest sizes. *C. cancellata*, *C. compta*, and *C. elevata* attain comparable maximum sizes.

Intraspecific Variation

Homogeneity of species data sets was tested with relative warp analysis of partial and uniform warp scores. All species exhibited some degree of intraspecific variation along the most significant relative warp, relative warp one. However, much of this variation can be interpreted as intraspecific allometry. For example, there is significant intraspecific sample differentiation of *C. elevata* along the first relative warp (Fig. 8.5). However, this axis covaries significantly with centroid size ($r^2 = -0.74$, $P < 0.0001$) (Fig. 8.6), and sample means also covary significantly with size ($r^2 = -0.95$, $P = 0.0013$). There is a clear trend of increasing maximum size from south to north (Texas to North Carolina), and scores on relative warp one decrease accordingly. Therefore, much of the intraspecific variation in this species can be explained by geographically mediated allometry. This statement holds true for every species in which significant intraspecific variation was detected (Table 8.2). In no case is there evidence of non-size-related differentiation.

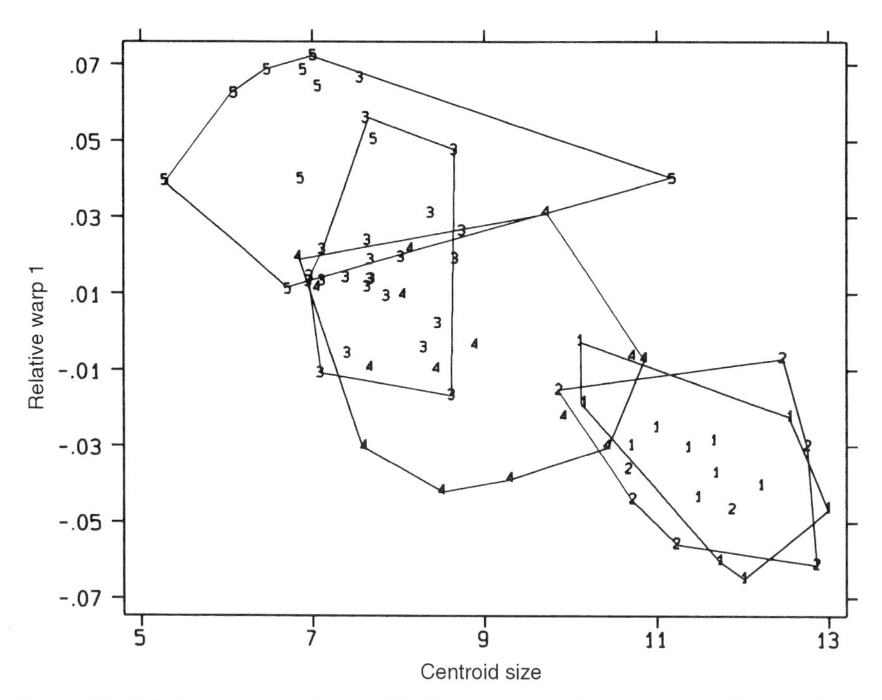

Figure 8.6. Relative warp 1, as illustrated in Figure 8.5, plotted against centroid size. Latitudinal location and centroid size are correlated, as indicated by the significant correlation between sample mean values and centroid size ($r^2 = -0.95$, $P = 0.0013$) (samples were ordered latitudinally, see legend of Fig. 8.5). There is an overall significant correlation between relative warp 1 and centroid size ($r^2 = -0.74$, $P < 0.0001$), indicating the presence of shape allometry.

Developmental Maps

The regressions of x and y shape coordinates on centroid size yield ontogenetic vector maps for each species (Fig. 8.7). These maps agree very well with thin-plate spline visualizations of development derived from relative warp analysis (Fig. 8.8) and represent an alternative method for displaying morphometric

Table 8.2. Correlation between specimen scores on relative warp one (analyses of individual species) and centroid size; relationship is not significant in *Chione cancellata*, but there is no evidence of geographic intraspecific variation in that species

Species	r^2	P Value
Chione californiensis	−0.7161	<0.0001
C. cancellata	0.1733	0.0616
C. chipolana	0.5436	0.0002
C. compta	0.6355	<0.0001
C. elevata	−0.7370	<0.0001
C. undatella	−0.8421	<0.0001

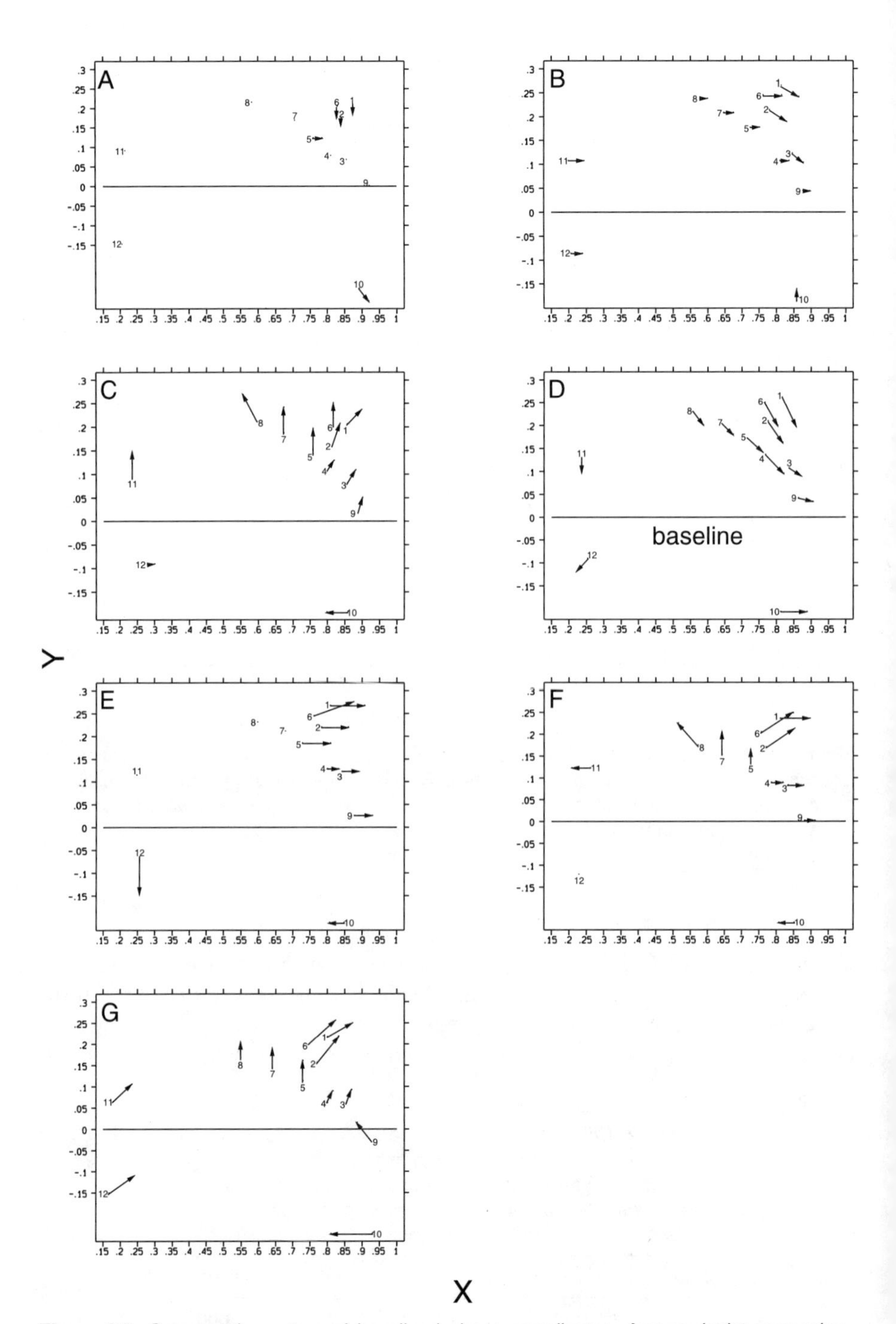

Figure 8.7. Ontogenetic vectors of baselined shape coordinates. Arrows depict regression-based vectors describing the change of relative landmark locations with increasing centroid size. The landmark numbers follow the sequence outlined in Figure 8.3, but landmarks 2 and 14 serve as the baseline anchors and are therefore missing. Landmark numbers are adjusted accordingly. A: *Puberella cribraria*. B: *Chione chipolana*. C: *C. californiensis*. D: *C. cancellata*. E: *C. compta*. F: *C. elevata*. G: *C. undatella*.

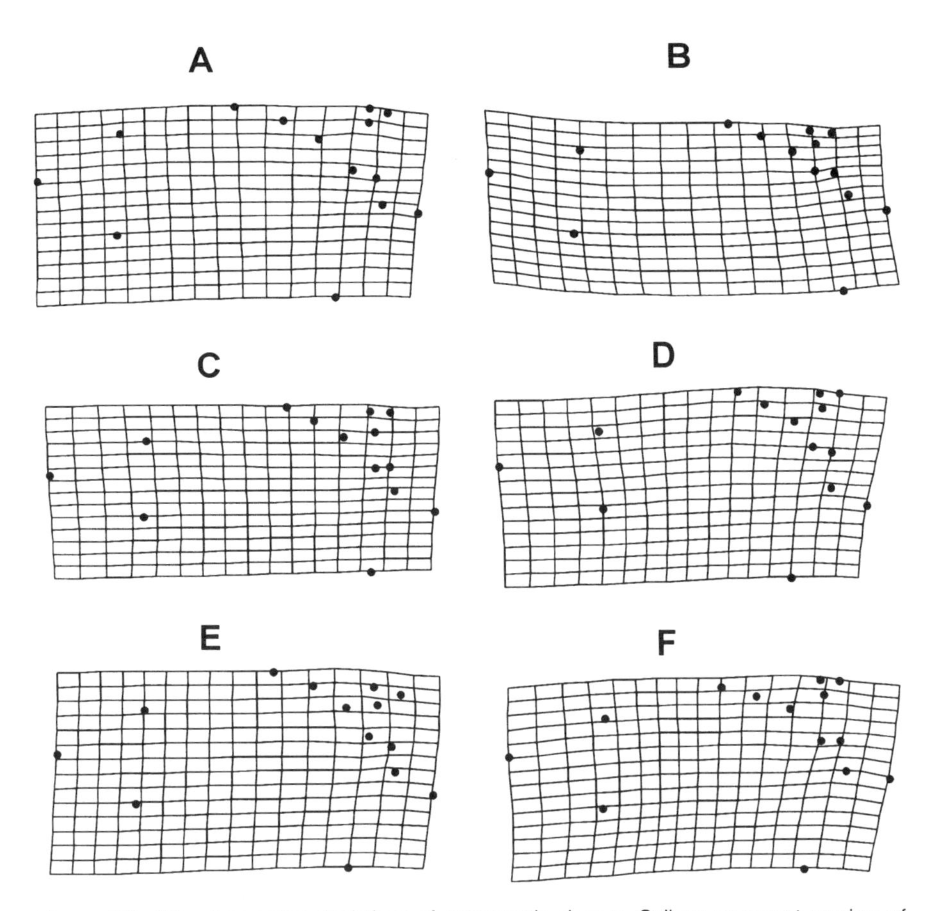

Figure 8.8. Thin plate spline depictions of ontogenetic change. Splines represent maxima of each species, reconstructed from intraspecific relative warp analyses. Landmark numbers are as shown in Figure 8.3. A: *C. californiensis*. B: *C. cancellata*. C: *C. chipolana*. D: *C. compta*. E: *C. elevata*. F: *C. undatella*.

developmental change (Zelditch and Fink, 1996). Viewing all the maps (and splines), it seems that each species can be placed in one of two categories on the basis of vectors associated with the umbo and cardinal teeth. The first set comprises the species *C. cancellata*, *C. chipolana*, and *C. compta*, where most of the vectors are directed anteriorwise, sometimes with a ventral component. In fact, the similarity between *C. chipolana* and *C. compta* is striking, given that the former is an Early Miocene, western Atlantic taxon and the other is an extant eastern Pacific species (with a fossil record extending only to the Late Pleistocene). The second set comprises *C. elevata*, *C. californiensis*, and *C. undatella*. Hinge landmarks of these species tend to migrate dorsally relative to the baseline during growth. *Puberella cribraria* bears no resemblance to any member of the ingroup, and its lack of landmark allometry is unusual.

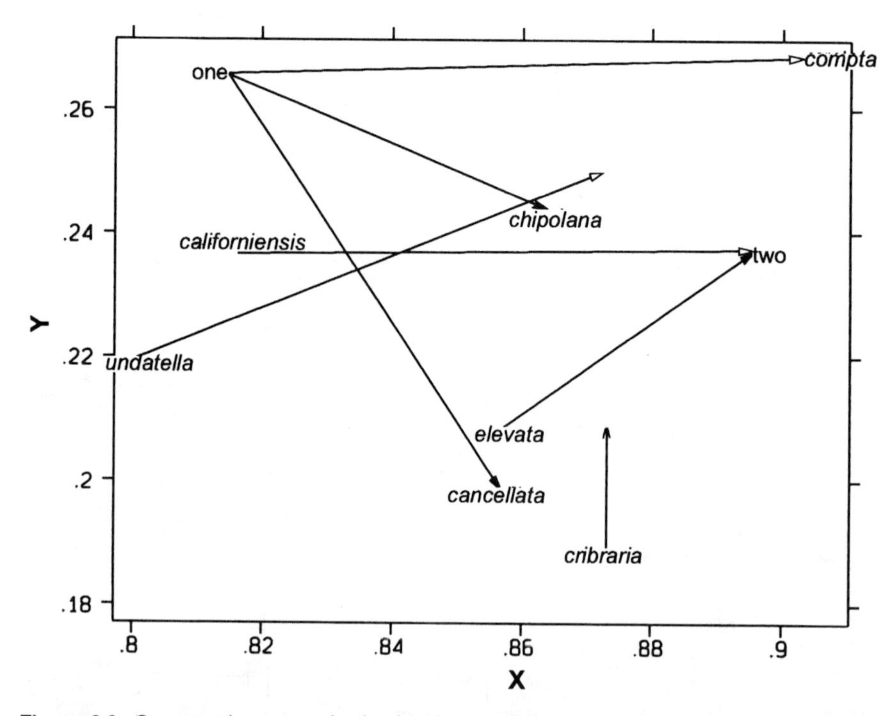

Figure 8.9. Ontogenetic vectors for landmark 1 (umbo). Vectors travel from size minima to maxima; *x* axis represents the baseline. Numbers, given instead of species names, represent points at which species vectors overlap or converge. One: *C. cancellata, C. chipolana,* and *C. compta.* Two: *C. californiensis* and *C. elevata.*

It is not obvious from the collection of maps, however, which vectors may represent heterochronic or heterotopic relationships as outlined in Figure 8.1. In phylogenetic interpretations of the maps, the trap of overall phenetic similarity must be avoided and instead be based on shared derived similarities. Therefore, although the vector maps may hint at developmental relationships among various species, determining actual morphometric developmental relationships requires closer inspection of the characters themselves, that is, the landmarks. To determine whether such relationships exist among any of the species, it is necessary to consider each landmark individually from allometric and developmental perspectives.

For example, Figure 8.9 illustrates the vectors of landmark 1 (umbo) for all the species. At this reduced geometric scale, similarities among some of the vectors are immediately apparent. *C. cancellata*, *C. chipolana*, and *C. compta* all originate from the same location; they are morphometrically similar at small sizes. Conversely, the *C. californiensis* and *C. elevata* vectors converge, resulting in these species being morphometrically similar at large sizes. Finally, small individuals of *C. elevata* are similar to large *C. cancellata*. These observations can be viewed quantitatively by comparing the distributions of the vector origins and terminuses using the expected variances of the fitted x and y values and testing the apparent differences and similarities of the vectors.

Applying regression confidence ranges to landmark 1 results in the collapse of the vector origins for *C. cancellata*, *C. chipolana*, and *C. compta* to a single point. Relative umbo location for these three species is geometrically and statistically indistinguishable at small sizes. Their divergence during development is an example of heterotopic alteration. Similarly, again with regards to the relative position of the umbo, *C. californiensis* and *C. elevata* are indistinguishable at large sizes. This could represent the retention of a plesiomorphic developmental pattern, or it could be convergent; there is no way to discriminate these two hypotheses on the basis of size and shape data alone. Finally, despite their proximity, the vector terminus of *C. cancellata* and origin of *C. elevata* are statistically different.

The final step to this landmark-by-landmark analysis is the comparison of relative sizes. The size distributions of *C. chipolana* and *C. cancellata* are comparable (Fig. 8.4), although *C. cancellata* does reach a greater size (maximum centroid size = 11.65 vs. 8.61 for *C. chipolana*). *C. compta* does comprise much smaller individuals in this exercise (minimum size = 1.66), but the allometric slopes are so gentle that coordinate values are still comparable at the *C. chipolana*/*C. cancellata* minimum size. The convergence between *C. californiensis* and *C. elevata*, on the other hand, occurs as *C. elevata* approaches its maximum size. Although *C. californiensis* attains larger size, allometry of landmark 1 occurs almost entirely in the x direction, resulting in the interception of *C. californiensis*' trajectory by *C. elevata*'s (Fig. 8.10).

There are several other landmarks that exhibit statistical similarity among species. These include landmarks 3–8, all of which are cardinal tooth landmarks (Fig. 8.11). Some of the similarities are large-size vector convergences, as may be expected of such a functionally significant region. Examples are *C. californiensis* and *C. undatella* at landmarks 5 and 6 and *C. elevata* and *C. undatella* at landmarks 4 and 7. There are, however, several unequivocal and interesting developmental relationships. *C. chipolana* and *C. compta* are linked at small size for landmarks 1, 3, 4, and 6 and *C. cancellata*, likewise, with the exception of landmark 4. Small individuals of *C. elevata* are again similar to large individuals of *C. cancellata* at landmarks 3, 6, and 8, and this time the similarities are statistically significant. There is no clear pattern of relationship among any species with regard to muscle scar landmarks. Although all the species exhibit a significant developmental change at these landmarks, it is likely that functional factors override phylogenetic ones.

The relationships among *C. cancellata*, *C. chipolana*, and *C. compta* are clearly heterotopic. There is a developmental relationship among the species, reflecting the close phylogenetic relationship between *C. cancellata* and *C. compta* and their apparent retention of the plesiomorphic condition represented

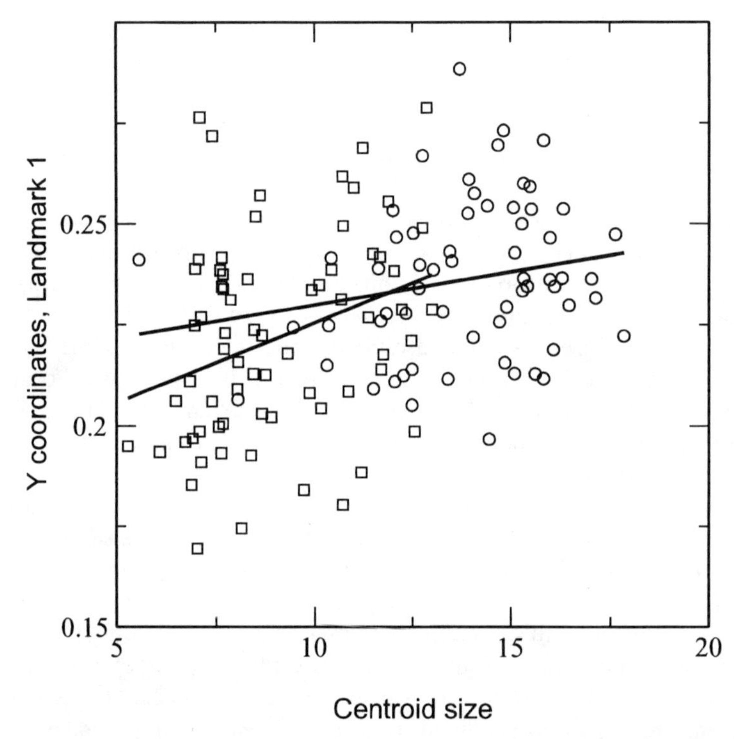

Figure 8.10. Plot illustrating allometric convergence of *C. elevata*'s landmark 1 vector with that of *C. californiensis* along the *y* axis.

by *C. chipolana*. Nevertheless, the relationships hold only at small size (that is, young age), after which morphologies become divergent. An exception is landmark 8 (posteroventral end of middle cardinal tooth), for which *C. compta* exhibits no ontogenetic change in landmark location; that location is identical to the terminus of *C. chipolana*'s trajectory. The relationship between *C. cancellata* and *C. elevata* is more puzzling. There is a clear implication of heterochrony, with small individuals of *C. elevata* exhibiting the adult morphology of *C. cancellata*. The subsequent development of *C. elevata* is not merely an extension of the *C. cancellata* vectors, however, as would be expected of a hypermorhpic or extended trajectory. Instead, subsequent *C. elevata* development is novel, pointing to a heterotopic alteration.

A final interesting observation concerns the locations of vector origins, at the above landmarks, for *C. californiensis* and *C. undatella* (Fig. 8.11). In all cases, they are displaced sequentially from the *C. cancellata-chipolana-compta* origin in the negative *y* direction, first *C. californiensis* and then *C. undatella*. This regularity hints at an underlying evolutionary transition or branching pattern.

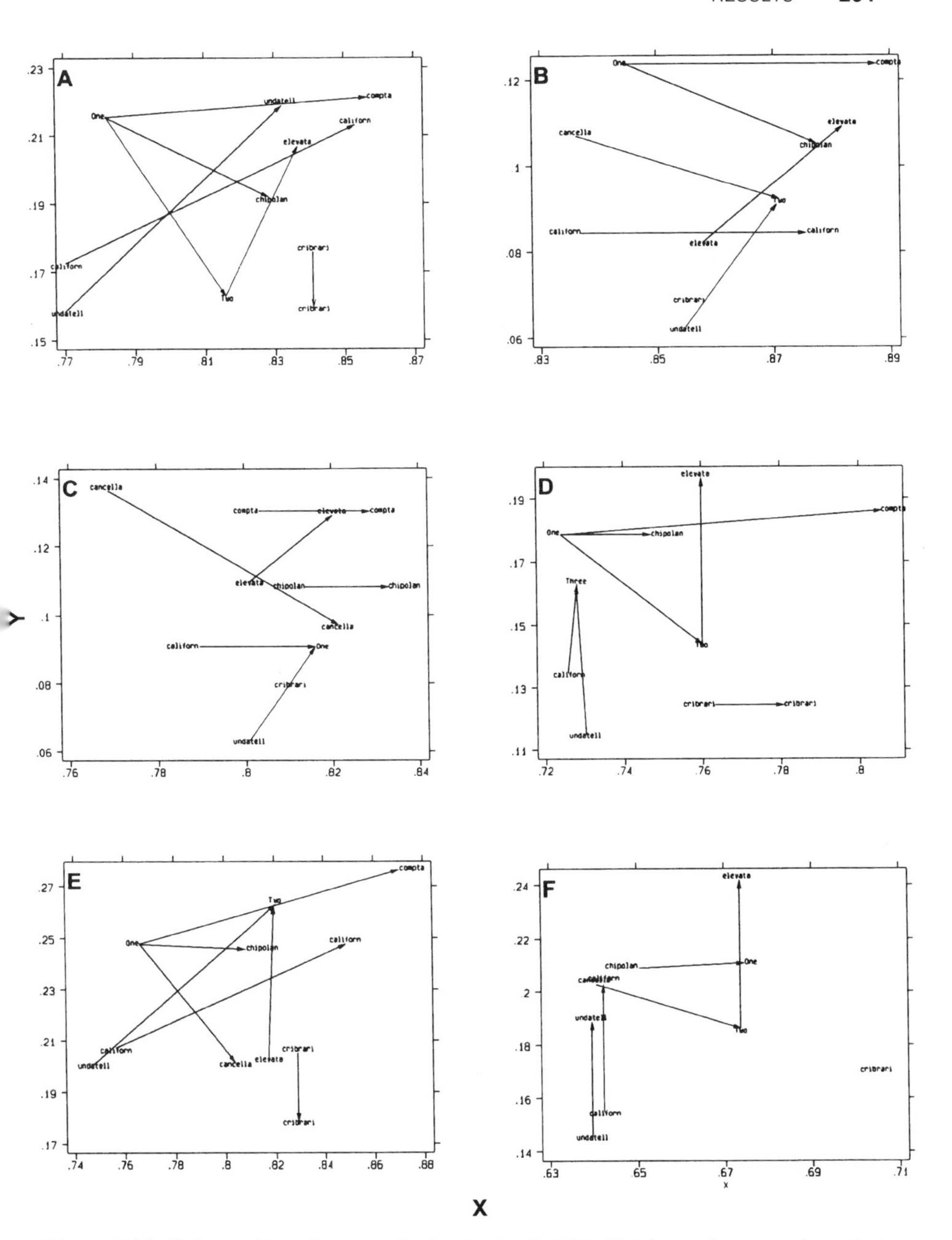

Figure 8.11. Ontogenetic vector maps for landmarks 3–8 (A–F). The vectors are interpreted as in Figure 8.9, but number points represent different groupings of taxa. **A:** one = *C. cancellata*, *C. chipolana*, and *C. compta*; two = *C. cancellata* and *C. elevata*. **B:** one = *C. chipolana* and *C. compta*; two = *C. cancellata* and *C. undatella*. **C:** one = *C. californiensis* and *C. undatella*. **D:** one = *C. cancellata*, *C. chipolana*, and *C. compta*; two = *C. cancellata* and *C. elevata*. **E:** one = *C. cancellata*, *C. chipolana*, and *C. compta*; two = *C. elevata* and *C. undatella*. **F:** one = *C. chipolana* and *C. compta*; two = *C. cancellata* and *C. elevata*.

Table 8.3. Multivariate pairwise comparison of ontogenetic trajectories between species (see Preface)

	Angle			
Species Comparison	Between	Within 1	Within 2	Comment
chipolana-compta	39.6	44.6	43.9	Not significant
chipolana-cancellata	51.5	48.2	34.3	Marginally significant
chipolana-elevata	92.7	38.7	48.1	
chipolana-undatella	71.2	26.3	36.4	
californiensis-chipolana	69.9	40.4	41.1	
californiensis-compta	45	42.6	40.4	
californiensis-elevata	39.9	34.4	30.7	
californiensis-undatella	58.3	36.9	23.5	
californiensis-cancellata	87.1	40.5	39.1	
compta-elevata	71.9	54.1	41.7	
compta-undatella	67.1	25.9	45.3	
compta-cancellata	63.2	47.4	40.2	
elevata-undatella	51.4	22.7	40	
elevata-cancellata	113	24.1	36.7	Negative correlation?
undatella-cancellata	117.3	48.8	22.2	Negative correlation?

The angle (correlation) between species' ontogenetic trajectories (vectors), derived from regression of Booksteinshape coordinates on centroid size, are compared relative to within-species variation (determined by resampling). A hypothesis of significant differences between the vectors is supported when the interspecific angle exceeds the intraspecific range of variation. The hypothesis is rejected in the *Chione chipolana-compta* comparison and is supported equivocally in the *C. chipolana-cancellata* comparison. Significant negative correlations are suggested by the last two comparison, *C. elevata-cancellata* and *C. undatella-cancellata*. The interpretation of these results is unclear, but the *C. elevata-cancellata* relationship is a mixed heterochronic-heterotopic one (see text).

An integrated analysis of the landmarks, carried out with a multivariate comparison of the shape coordinate vectors (see Preface), produced an overall similar picture (Table 8.3). There is only one failure to reject the null hypothesis of no interspecific similarity (that is, a working hypothesis that the ontogenetic vectors are indeed similar), and that is between *C. chipolana* and *C. compta*. A similar relationship between *C. chipolana* and *C. cancellata*, as suggested by the analysis of individual landmarks, is not supported, although the rejection is marginal. Also interesting is the large angle between *C. cancellata* and *C. elevata*, possibly suggesting a negative correlation between these two trajectories. This is certainly congruent with the earlier observation that small-size *C. elevata* is similar to large-size *C. cancellata* but diverges during ontogeny. Whether this type of developmental relationship may be represented by a negative angle between trajectories requires further investigation.

INTERPRETATION

Intraspecific Variation

The nature and basis of intraspecific variation must be assessed in this type of study before any conclusions can be made concerning phylogenetic patterns and relationships. The most appropriate method for dealing with this when using extant taxa is to use broad geographic coverage (and environmental, when appropriate) of single species. It is important to detect reifiable morphometric variation, which may indicate phylogenetic differentiation before the calculation of ontogenetic vectors. All the extant species examined in this chapter exhibited some level of intraspecific variation based on geographic samples, but simple size allometry explains the differentiation in every case. For example, the relative warp analysis of *Chione elevata* uncovers intersample discrimination along the first relative warp. The relative warp, however, covaries significantly with centroid size and represents a geographical cline in this species.

Heterochrony, Heterotopy, and Phylogeny

The ontogenetic vector maps of each landmark reveal morphometric developmental relationships among a number of the species. *Chione chipolana*, *C. cancellata*, and *C. compta* are united by sharing a common early developmental "stage" for five landmarks (four in the case of *C. cancellata*; integrated support for this relationship is equivocal). Mapping this set onto the phylogeny suggests that *C. chipolana* represents the plesiomorphic condition of these landmarks, and the other two species retain that plesiomorphic condition. The situation holds only for the early ontogenetic stage, however, as the two extant taxa diverge almost immediately from *C. chipolana*'s trajectory. This type of pattern qualifies as heterotopic morphometric change because there is both geometric and developmental homology among the species, yet ontogeny does not reflect phylogeny. There is spatial repatterning, and the information borne by sharing of an early developmental stage is symplesiomorphic; it bears on the monophyly of the species but offers no resolution of the pattern of relationships. Moreover, the resulting trajectories may be considered autapomorphic or novel for each species.

There are two situations in the present example in which overlapping vectors may fit a heterochronic framework and convey phylogenetic information. First, *C. elevata* is linked significantly to *C. cancellata* at landmarks 3, 6, and 8 and equivocally at landmark 1. In all cases, the link is the same; small individuals of *C. elevata* are geometrically similar to large individuals of *C. cancellata*. This is a very real pattern, as the geographic ranges of both species are covered extensively, and there is considerable overlap between their size ranges. Moreover, it is possible that *C. elevata*, which first appears suddenly in the Early Pleistocene of the southeast United States and replaces the extinct *C. erosa* (Early-Late Pliocene), is an evolutionary offshoot of the more ancient *C. can-*

cellata lineage (Roopnarine, 1996), which is endemic to more southern Caribbean waters. A plausible developmental hypothesis, although not testable in the present context, is that *C. elevata* is derived from an extreme acceleration of the rate of morphometric differentiation normal to *C. cancellata*. This is not a purely heterochronic change, however, because *C. elevata* subsequently proceeds on a unique developmental path. The subsequent relationship is not hypermorphic because it is not an extension, nor is it predictable on the basis of *C. cancellata*'s trajectory. The trajectories of *C. elevata* captured here for landmarks 3, 6, and 8 must, therefore, according to this hypothesis, be combined heterochronic and heterotopic derivatives of the homologous *C. cancellata* trajectories. Consideration of all the landmarks together, based on measurement of the angle between the ontogenetic trajectories, does not support a heterochronic developmental link between these two species. However, the very large size of the angle does suggest a negative correlation, and consideration should be given as to how the diversity of heterochronic and perhaps heterotopic patterns may be represented by such trajectory comparisons in morphological space.

Second, in addition to sharing apparent plesiomorphic conditions for several landmarks, *C. chipolana* and *C. compta* are also linked at landmark 7 (dorsal end of middle cardinal tooth). In this case though, the relationship is somewhat similar to the one described above between *C. cancellata* and *C. elevata*. Individuals of *C. compta* are indistinguishable, at all sizes, from large individuals of *C. chipolana*. Again, we have evidence of accelerated development, but there is no evidence of a subsequent heterotopic trajectory. If one supposes that, at this landmark, *C. compta* begins at the same point as *C. chipolana*, as it does for almost every other landmark of the hinge area, then the trajectory is simply missing from these data because it is accelerated (time compressed). This is the sole suggestion of pure heterochrony in the entire analysis.

Phylogenetic Revision

The implications of phylogenetic information outlined above were used as the basis for a reanalysis of *Chione* phylogeny. The analysis is contingent on the validity of several observations:

1. The landmark ontogenetic trajectories are themselves character state trees and can inform us of character polarity when viewed in the context of a phylogenetic hypothesis. Such a hypothesis already exists for *Chione*.
2. Coincidence with any segment of a plesiomorphic trajectory is indicative of a heterochronic relationship (for example, *C. chipolana* and *C. compta*), although this may be followed by heterotopic divergence (for example, *C. cancellata* and *C. elevata*). These coincidences are phylogenetically informative and represent shared, derived trajectories.
3. Convergent trajectories may result from exactly that, evolutionary convergence. An existing phylogeny can be used to indicate the likelihood

that a vector convergence resulted from anything other than evolutionary convergence. Although these could be coded as characters and subjected to the usual character congruency test (along with all other characters), the lack of a morphometric-ontogenetic connection between the trajectories dictates that they should be treated with less confidence than ontogenetically linked characters.

Following these guidelines, two approaches were adopted for analysis of the ontogenetic data (Mabee, 2000). First, the four instances in which *C. cancellata* and *C. elevata* and the single instance in which *C. chipolana* and *C. compta* are heterochronically linked were coded as separate binary characters (Table 8.4). One additional observation should be added to the above list when this coding is employed: shared trajectory origins cannot be used if they coincide only at an origin. This simply indicates that the trajectories retain plesiomorphic early stages of shell development, but subsequent developmental stages are autapomorphic. For example, *C. cancellata* and *C. elevata* are indicated as sharing a derived condition, whereas all other taxa are coded as character absent or in-

Table 8.4. Character data set, including adult conchological characters (Roopnarine and Vermeij, 2000; Roopnarine, 2001) and ontogenetic morphometric characters (final six characters, this chapter)

Species	Character Data Set
Chionopsis procancellata	1111311101011101110011l??????
Chionopsis amathusia	11113100010111011100111??????
Puberella cribraria	11121101010111011100001007777
Lirophora latilirata	01111000100110100011?11??????
Panchione ulocyma	01111000100010100111?11??????
Mercenaria mercenaria	01110101100021010111?11??????
Chione elevata	10112211111000101100111014444
C. californiensis	02002211111000101100111005555
C. undatella	01102211111000101100111006666
C. tumens	1033021000100010100?11??????
C. compta	12103211111000101100111102222
C. cancellata	10103210111000101100111013333
C. guatulcoensis	10112210101000101100111??????
C. mazyckii	11112210101000101100111??????
C. chipolana	12121210100000101100111101111

The first two ontogenetic characters are binary encodings of the implied heterochronic relationships between *C. cancellata-C. elevata* and *C. chipolana-C. compta*. The remaining four are the character states used for step matrix transformation (see Table 8.5). Although these latter characters appear to be autapomorphic (every taxon possesses a unique state or the state is missing), this is reflective of the fact that every taxon has unique ontogenetic landmark vectors. Synapomorphic relationship among the vectors is captured by the transformation matrices. Analysis of the data set comprised three separate subsets of the character matrix: adult conchological characters only (Fig. 8.2), adult characters plus binary ontogenetic characters (Fig. 8.12), and adult characters plus ontogenetic transformational characters (Fig. 8.13).

Table 8.5. Step transformation matrices of characters 26–29, representing landmarks 3–8 (Fig. 3)

	Landmarks	
	3 and 6	**4**
Character states	1 2 3 4 5 6 7	1 2 3 4 5 6 7
Transformation costs	– 1 1 2 9 9 9	– 1 9 9 9 9 9
	1 – 1 2 9 9 9	1 – 9 9 9 9 9
	1 1 – 1 9 9 9	9 9 – 9 9 9 9
	2 2 1 – 9 9 9	9 9 9 – 9 9 9
	9 9 9 9 – 9 9	9 9 9 9 – 9 9
	9 9 9 9 9 – 9	9 9 9 9 9 – 9
	9 9 9 9 9 9 –	9 9 9 9 9 9 –
	7	**8**
	– 1 1 9 9 9 9	– 1 9 9 9 9 9
	1 – 1 9 9 9 9	1 – 9 9 9 9 9
	1 1 – 9 9 9 9	9 9 – 1 1 9 9
	9 9 9 – 9 9 9	9 9 1 – 2 9 9
	9 9 9 9 – 9 9	9 9 1 2 – 9 9
	9 9 9 9 9 – 9	9 9 9 9 9 – 9
	9 9 9 9 9 9 –	9 9 9 9 9 9 –

Matrix elements represent the cost of transformation between two character states. A value of nine indicates that there is no morphometric evidence to support the transformation, that its cost is very high and improbable.

formation missing. Likewise, *C. chipolana* and *C. compta* were coded as sharing a derived condition for a separate character. Binary coding does not capture the full complexity of the situation, but coding-shared states as 1, but unique states as 0 (that is, *Chione* species with unique trajectories) and all others as missing ("?"), serve to highlight the known synapomorphies.

A second approach was to code the ontogenetic transformations themselves using the step matrix option in PAUP (Mabee and Humphries, 1993; Mabee, 2000) (Table 8.5). Landmarks with meaningful ontogenetic vectors (Fig. 8.11) were selected as characters, and each vector was coded as a unique character state (because no taxa shared the same vectors). Transitions between states were permitted if they shared a point of origin or one vector could be traced backward in development to another. Transition costs were calculated as the number of steps required to move from one vector to the other. Convergences were ignored. The modified data matrices of both coding schemes were analyzed with a branch and bound algorithm, using PAUP4.0b4a for Linux (Swofford, 1998).

Results of the original analysis comprised 20 most-parsimonious cladograms (MPCs) (consistency index = 0.5778, length = 49) (Fig. 8.2) and are discussed in detail elsewhere (Roopnarine and Vermeij, 2000; Roopnarine, 2001). The analysis here included the same taxa (6 outgroup species with the addition of Mercenaria mercenaria, 15 taxa total) and two binary ontogenetic characters, as well as four ordered characters with transformation matrices. Analysis of the

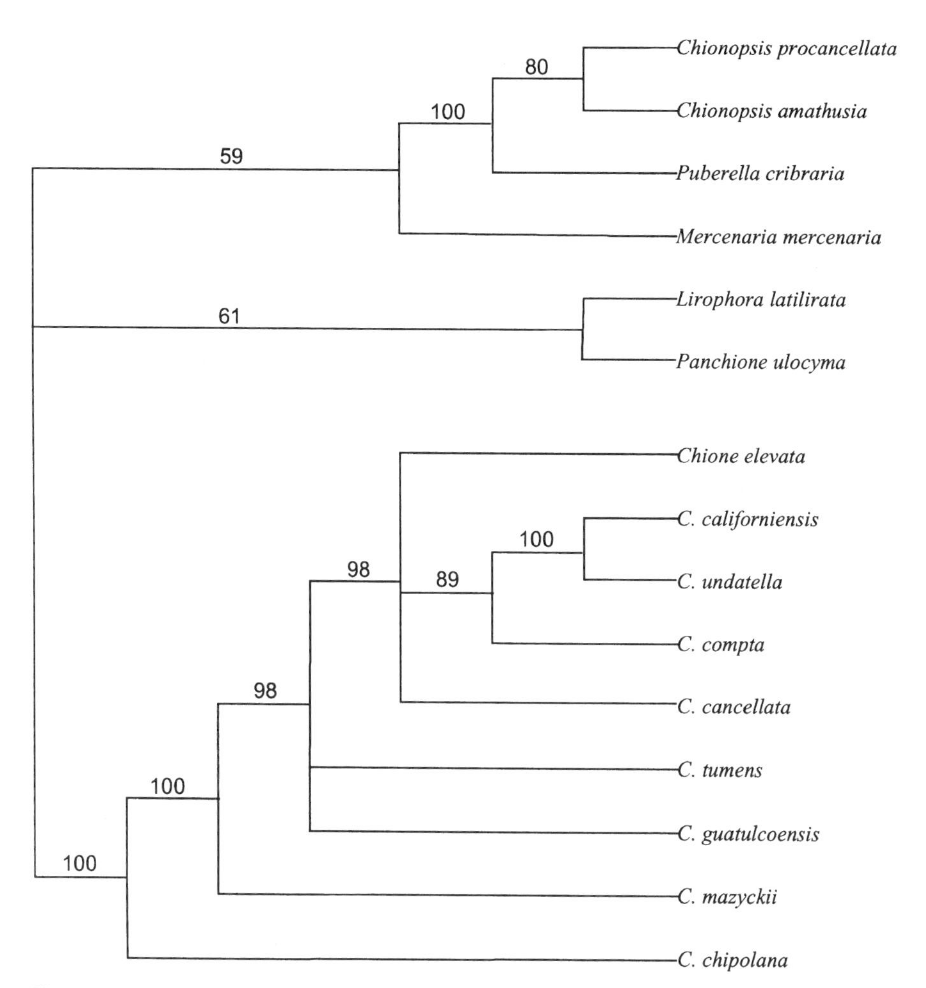

Figure 8.12. A 50% majority-rule consensus tree of 46 equally parsimonious cladograms (consistency index = 0.5714, length = 52). Character data comprise adult conchological and binary encoded ontogenetic morphometric characters. The topology is unchanged relative to analysis of the adult characters only (Fig. 8.2), with only a minor increase in tree length.

original character data plus the two binary characters resulted in 46 MPCs (consistency index = 0.5714, length = 52). A 50% majority-rule consensus tree shows that the new results do not differ significantly from previous results, with the exception of slightly less support for several outgroup nodes and ingroup nodes supporting monophyly of the large, shallow water taxa (Fig. 8.12). Support of the eastern Pacific subset of this latter group, however, is improved; this indicates that the analysis recognized the heterochronic link between *C. compta* and *C. chipolana* as retention of a plesiomorphic condition. Inclusion of the character indicating a similar relationship between *C. cancellata* and *C. elevata* fails to alter the overall topology of the results (that is, the relationship is not a ubiquitous feature in the resultant MPCs). In general, use of the binary coding scheme fails to alter the phylogeny. This most likely results from both inadequate representation of the developmental characters by the binary coding and the small number of developmental characters vs. adult conchological characters in the analysis.

Analysis of the original data set plus the four ordered characters (with underlying transformation matrices) resulted in 80 MPCs (consistency index = 0.7273, length = 195) and an altered topology; note the significant increase in both the consistency index and tree length. The MPCs were again summarized with a 50% majority-rule consensus tree (Fig. 8.13). The major

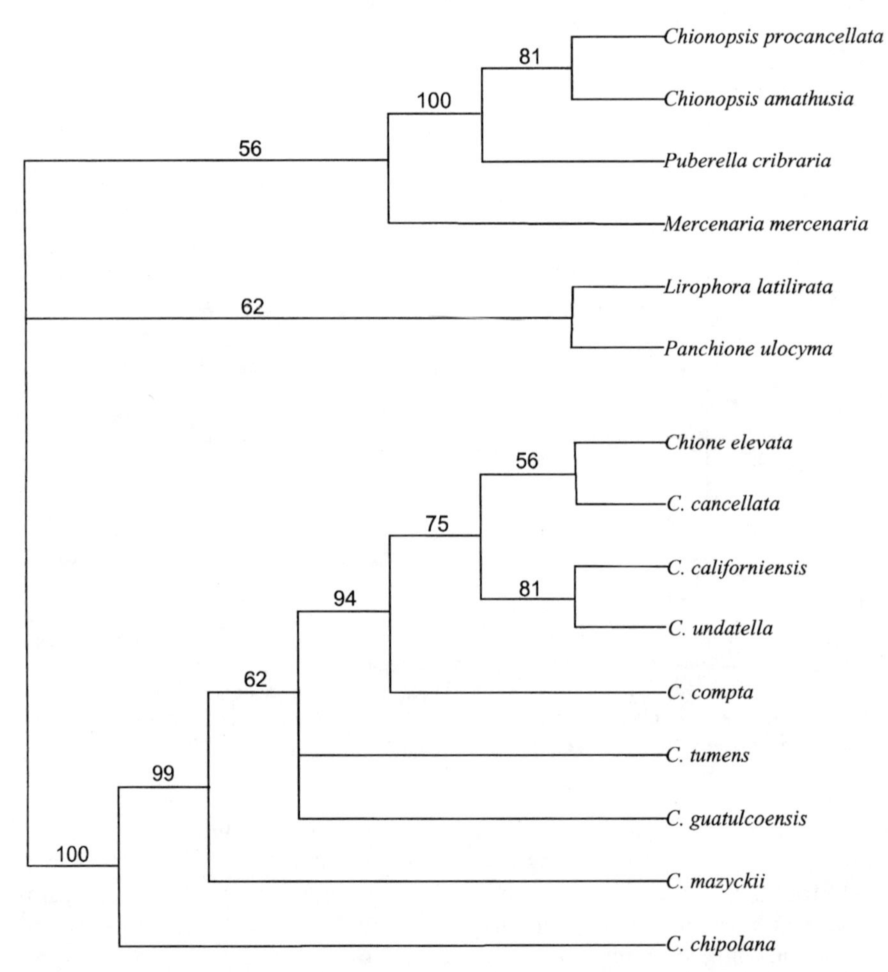

Figure 8.13. A 50% majority-rule consensus tree of 80 equally parsimonious cladograms (consistency index = 0.7273, length = 195). Character data comprise adult conchological and ordered ontogenetic morphometric characters (using step matrices). The topology differs from the original, based on adult characters only, in placing *C. cancellata* and *C. elevata* as sister taxa on the basis of their apparent heterochronic-heterotopic relationship. The significant increase in length results from coding transition costs in the step matrices (Table 8.5).

difference between this tree and the original was again lessened majority support of ingroup branches, although monophylies of the genus as well as the clade of extant shallow water taxa are maintained. Topology of this latter clade is altered, however, with 56% of the MPCs supporting a sister-taxon relationship between *C. cancellata* and *C. elevata*. This is of course a direct result of the ontogenetic data. The ontogenetic links between *C. chipolana* and the extant *C. cancellata* and *C. compta* are recognized correctly as symplesiomorphic.

There is a danger inherent to the interpretation of consensus trees, but they represent the data common to a set of equally parsimonious hypotheses. Advancing from the use of consensus trees to the comparison of actual cladograms requires the inclusion of several taxa absent from the current analyses, namely, extinct fossil species. These taxa may be critical to our understanding of ancestor-descendant ontogenetic transformations via heterochrony and heterotopy, thereby testing the overturning of adult characters-only phylogenies by phylogenies, which include ontogenetic data. The *Chione* taxa required for such an analysis are available, and, although exploration of this issue is beyond the scope of this chapter, a phylogeny based only on adult discrete characters has been completed (Roopnarine, 2001) (Fig.'8.2).

Evolutionary History

Although addition of the developmental characters did not revise the previous tree radically, the slight alterations of topology lend themselves to an interesting interpretation of the history of this genus. Given its probable tropical western Atlantic origin (Roopnarine, 1996), *Chione* has apparently invaded the eastern Pacific at least twice, one event leading to the *C. tumens* group, which may include *C. guatulcoensis*, and another invasion involving shallow-water, larger taxa, leading to *C. compta*, *C. californiensis*, and *C. undatella*. When the ontogenetic data are coded as binary presence/absence characters, the sister subclade of this latter subclade is the western Atlantic *C. cancellata-C. elevata* pair. Coding the ontogenetic data as ordered characters with transformation costs, however, suggests several alternative scenarios:

1. at least three invasions of the eastern Pacific, with separate dispersal events resulting in *C. compta* and the sister-taxon *C. californiensis* and *C. undatella*;
2. a single eastern Pacific origin of those taxa, with subsequent reinvasion of the western Atlantic by the common ancestor of *C. cancellata* and *C. elevata*; and
3. vicariant fragmentation of an ancestral eastern Pacific and western Atlantic shallow-water ancestor, resulting in the current species.

Examination of fossil taxa may eventually assist in testing these alternative hypotheses. Regardless of which hypothesis is supported, this radiation of shallow-water, large taxa was accomplished by an apparent combination of heterotopy and heterochrony. Both *C. cancellata* and *C. compta* retain some plesiomorphic features at small shell size, as shown in *C. chipolana*, but both taxa are also obvious heterotopic derivatives of the ancestral condition. Evolution of the other taxa within the subclade is less clear. *C. californiensis* and *C. undatella* may also be heterotopic descendants of the ancestral condition, but the information necessary to test this either is lost because of the loss of stages of the ontogenetic trajectories or is present in unexamined fossil taxa. *C. elevata* may be a cladogenetic descendant of *C. cancellata* or a common ancestor that was significantly more similar to *C. cancellata*. *C. elevata*'s development can be traced back to the plesiomorphic conditions of the clade but is removed from it as a heterochronically and heterotopically altered version of *C. cancellata*.

This type of interpretation is impossible, given solely a standard cladistic analysis and cladogram topology. Parsimony forbids us from speculating on the direct ancestor-descendant relationships among taxa, which would constitute construction of evolutionary trees rather than cladistic hypotheses. To make the transition from cladogram to tree a testable one, it is generally considered necessary to employ data independent of the derivation of the cladogram, such as ecological or stratigraphic information. Nevertheless, at the character level, ontogenetic pathways that are linked heterotopically and particularly heterochronically predict the nature of intervening and hypothetical taxa. For example, given the knowledge that plesiomorphic states are retained by some upper branches of the clade, that heterotopic alterations of these states during development serve to distinguish derived taxa, and that heterochronic alterations connect sister taxa, we can make predictions regarding the morphometric development of common ancestors and intervening taxa.

The complete tree topology is likely to increase in complexity with the addition of fossil taxa. Before this study, such an addition would probably have resulted in a significant reduction of phylogenetic resolution. This would be due to the low number of phylogenetically informative characters and the large number of fossil taxa within a clade such as this one. The ability to use morphometric developmental data, and discriminate symplesiomorphies, synapomorphies, and autapomorphies on the basis of heterochrony and heterotopy, greatly increases the number of characters available for analysis.

SUMMARY

Testing the null hypothesis of no heterochrony is an essential first step that must be taken when conducting morphometric studies of heterochrony. Appropriate testing requires the independent quantification of size and shape as is possible in the geometrically homologous landmark approach. Intertaxon intersection of shape distributions is an essential precursor to any explanations based on heterotopy or heterochrony. Ordination of shape in accordance with developmental time is necessary to discriminate between the two classes of process.

Developmental time can be represented adequately in most cases by body (centroid) size, making chronological time irrelevant at this stage of analysis. Overlap, even partially, of segments of the resulting ontogenetic vectors is indicative of heterochrony, whereas the sharing of points on vectors, followed by subsequent divergence, is indicative of heterotopy. The two processes however are not exclusive. Finally, determination of specific types of heterochrony will in most cases require knowledge of chronological time.

In summary, consideration of a null hypothesis of no heterochrony, or no overlap of ontogenetic trajectories when projected onto shape dimensions, clarifies the frequency of heterochrony as a plausible mechanism for evolutionary change. In the present example, of the six ingroup species considered (and the 12 landmarks examined for each), there are only two supportable instances of character heterochrony. And one of those instances is a probable combination of heterochrony and heterotopy. There is significant evidence for a higher frequency of heterotopy, and gaps in the phylogenetic scenarios highlight the ways in which the extinction of taxa or loss/compression of developmental stages may limit our ability to recognize heterochrony and heterotopy in morphometric data.

ACKNOWLEDGMENTS

I thank Miriam Zelditch for kindly extending an invitation to participate in this volume, assisting with the analysis and interpretation of data, and adding significant insight to my ideas. I am also very grateful to the numerous colleagues with whom I have discussed these issues over the years, including Sandy Carlson, Glenn Jaecks, Lindsey Leighton, Rich Mooi, Ross Nehm, Gary Vermeij, and Miriam Zelditch. I particularly thank the students of Sandy Carlson's Heterochrony Graduate Seminar at UC Davis for stimulating discussions.

REFERENCES

Alberch P, Gould SJ, Oster GF, Wake DB (1979): Size and shape in ontogeny and phylogeny. Paleobiology 5: 296–317.

Berry WBN, Barker RM (1975): Growth increments in fossil and modern bivalves. In: Rosenberg GD, Runcorn SD (eds), *Growth Rhythms and the History of the Earth's Rotation*. New York: John Wiley and Sons, p. 9–25.

Blackstone NW, Yund PO (1989): Morphological variation in a colonial marine hydroid; a comparison of size-based and age-based heterochrony. *Paleobiology* 15: 1–10.

Bookstein FL (1991): *Morphometric Tools for Landmark Data*. New York: Cambridge University Press.

Bookstein FL (1996): Combining the tools of geometric morphometrics. In: Marcus LF, Corti M, Loy A, Naylor GJP, Slice DE (eds), *Advances in Morphometrics*. New York and London: Plenum Press.

de Beer G (1958): *Embryos and Ancestors*. Oxford, UK: Oxford University Press.

Dettman DL, Roopnarine PD, Flessa KW (1998): Hydrology on the half-shell; estimating ancient Colorado River discharge from isotopic variation in fossil bivalve mollusks. Geol Soc Am Abstracts Programs 30: 118.

Fink WL (1982): The conceptual relationship between ontogeny and phylogeny. Paleobiology 8: 254–264.

Goodwin DH (2000): Almost weather: high-resolution sclerochronological calibration of isotopic variation in a bivalve mollusk. Geol Soc Am Abstracts Programs 32: A-298.

Gould SJ (1977): *Ontogeny and Phylogeny*. Cambridge, MA: The Belknap Press of Harvard University Press.

Gould SJ (1988): The uses of heterochrony. In: McKinney ML (ed), *Heterochrony in Evolution: A Multidisciplinary Approach*. New York: Plenum Press, p. 1–13.

Jablonski D (1997): Body-size evolution in Cretaceous molluscs and the status of Cope's rule. Nature 385: 250–252.

Jackson JBC (1973): The ecology of molluscs of *Thalassia* communities, Jamaica, West Indies. I. Distribution, environmental physiology, and ecology of common shallow-water species. Bull Mar Sci 23: 313–350.

Jones CC (1976): *The Biology and Evolution of Textit Chione (Bivalvia; Veneridae) in the Eastern Coastal Plains of North America*. Dissertation. Cambridge, MA: Harvard University.

Jones DS (1991): Habitat-specific growth of hard clams textit Mercenaria mercenaria (L.) from the Indian River, Florida. J Exp Mar Biol Ecol 147: 245–265.

Klingenberg CP (1996): Multivariate allometry. In: Marcus LF, Corti M, Loy A, Naylor GJP, Slice DE (eds), *Advances in Morphometrics*. New York and London: Plenum Press, p. 23–49.

Mabee PM (2000): The usefulness of ontogeny in interpreting morphological characters. In: Wiens JJ (ed), *Phylogenetic Analysis of Morphological Data*. Smithsonian Institution Press, p. 84–114.

Mabee PM, Humphries J (1993): Coding polymorphic data: examples from allozymes and ontogeny. Syst Biol 42: 166–181.

McKinney ML (1988): Classifying heterochrony: allometry, size, and time. In: McKinney ML (ed), *Heterochrony in Evolution: A Multidisciplinary Approach*. New York: Plenum Press, p. 17–34.

McKinney MC, McNamara KJ (1991): *Heterochrony: The Evolution of Ontogeny*. New York: Plenum Press.

Raff RA (1996): *The Shape of Life. Genes, Development, and the Evolution of Animal Form*. Chicago, IL: The University of Chicago Press.

Rice SH (1997): The analysis of ontogenetic trajectories: when a change in size or shape is not heterochrony. Proc Natl Acad Sci USA 94: 907–912.

Rohlf FJ (1996): Morphometric spaces, shape components and the effects of linear transformation. In: Marcus LF, Corti M, Loy A, Naylor GJP, Slice DE (eds), *Advances in Morphometrics*. New York and London: Plenum Press, p. 117–129.

Rohlf FJ (1999): *tpsRelw-Thin-plate spline relative warp*. Stony Brook, NY: Department of Ecology and Evolution, State University of New York.

Roopnarine PD (1995): A re-evaluation of evolutionary stasis between the bivalve species *Chione erosa* and *Chione cancellata* (Bivalvia: Veneridae). J Paleontol 69: 280–287.

Roopnarine PD (1996): Systematics, biogeography and extinction of chionine bivalves (Early Oligocene-Recent) in the Late Neogene of tropical America. Malacologia 38: 103–142.

Roopnarine PD (2001): A history of diversification, extinction, and invasion in tropical America as derived from species-level phylogenies of chionine genera (Family Veneridae). J Paleontol 75: 644–658.

Roopnarine PD, Beussink A (1999): Extinction, geographic replacement, and escalation of the bivalve *Chione* in the Late Neogene of Florida. Paleontologia Electronica vol. 2.

Roopnarine PD, Vermeij GJ (2000): One species becomes two: the case of *Chione cancellata*, the resurrected *C. elevata*, and a phylogenetic analysis of *Chione*. J Molluscan Stud 66: 517–534.

Schoene BR (2000): Effects of river discharge on growth rates in marine bivalve mollusks. Geol Soc Am Abstracts Programs 32: A-298.

Swofford DL (1998): *PAUP* (Phylogenetic Analysis Using Parsimony (*and Other Methods) Version 4.0*. Sunderland, MA: Sinauer Associates.

Thompson, D'AW (1942): On Growth and Form, abridged edition, JT Bonner (ed.). Cambridge University Press, p. 346.

Zelditch ML, Fink WL (1996): Heterochrony and heterotopy: stability and innovation in the evolution of form. Paleobiology 22: 241–254.

Zelditch, ML, Swiderski DL, Fink, WL (2000): Discovery of phylogenetic characters in morphometric data. In Phylogenetic Analysis of Morphological Data. JJ Wiens (eds.). Smithsonian Institution Press, pp. 37–83.

9

TESTING MODULARITY AND DISSOCIATION: THE EVOLUTION OF REGIONAL PROPORTIONS IN SNAKES

P. David Polly

Molecular and Cellular Biology Section, Biomedical Science Division, Queen Mary & Westfield College, London, United Kingdom

Department of Palaeontology, The Natural History Museum, London, United Kingdom

Jason J. Head

Department of Geological Sciences, Southern Methodist University, Dallas, Texas

Martin J. Cohn

Division of Zoology, School of Animal and Microbial Sciences, University of Reading, Whiteknights, Reading, United Kingdom

INTRODUCTION

In this chapter, we test a hypothesis about dissociation in the evolution of body and tail proportions in snakes and examine whether that dissociation represents heterochronic change. Heterochrony can be defined as a dissociation in timing between two developmental modules. This definition—subtly different from the usual "evolutionary change in developmental timing"—was used by Raff

Beyond Heterochrony: The Evolution of Development, Edited by Miriam Leah Zelditch
ISBN 0-471-37973-5

(1996) after he explored the concepts of dissociation and modularity in relation to the evolution of development. Raff defined *modularity* as the division of an organism into units, each with its own genetic specification, hierarchical organization, and interactions with other similar units. Through the homeostatic influences of functional and developmental integration, modules are usually linked. Raff defined *dissociation* as the process of unlinking. A heterochronic change thus requires dissociation between two dissociable modules. Distinguishing heterochronic change from its alternatives therefore requires analysis of at least two key elements: that evolutionary change can be attributed to shifts in the timing between modules and that those modules are dissociable. Attention is now being paid to the former (the chapters in this volume are a notable example) but less so the latter. In this chapter, we test two possible developmental modules—those controlling the number of vertebrae in the body region and in the tail region of snakes—for dissociation.

Despite the simplicity of the concepts, the empirical recognition of modularity and dissociation can be difficult. One reason is that developmental modules often do not correspond with obvious morphological boundaries. Lovejoy et al. (2000) recently identified an unexpected module in the mammalian forelimb. By analyzing growth patterns in the forearm of several primates, they identified a surprising suite of skeletal elements that scale together during growth. The bones affected by the scalar are the distal radius and the digits II–V, which have a linear relationship and behave as a developmental module, but not the proximal radius or digit I. Interestingly, the *Hoxd11* gene is expressed precisely in this region (Lovejoy et al., 2000) and is known to play a role in growth of long bones (Goff and Tabin, 1997). A second complication is that developmental processes may have overlapping physical domains. For example, several topographically disjunct skeletal elements—notably membranous bones—are linked through the common expression of *CBFA1*, a transcription factor that plays multiple roles in bone formation (Mundlos, 1999). Humans with mutations in CBFA1 may have open skull sutures, supernumerary teeth, absent or reduced clavicles, and unfused pubic symphyses—symptoms clinically diagnosed as "cleidocranial dysplasia." The combined targets of CBFA1 mutational effects do not function as a module (they do not form a hierarchical unit), but the pattern of *CBFA1* expression *links* modules, making complete dissociation between them difficult. Dissociation may thus be more a matter of degree than an absolute phenomenon.

Furthermore, dissociation is an *evolutionary* phenomenon. Unlike modularity, which is a property of the individual organism created by genetic and developmental correlation, dissociation is a process of unlinking over evolutionary time. Modules become dissociated along branches of a phylogenetic tree. Because modules may be linked by many independent interactions—some direct (like induction), others less so (like the influence of CBFA1)—dissociation may be subtle enough to require a statistically significant sample of comparisons. The scalar identified by Lovejoy et al. (2000) was apparent only after several species were compared. A comparison of distal limb elements in only two species reveals simply that each species has distal limb elements that differ in size and shape from those in the other species. Only when a large enough sample of species is studied does it become apparent that the distal radius and digits II–V are correlated with one another relative to the proximal radius and digit I. It is also possible that links between modules may not be linear, but the scalar may cause them to change allometrically. A single between-species comparison might reveal that the relationship between two modules has shifted, but many such comparisons may reveal that those shifts are correlated. The identification of modules and dissociation is thus a type of morphological integration study (Olson and Miller, 1959; Cheverud, 1982; Atchley, 1991; Zelditch, 1995; Leamy et al., 1999).

Because of the variation in vertebral segment number, snakes are ideal for studying the evolutionary interactions between developmental processes determining segment number and those determining regional segment identity. Unlike in limbed vertebrates, locomotory and ecological diversity in snakes have been attained exclusively through specialization of the vertebral column and axial muscular system. The range of snake ecomorphology has been realized through modification of vertebral proportions (Gasc, 1976), vertebral number (Lindell, 1996), body length (Lindell, 1996), body mass (Shine, 1986), and muscular arrangements (Jayne, 1982, 1988). The extremes of snake ecomorphology—arboreal vs. terrestrial vs. fossorial vs. aquatic forms—can often be recognized simply by the proportion of body to tail length. Many arboreal snakes, for example, have long bodies with proportionally long tails, terrestrial taxa have long bodies with shorter tails, and fossorial snakes are short with extremely short tails (Fig. 9.1). These proportional differences are found in both the linear lengths of these regions and the number of vertebral segments in each. Specifically, we test the hypothesis that control of the number of body segments is "modularized" and can be dissociated from the tail. As reviewed below, patterns of development in the two regions suggest that they are modularized; however, several developmental processes are common to both and may prevent them from being dissociated. Our hypothesis is thus based on known developmental mechanisms; we test that hypothesis against adult comparative data in a phylogenetic context. Our test thus goes further than looking at simple correlations between the number of body and tail segments in adult snakes; we examine their correlation as evolutionary *transformations*.

MORPHOLOGY AND DEVELOPMENT OF THE SNAKE VERTEBRAL COLUMN

Vertebral morphology in snakes is highly derived, and the primary regions of the vertebral column can be recognized based on discrete morphological features. All snakes possess zygosphene-zygantral articular processes along the dorsolateral margins of the neural canal throughout the column, as well as synapophyses (paired dia- and parapophyseal articulations for the ribs), corresponding to the absence of vertebral transverse processes. The primary division in the snake vertebral column is the recognition of pre- and postcloacal (that is,

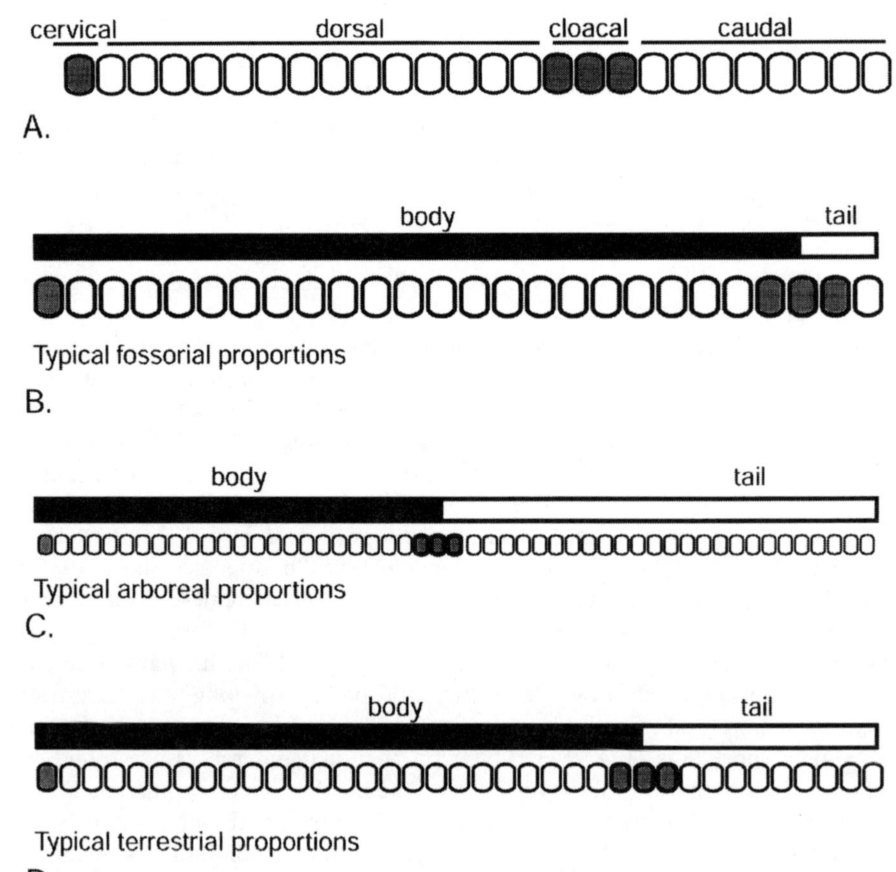

Figure 9.1. Differences in regional proportions in terrestrial, arboreal, and fossorial snakes. A: morphologically defined axial regions within snakes. Snakes have a cervical vertebra, which articulates with the back of the skull, a large number of rib-bearing dorsal vertebrae (homologous to thoracic and lumbar vertebrae in mammals and birds), a small number of cloacal vertebrae (homologous to sacral vertebrae in other tetrapods), and a series of caudal vertebrae. B: typical proportion between body and tail (pre- and postcloacal) segment numbers in a typical fossorial snake (e.g., *Typhlops*). There are few vertebrae overall, and the proportional number in the tail is small. C: proportion of body and tail in a typical arboreal snake (e.g., *Ahaetulla*). There are many vertebrae, almost half of which are found in the tail. D: proportion of body and tail in a typical terrestrial snake (e.g., *Python*). There are a large number of vertebrae, but the proportion found in the tail is considerably less than in arboreal snakes.

caudal) regions, which are separated by a series of three to five cloacal vertebrae. These regions are differentiated morphologically by the transition from precloacal synapophyses to forked cloacal lymphapophyses to caudal pleurapophyses (sensu Hoffstetter and Gasc, 1969), the transition from a single ventral hypapophysis to paired hemapophyses at the cloaca, and an abrupt shortening of the centrum at the cloaca (Thireau, 1967). Within the precloacal column, regions have been recognized for some taxa, including "cervical" or anterior trunk, midtrunk, and posterior trunk regions (e.g., LaDuke, 1991a). Prominent hypapophyses are restricted to the "cervical" region and subcentral paramedian lymphatic fossae to the "posterior trunk" region. However, these characteristics are restricted to inclusive clades of snakes and are not present in the majority of taxa. Among all snakes, the division of the precaudal vertebral column into discrete regions can only be achieved by recognition of relative placement. The only universally recognizable regions are pre- and postcloacal (Hoffstetter and Gasc, 1969).

The development of the snake axial skeleton can be thought of as three conceptually separate processes: segmentation, regionalization, and skeletogenesis. During early development, the embryo's paraxial mesoderm is partitioned into segments, the somites, which then differentiate into axial bones, muscles, and dermis. As development progresses, the structures within each segment continue to grow, accentuating regional differences along the axis. The degree to which these three processes are integrated is an open question, and at least two of them, segmentation and skeletogenesis, are common to both body and tail. In this chapter, we ask whether the number of segments—and therefore the number of vertebrae—in the body and tail regions can be dissociated evolutionarily.

Most of the work on somitogenesis has been carried out using the chick embryo, but similar studies of other vertebrate embryos suggest that mechanisms are generally conserved. Cells that emerge from the primitive streak and Hensen's node during gastrulation form the embryonic mesodermal layer. The streak deepens into a *primitive groove* with a thickening known as Hensen's node at the anterior end (Fig. 9.2, A and B). Premesodermal epiblast cells move toward the midline of the embryo and ingress through the streak. After ingression, cells move laterally and anteriorly, giving rise to the mesodermal layer of the embryo. Fate mapping of the streak and node has identified regional subpopulations of cells that act as a source of somitic mesoderm (Tam and Tan, 1992). After the anterior (prelumbar) part of the body is laid down (LeDouarin et al., 1996), the primitive streak and node regress posteriorly, leaving the notochord in their wake, and are replaced by a bulb of mesenchymal cells in the tail bud (Fig. 9.2C).

In the tail bud, cells continue to ingress from the surface, and many genes associated with gastrulation continue to be expressed (Knezevic et al., 1998). The tail bud continues to produce somites that are added onto the posterior end of the segmental plate (Fig. 9.2D) beginning with the first lumbar vertebra (LeDouarin et al., 1996). The embryonic distinction between Hensen's node and the primitive groove on the one hand and the tail bud on the other thus presumably corresponds to the two primary regions into which adult snake axial morphology is divided—precloacal and postcloacal, or body and tail. Experimental manipulations of embryos provide evidence for the idea that the tail bud retains competence to generate somites beyond the stage at which the

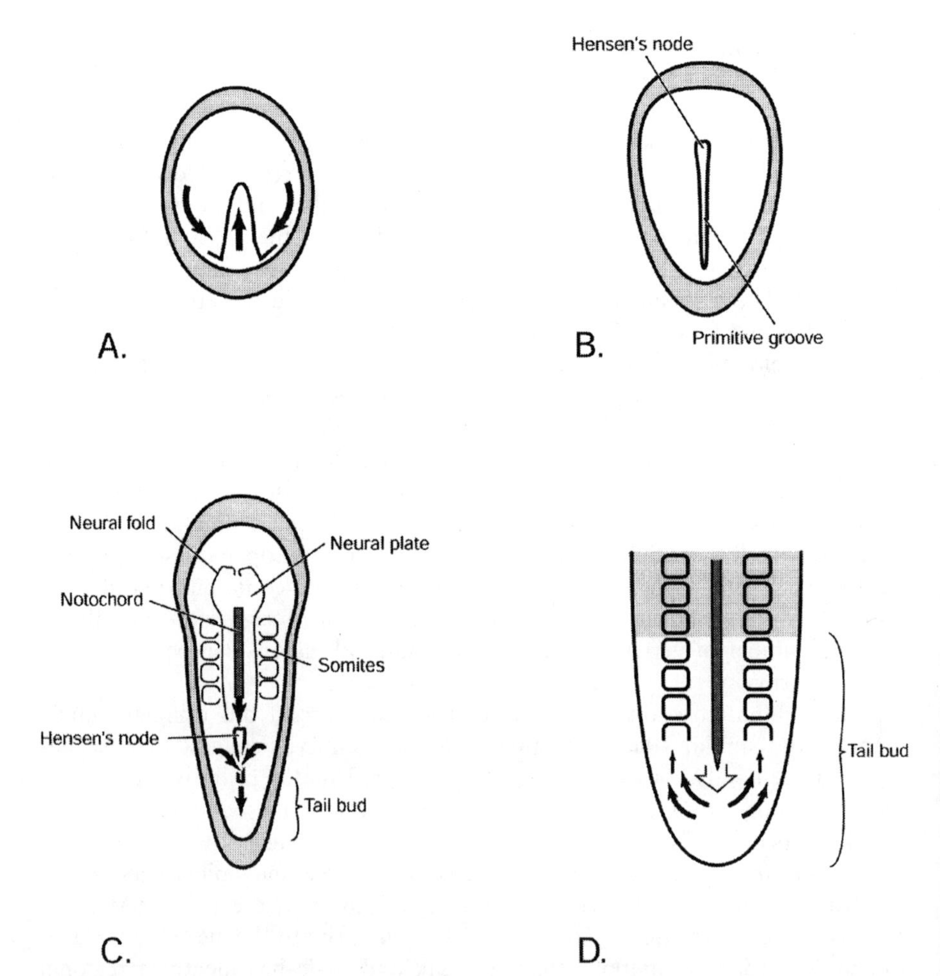

Figure 9.2. Early development and segmentation. A: early gastrulation. Primitive streak forms and extends anteriorly as cells migrate through it to form mesodermal tissue. B: streak extends and deepens to become the primitive groove, with Hensen's node at the anterior end. C: further differentiation takes place as Hensen's node and the primitive groove regress posteriorly. Precloacal body segments are formed in the paraxial mesoderm lateral to the full length of the groove. D: gastrulation-like processes continue in the tail bud behind the posterior end of the primitive groove. Cells are added to the end of the growing bud, allowing the segmentation process to continue posteriorly. See text for details.

expected number of somites are laid down. For example, transplanting the tail bud of a 13.5-day mouse embryo into an 8.5-day embryo shows that tail bud cells can "continue to participate in somitogenesis well beyond their expected developmental life span," although somitogenesis ceases when the appropriate numbers of host somites have formed (Tam and Tan, 1992, p. 714). These results suggest that posterior segment number is not solely determined by the competence of tail bud cells to form mesoderm but rather by extrinsic cues acting on those cells.

The process of neurulation begins on the ectodermal surface as gastrulation and segmentation proceed posteriorly. Above the notochord and anterior to the regressing Hensen's node, two lateral folds develop around the open neural plate (Fig. 9.2C). These arch medially over the plate to form an enclosed neural tube, which sinks into the embryo over the notochord. The mesodermal layer is now divided by the tube and notochord into two lateral halves, known as paraxial mesoderm. After neurulation, the paraxial mesoderm is segmented beginning at the anterior end, as somites condense within the segmental plate. The timing of individual somitogenesis is foreshadowed by the *hairy1* and *lunatic fringe* genes, which are expressed in a wave-like pattern spreading anteriorly from the caudal end of the presegmental plate (Palmeirim et al., 1998). As the wave reaches its anterior-most point, an expression pathway involving the *Delta-1* and *Notch-1* genes polarizes a package of mesoderm one segment in length into rostral and caudal halves. A new somite is thus added to the posterior end of the growing somitic chain (Palmeirim et al., 1997). The process continues down the axis and into the tail bud following the addition of new mesoderm and somites at the caudal end (LeDouarin et al., 1996). Once formed, the somites differentiate into dermatome, which becomes the connective tissues of the skin, sclerotome, which condenses around the notochord and neural tube as vertebral elements and ribs, and myotome, which migrates ventrolaterally to become the axial and appendicular muscles.

The number of vertebrae does not change after early development, and, at hatching, precloacal, cloacal, and postcloacal vertebrae are clearly distinguishable. Subsequent differentiation within these regions can be quite complex (LaDuke, 1991b) but does not seem to be mediated by sharp Hox boundaries. This may not be surprising given the complexity and number of gene expression pathways contributing to vertebral development (Monsoro-Burq et al., 1996). Unlike vertebrates with forelimbs, Hox gene expression is not regionalized in the precloacal segments (Cohn and Tickle, 1999). In the chicken, for example, the expression boundaries of *Hoxb5*, *Hoxc8*, and *Hoxc6* correspond to the boundaries of the adult cervical, thoracic, and lumbar regions, respectively, whereas, in snakes, the expression of these genes is coextensive along the entire precloacal region. The vertebrae are at first also uniform in size. In *Python molurus* at 10 days posthatching the first vertebra is the largest in both length and width. Posteriorly, all segments are approximately the same length, but their width tapers very gradually along the body, dropping sharply about halfway down the tail. Snakes grow not by addition of additional segments but by growth in segments. In amniotes, the number of vertebrae in an individual is constant, once established during early embryogenesis, except in species capable of regenerating their tails. At hatching, the vertebrae of snakes (and other limbless vertebrates) are subequal in size (see Wake, 1980, for data on caecilian axial skeleton growth). As the snake grows to maturity, each of the segments increases in both length and width, although not all segments increase at the

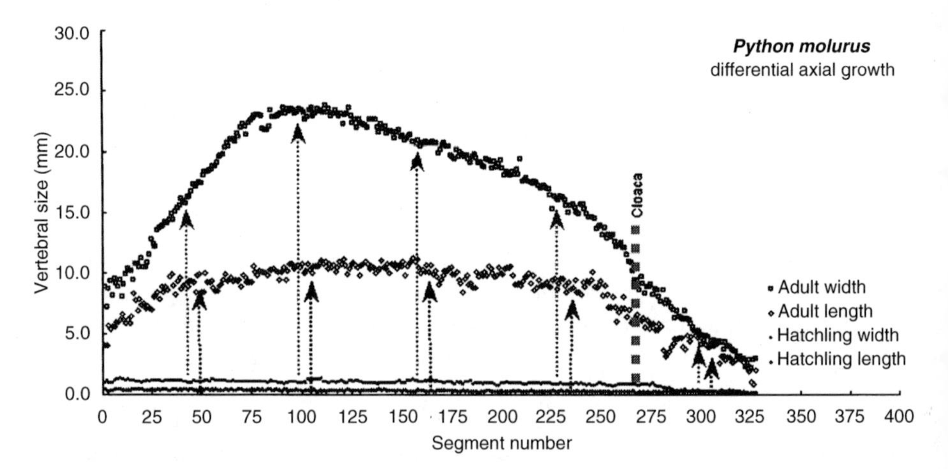

Figure 9.3. Differential growth in length and width of vertebrae in *Python molurus*. Horizontal axis shows vertebral position from head to tail; vertical axis shows linear dimension (length or width) in millimeters. Two individuals are shown, a 10-day-old hatchling (parallel lines with solid shapes along the bottom of the graph) and an adult (two curved lines with open shapes). The length (diamonds) and width (squares) of each vertebrae of each snake are shown. The position of the cloacal vertebrae is shown by the broken gray line. The amount of posthatching growth in each segment is indicated by the difference in adult and hatching sizes (indicated by the broken vertical arrows). Growth in tail vertebrae is less than in body vertebrae, and growth in width is greater than growth in length. The region of maximum growth is in the body, about one-third back from the head. The hatchling snake had more vertebrae (349) than the adult (327), so some were removed to make the two data series equal length for the purposes of this figure.

same rate or to the same final size (Fig. 9.3). Both rate of growth of an individual snake and its maximum adult size are correlated with the total number of vertebrae (Lindell, 1996).

Because of the variation in both segment number and in regional proportions, snakes are ideal for studying the evolutionary interactions between developmental processes determining segment number and those determining regional segment identity. Snakes have greater than average among-species variation in vertebral number, including the proportion found in the pre- and postcloacal regions (Lindell, 1994). They also have unusually high within-species variation, a phenomenon that is common in elongate vertebrates (Jockusch, 1997; Lindell, 1996). *Vipera berus*, for example, has vertebral counts ranging from 139 to 157—about plus or minus about 6% of the total number—but extreme variation is apparently not common because the standard deviation is only about 3.0 (Lindell, 1996). Thus, among-species variation in vertebral counts is much greater than within-species. This is an important point for our analysis because it implies that we are actually measuring evolutionary changes in vertebral number rather than sampling an artifact of within-species variation.

Related to this is the question of intraspecific variation in body and tail proportions, including sexual dimorphism. For example, male snakes are generally smaller than females and often have longer tails (Shine, 1993; Lindell, 1996). The latter is possibly because males store their hemipenes in the proximal part of the tail. Snake dimorphism may be more due to variation in growth rather than to variation in vertebral number. Lindell (1996) found small but statistically significant differences in the vertebral counts of *Vipera berus*, the European adder. On average, males had 144.9 vertebrae [standard deviation (SD): 2.21] and females 148.8 (SD: 3.14). Within- and between-sex variance in snout-vent length, however, was much greater, on average 500.9 mm (SD: 75.5) in males and 523.2 mm (SD: 105.7) in females. Like axial regionalization, dimorphism seems to develop during maturation because it is rarely apparent at hatching (Shine, 1993). The time to maturation is usually longer in females, leading to an overall greater size. Because growth of midbody vertebrae is greater than that of tail vertebrae, a longer period of growth should result in a proportionally shorter tail (Fig. 9.3). The snout-vent length of females should, therefore, increase more than it does in males during maturational growth. Thus dimorphism in growth pattern and overall size may also be factors explaining the between-sex differences in regional proportion (King et al., 1999).

TESTING FOR DISSOCIATION

We hypothesize that—in regards to the number of vertebrae—the body and tail of snake are modules that can be dissociated during evolution (Fig. 9.4). The modularity of the two regions is suggested by the differences in the formation of their mesoderm and by the ecomorphological differences among snake species, shown in Figure 9.1. However, modularity does not necessarily imply dissociability. Because segmentation and somitogenesis are continuous across the body/tail bud boundary, differences in regional proportion may be an allo-

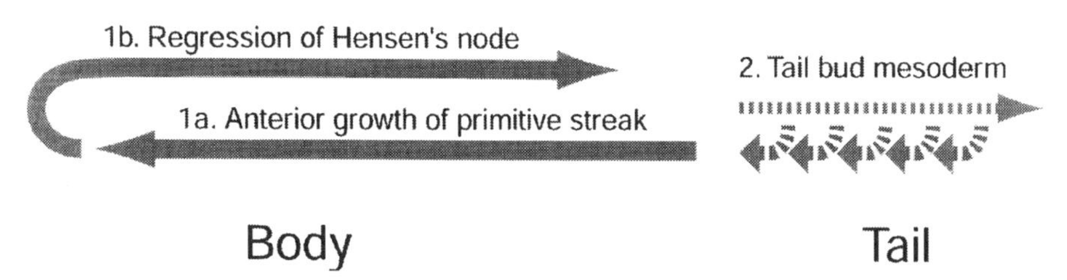

Figure 9.4. Two modules in snake axial development. The precloacal vertebrae are derived from somites that formed in paraxial mesoderm along the margins of the primitive streak. Postcloacal vertebrae are derived from somites laid down in segmented mesoderm that is added to the end of a growing tail bud. The adult cloacal region is thus located at the embryonic junction between tail bud and posterior-most streak. See text for details.

metric function of the total number of segments. In other words, when the total number of segments is changed, vertebrae may be added or removed from the two regions in some constant proportion. Such a situation could explain the difference between the fossorial and arboreal snakes in Figure 9.1 without invoking dissociation. The fossorial snake has few total segments and a very short tail. If evolutionary changes in the two regions are correlated and changes in the tail are proportionally greater than in the body, then one would expect a snake with lots of vertebrae to have a proportionally longer tail. Thus, inability to dissociate the two regions might imply that the long *tail* of arboreal snakes is not related to its habitat so much as a long body is; however, dissociability would allow body and tail to evolve independently, permitting a wider variety of ecomorphologies. The fact that some taxa (like the one shown in Fig. 9.1B) have lots of vertebrae but relatively few in the tail hints at such a possibility.

We test this hypothesis by looking at evolutionary changes in the number of body vertebrae relative to the number in the tail. We ask (1) whether a simple change in the number of segments is responsible for the differences among snakes both in segment number and relative length of body and tail regions and (2) whether there is evidence for independence in the number of segments and the relative proportion of body and tail. We first investigate the relationship between number of body vertebrae, the number of tail vertebrae, and the total number of vertebrae to determine the extent to which they are correlated. This is used to determine whether simple allometry explains the diversity of body-to-tail proportion among snakes. We also look at changes in vertebral number along the branches of snake phylogeny. In particular, we look for a correlation between change in the number of body vertebrae and change in the number of tail vertebrae. The relationship should be constant if there is no dissociation. We identify exceptions as cases of dissociation.

In this study, dissociation is therefore a departure from the usual relationship between change in the number of body vertebrae and in the number of tail vertebrae. Dissociation could manifest itself in three ways: (1) segments are added or removed to the tail without affecting the number in the body region, (2) segments are added or removed in the body region without affecting the tail, or (3) segments are added or removed from one region and the opposite done in the other. We do not invoke dissociation when segments are added or deleted from both regions, even if in different proportions. After identifying dissociations, we return to the question of whether they represent heterochronic change in **CONCLUSIONS**.

MATERIALS AND METHODS

We examined the skeletons of 32 species of snakes, recording the number of precloacal (body) and postcloacal (tail) vertebrae in each (Table 9.1). Data were collected from articulated skeletons. Not all specimens were fully mature individuals, but all specimens were at least 28% of reported adult body lengths. Posthatchling vertebral counts reflect the number of segments produced during early embryonic development. Neither the number of vertebrae nor their regional identities change after being established in the embryo. We also examined embryos of *Python molurus* and *Thamnophis ordinoides*. The species considered in this study represent all of the higher orders of living snakes, as well as the range of their ecological diversity.

The ratio of tail-to-body vertebrae (TtB) and the percentage of body vertebrae relative to the total were calculated from the raw data. These are reported as "odds ratios" of the number of tail to body vertebrae and as pie graphs showing the percentage of body (white circles) and tail (black wedges) (Table 9.1). Odds ratios are useful for clearly representing differences between the number of segments in the body and tail among snakes. A ratio of 0.5:1 indicates both that there is one-half tail vertebra per body vertebra and that the tail is one-half as many segments long as is the body. The ratios provide a more intuitive summary of differences in the number of vertebrae in each region than do equivalent percentages of the total number of segments (Sokal and Rohlf, 1995).

The phylogeny we employ in this chapter is derived from analyses by Dowling et al. (1983), Ashe and Marx (1987), Cadle (1988), Cundall et al. (1993), Kluge (1991, 1993), and Keogh (1998). Morphological and molecular data are generally congruent in their expression of snake relationships (Kluge, 1989; Cundall et al., 1993; Heise et al., 1995): Serpentes consists of a sister-taxon relationship between Scolecophida (blind snakes) and Alethinophidia (all other snakes), with Alethinophida composed of a gradation of aniliod (pipe snakes), xenopeltid (sunbeam snakes), "henophidian" (boas, pythons), and acrochordid (file snakes) taxa, culminating in Colubroidea, which is composed of Viperidae (vipers, rattlesnakes), Colubridae (racers, grass snakes), Elapidae (cobras, kraits, sea snakes), and Actraspidae (stiletto snakes). Differences in phylogenetic hypotheses include monophyly vs. paraphyly of Anilioidea and Boidae (Rieppel, 1988; Cundall et al., 1993), as well as the relative relationships of various "henophidian" taxa with respect to Colubroidea (Cundall et al., 1993; Heise et al., 1995). Additionally, interrelationships of many colubrid taxa are poorly known.

Because of uncertainty about certain aspects of snake phylogeny, we considered two alternate trees (Fig. 9.5). The first makes a minimum of assumptions about those colubrines whose relationships have not been studied or are controversial, and it supports paraphyly of *Cylindrophis* and *Anilius* (Cundall et al., 1993). The second tree assumes both relationships among Colubrinae based on continental-scale geographic provenance and the monophyly of Aniloidea (*Cylindrophis* + *Anilius*; Rieppel, 1988; Kluge, 1991). Divergence times for key clades were determined using oldest known occurrences (e.g., Rage, 1984; Rage and Richter, 1994; Gardner and Cifelli, 1998). Dates for other clades were extrapolated linearly from these. For example, if the parent clade originated

Table 9.1. Pre- and postcloacal proportions in selected snake species

Species	Precloacal	Postcloacal	Total	Tail: Body and Percent Tail	Estimated SVL, cm	Typical Adult SVL, cm
Acrochordus javanicus	204	67	271	0.3:1	98	130
Agkistrodon piscivorous	138	41	179	0.3:1	78	120
Ahaetulla nasuta	181	134	315	0.7:1	50	160
Anilius scytale	217	22	239	0.1:1	70	70
Arizona elegans	225	57	282	0.3:1	99	90
Bitis arietans	143	27	170	0.2:1	98	150
Crotalus molossus	177	28	205	0.2:1	99	105
Cylindrophis rufus	203	20	223	0.1:1	44	
Dendroaspis viridis	226	122	348	0.5:1	181	220
Epicrates chenchria	265	64	329	0.2:1	163	200
Erythrolamprus aesculapii	198	46	244	0.2:1	98	100
Eunectes murinus	252	59	311	0.2:1	199	900
Hydrophis fasciatus	228	27	255	0.1:1	92	90
Lachesis muta	232	45	277	0.2:1	273	225
Laticaudata laticaudata	245	37	282	0.2:1	89	110
Masticophis flagellum	237	53	290	0.2:1	62	225
Micrurus fulvius	213	49	262	0.2:1	78	75
Morelia spilota	286	75	361	0.3:1	186	245
Naja nigricollis	203	63	266	0.3:1	144	220
Oxybelis fulgidus	206	142	348	0.7:1	140	130
Pareas carinatus	173	64	237	0.4:1	48	90
Python molurus	259	68	327	0.3:1	281	400
Python regius	215	33	248	0.2:1	116	155
Storeria dekayi	114	35	149	0.3:1	28	35
Thamnophis ordinoides	142	60	202	0.4:1	43	95
Typhlops jamanicus	200	15	215	0.1:1	22	25

Daboia russelli	175	48	223	0.3:1	91	135
Xenodon severus	143	36	179	0.3:1	155	130
Xenopeltis unicolor	188	30	218	0.2:1	64	90

The number of precloacal vertebrae, postcloacal vertebrae, and the total number of vertebrae in the axial skeleton are reported for each species considered in this study. The proportion of postcloacal to precloacal vertebrae is reported both as an odds ratio and as a pie graph showing the percentage of postcloacal vertebrae relative to the total. An estimated snout-vent length (SVL) for the specimen used to represent each species is reported with an adult average for that species.

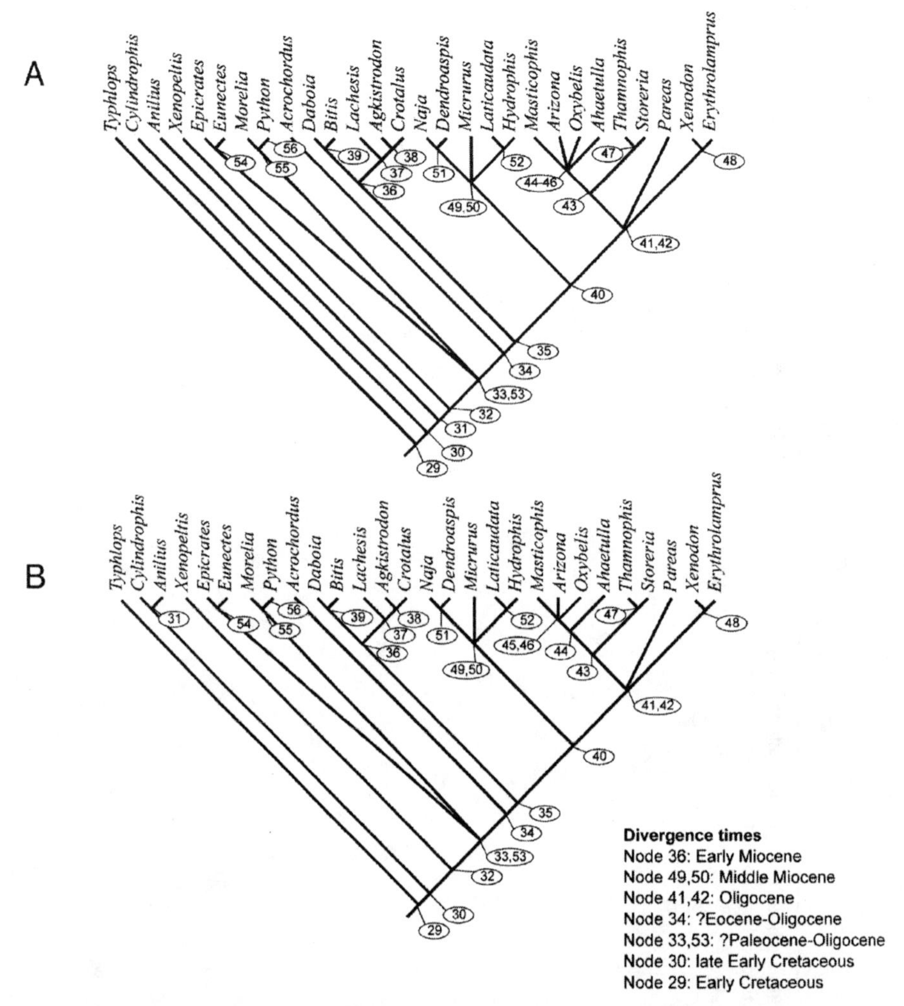

Figure 9.5. Cladograms of the snakes considered in this study. Because of uncertainty about the relationship of some of the snakes considered in this study, we considered two trees (A and B). The nodes are numbered as discussed in the text. Divergence times for seven nodes were estimated from paleontologic data. See text for details.

15 million years ago and has two nested subclades, the latter were estimated to have originated 10 and 5 million years ago, respectively. These dates were used to estimate branch lengths for the tree in millions of years (see Table 9.3). The phylogeny-based analyses below require fully dichotomous trees; polytomies were accommodated by assigning very small branch lengths (0.0001 in this case) to branches connecting nodes forming the polytomy as suggested by Martins (1995). For repeatability, the two trees are presented here with branch lengths in New Hampshire format:

Tree A

(Typhlops:140.0,((Anilius:80.0,(Xenopeltis:65.0,((Acrochordus:34.0,(((La
chesis:12.0,(Agkirstrodon:7.0,Crotalus:7.0):5.0):8.0,(Bitis:10.0,Daboia:10.
0):10.0):13.0,((Pareas:30.0,(((Oxybelis:5.0,(Ahaetulla:5.0,(Arizona:5.0,M
asticophis:5.0):0.001):0.001):15.0,(Storeria:10.0,Thamnophis:10.0):10.0):
10.0,(Erythrolamprus:15.0,Xenodon:15.0):15.0):0.001):1.0,(Micrurus_:15.
0,((Dendroaspis:7.0,Naja:7.0):7.99,(Hydrophis:7.0,Laticaudata:7.0):7.99):
0.01):16.0):2.0):1.0):6.0,((Epicrates_:15.0,Eunectes:15.0):25.0,(Morelia:1
5.0,(Python_m:10.0,Python_r:10.0):5.0):25.0):0.01):25.0):15.0):20.0,Cyli
ndrophis:100.0):40.0);

Tree B

(Typhlops:140.0,((Anilius:60.0,Cylindrophis:60.0):40.0,(Xenopeltis:65.0,(
(Acrochordus:34.0,(((Lachesis:10.0,(Agkirstrodon:5.0,Crotalus:5.0):5.0):1
0.0,(Bitis:5.0,Daboia:5.0):15.0):13.0,((Pareas:30.0,(((Oxybelis:5.0,(Ahaet
ulla:5.0,(Arizona:5.0,Masticophis:5.0):0.001):0.001):15.0,(Storeria:10.0,T
hamnophis:10.0):10.0):10.0,(Erythrolamprus:15.0,Xenodon:15.0):15.0):0.
0010):2.0,(Micrurus_:15.0,((Dendroaspis:7.0,Naja:7.0):7.99,(Hydrophis:7.
0,Laticaudata:7.0):7.99):0.01):17.0):1.0):1.0):6.0,((Epicrates_:15.0,Eunect
es:15.0):25.0,(Morelia:15.0,(Python_m:5.0,Python_r:5.0):10.0):25.0):0.00
1):25.0):35.0):40.0);

Correlations among the number of body vertebrae, the number of tail vertebrae, and the total number of vertebrae were assessed using model I regression. This is the appropriate regression because we wished to assess the degree to which each y variable (either pre- or postcloacal number) is predictable given a particular value of the x (either total number or precloacal number). Because our purpose was to determine whether evolutionary change in the number of axial segments results in change exclusively in the tail, either in the body or tail, or in both body and tail, we also looked at the correlation of changes in these traits along branches of the tree. To do this, we optimized the three traits—the number of body, tail, and total vertebrae—on both trees using Martins and Hansen's (1997) generalized linear model. Node values and their standard errors (or variances) are estimated by the method and are reported in Table 9.2. The node values are the most probable given the topology of the tree, the data at the tips, and certain assumptions about evolutionary change in the traits. Those assumptions are that gain and loss of segments are equally probable and that the direction of change at one point in time is dependent on neither preceding nor subsequent changes (i.e., a Brownian motion model of evolution). As performed here, this optimization is the same as a squared-change parsimony optimization (Maddison, 1991).

Changes along each branch were calculated from the optimized node values by subtracting the end value from the beginning value. These were standardized by dividing through by the length of each branch in millions of years, resulting in an estimate of change in each trait per million years per branch. Both standardized changes and branch lengths are reported in Table 9.3. Two sorts

Table 9.2. Tree node reconstructions and their standard errors

	Total Vertebrae				Precloacal Vertebrae				Postcloacal Vertebrae			
	Tree A	Tree B	Mean	Var	Tree A	Tree B	Mean	Var	Tree A	Tree B	Mean	Var
RSS	22.1	21.8			22.7	22.6			26.9	27.1		
sigma	191.4	189.8			117.9	121.2			41.8	46.1		
Node 29	250	254	252	9	187	191	189	7	63	64	63	0
SE	95	94	95	0	75	76	75	1	44	47	45	3
Node 30	244	249	247	16	182	187	184	12	62	63	62	0
SE	58	58	58	0	45	46	46	0	27	28	28	1
Node 31	254	248	251	16	188	171	180	146	65	77	71	65
SE	56	62	59	17	44	50	47	16	26	31	28	10
Node 32	250	246	248	6	194	197	196	4	55	49	52	18
SE	53	55	54	2	41	44	43	3	25	27	26	3
Node 33	247	247	247	0	196	197	196	0	51	50	51	0
SE	34	34	34	0	26	27	27	0	16	17	16	0
Node 34	261	261	261	0	206	207	206	0	55	54	55	0
SE	29	28	28	0	22	23	23	0	13	14	14	0
Node 35	263	263	263	0	208	208	208	0	56	55	55	0
SE	27	27	27	0	22	22	22	0	13	13	13	0
Node 36	262	261	262	1	206	205	206	0	56	55	56	0
SE	31	32	32	1	24	26	25	1	14	16	15	1
Node 37	221	215	218	21	184	180	182	7	37	34	36	4
SE	27	26	26	1	21	21	21	0	13	13	13	0
Node 38	217	214	216	4	188	187	187	0	29	27	28	2
SE	23	20	21	5	18	16	17	2	11	10	10	0
Node 39	313	327	320	102	232	240	236	27	81	88	84	24
SE	27	21	24	22	21	17	19	12	13	10	11	3
Node 40	267	265	266	2	210	209	210	1	57	56	56	1
SE	27	27	27	0	21	21	21	0	12	13	13	0
Node 41	269	268	268	0	212	213	213	0	57	55	56	1
SE	27	27	27	0	21	22	21	0	13	13	13	0

Node 42	269	268	268	0	212	213	213	0	57	55	56	1
SE	27	27	27	0	21	22	21	0	13	13	13	0
Node 43	274	273	273	1	223	224	223	0	51	49	50	2
SE	31	30	31	0	25	24	24	0	15	15	15	0
Node 44	281	275	278	16	238	233	236	12	43	42	42	0
SE	15	25	20	48	12	20	16	32	7	12	10	14
Node 45	281	286	283	15	238	241	239	3	43	46	44	5
SE	15	17	16	1	12	13	13	1	7	8	8	1
Node 46	281	286	283	15	238	241	239	3	43	46	44	5
SE	15	17	16	1	12	13	13	1	7	8	8	1
Node 47	275	275	275	0	224	225	224	0	51	50	51	0
SE	27	27	27	0	21	22	22	0	13	13	13	0
Node 48	299	298	298	0	234	234	234	0	65	64	65	0
SE	32	32	32	0	25	26	25	0	15	16	15	0
Node 49	269	269	269	0	198	197	198	0	72	72	72	0
SE	26	26	26	0	20	21	20	0	12	13	12	0
Node 50	269	269	269	0	198	197	198	0	72	72	72	0
SE	26	26	26	0	20	21	20	0	12	13	12	0
Node 51	278	278	278	0	210	210	210	0	68	68	68	0
SE	23	23	23	0	18	18	18	0	11	11	11	0
Node 52	220	220	220	0	175	175	175	0	45	45	45	0
SE	23	23	23	0	18	18	18	0	11	11	11	0
Node 53	247	247	247	0	196	197	196	0	51	50	51	0
SE	34	34	34	0	26	27	27	0	16	17	16	0
Node 54	217	217	217	0	177	177	177	0	41	40	41	0
SE	34	34	34	0	27	27	27	0	16	17	16	0
Node 55	213	218	215	9	173	176	174	3	40	42	41	2
SE	29	33	31	7	23	26	24	6	13	16	15	4
Node 56	203	202	203	1	168	168	168	0	35	35	35	0
SE	20	21	20	0	16	16	16	0	9	10	10	0

The estimated total number of vertebrae, precloacal vertebrae, and postcloacal vertebrae are reported for each node of the phylogenetic trees in Figure 9.5. Estimates were made using the topologies of both Tree A and Tree B. Martins and Hansen's (1997) generalized linear model was used to estimate the values from the species data in Table 9.1 and the phylogenetic branching pattern. Standard errors (SE) for each estimate are also reported. The mean and variance (Var) of the Tree A and Tree B node estimates are reported as an index of discrepancy between the two optimizations.

Table 9.3. Evolution of vertebral numbers

	Branch Length, MY		ΔTotal Vertebrae		ΔPrecloacal		ΔPostcloacal	
	Tree A	Tree B	Tree A	Tree B	Tree A	Tree B	Tree A	Tree B
Node 29 → node 30	40	40	−0.15	−0.13	−0.12	−0.10	−0.03	−0.02
Node 29 → Typhlops	140	140	−0.25	−0.28	0.09	0.07	−0.34	−0.35
Node 30 → Cylindrophis	100	N/A	−0.21	N/A	−0.41	N/A	−0.42	N/A
Node 30 → node 31	20	100	0.50	−0.01	0.50	−0.01	0.18	0.14
Node 30 → node 32	N/A	10	N/A	−0.30	N/A	−0.31	N/A	−1.34
Node 31 → Anilius	80	40	−0.19	−0.23	0.36	1.14	−0.54	−1.37
Node 31 → Cylindrophis	N/A	40	N/A	−0.60	N/A	−1.12	N/A	−1.42
Node 31 → node 32	15	N/A	−0.27	N/A	0.41	N/A	−0.68	N/A
Node 32 → node 33, 53	25	25	−0.12	0.04	0.06	−0.01	−0.17	0.04
Node 32 → Xenopeltis	65	65	−0.49	−0.48	−0.10	−0.14	−0.39	−0.29
Node 33, 53 → node 34	6	6	2.33	2.33	1.74	1.63	0.68	0.74
Node 33, 53 → node 54	25	25	−1.20	−1.20	−0.77	−0.80	−0.77	−0.80
Node 33, 53 → node 55	25	25	−1.36	−1.16	−0.91	−0.85	−0.43	−0.32
Node 34 → Acrochordus	33	33	0.30	0.30	−0.07	−0.08	0.36	0.38
Node 34 → node 35	1	1	2.00	2.00	1.19	1.00	0.62	0.67
Node 35 → node 36	13	13	−0.08	−0.15	−0.15	−0.23	0.05	0.03
Node 35 → node 40	1	1	4.00	2.00	2.60	1.28	1.15	0.64
Node 36 → node 37	8	8	−5.14	−5.75	−2.74	−3.10	−2.40	−2.64
Node 36 → node 39	10	10	5.07	6.66	2.63	3.44	2.44	3.21
Node 37 → Lachesis	12	12	4.64	5.18	3.99	4.29	0.65	0.89
Node 37 → node 38	5	5	−0.86	−0.11	0.69	1.27	−1.55	−1.38
Node 38 → Agkistrodon	7	7	−5.43	−5.04	−7.08	−6.98	1.65	1.94
Node 38 → Croatalus	7	7	−1.72	−1.33	−1.51	−1.40	−0.20	0.08
Node 39 → Bitis	10	10	−14.31	−15.74	−8.94	−9.68	−5.38	−6.06
Node 39 → Daboia	10	10	−9.01	−10.44	−5.74	−6.48	−3.28	−3.96
Node 40 → node 41, 42	2	2	1.00	1.50	1.03	1.96	−0.17	−0.29
Node 40 → node 49, 50	16	16	0.13	0.25	−0.77	−0.72	0.92	0.99
Node 41, 42 → node 43	10	10	0.50	0.50	1.10	1.06	−0.55	−0.60
Node 41, 42 → node 48	15	15	1.99	2.02	1.44	1.40	0.55	0.61

Node 41, 42 → Pareas	30	30	−1.06	−1.04	−1.31	−1.33	0.25	0.29
Node 43 → node 44	N/A	10	N/A	0.20	N/A	0.96	N/A	−0.73
Node 43 → node 44–46	15	N/A	0.47	N/A	0.99	N/A	−0.55	N/A
Node 43 → node 47	10	10	0.10	0.20	0.12	0.10	0.00	0.12
Node 44 → Ahaetulla	N/A	10	N/A	3.99	N/A	−5.22	N/A	9.13
Node 44 → node 45, 46	N/A	5	N/A	2.24	N/A	1.48	N/A	0.76
Node 44–46 → Ahaetulla	5	N/A	6.86	N/A	−11.40	N/A	18.26	N/A
Node 44–46 → Arizona	5	N/A	0.26	N/A	−2.60	N/A	2.86	N/A
Node 44–46 → Masticophis	5	N/A	1.86	N/A	−0.20	N/A	2.06	N/A
Node 44–46 → Oxybelis	5	N/A	13.46	N/A	−6.40	N/A	19.86	N/A
Node 45, 46 → Arizona	N/A	5	N/A	−0.85	N/A	−3.11	N/A	2.25
Node 45, 46 → Masticophis		5	N/A	0.75	N/A	−0.71	N/A	1.45
Node 45, 46 → Oxybelis	N/A	5	N/A	12.35	N/A	−6.91	N/A	19.26
Node 47 → Storeria	10	10	−12.64	−12.59	−16.14	−11.05	−1.60	−1.54
Node 47 → Thamnophis	10	10	−7.34	−7.29	−8.24	−8.25	0.90	0.96
Node 48 → Erythrolampris	15	15	−3.64	−3.63	−2.38	−2.40	−1.26	−1.23
Node 48 → Xenodon	15	15	−7.97	−7.96	−6.05	−6.07	−1.92	−1.89
Node 49, 50 → Micrurus	15	15	−0.50	−0.47	1.01	1.04	−1.51	−1.51
Node 49, 50 → node 51	7	7	1.29	1.29	1.81	1.84	−0.56	−0.56
Node 49, 50 → node 52	7	7	−7.05	−7.00	−3.31	−3.27	−3.74	−3.73
Node 51 → Dendroaspis	7	7	9.98	10.00	2.22	2.24	7.76	7.76
Node 51 → Naja	7	7	−1.74	−1.71	−1.07	−1.00	−0.67	−0.71
Node 52 → Hydrophis	7	7	4.98	5.00	7.62	7.57	−2.64	−2.64
Node 52 → Laticaudata	7	7	8.86	8.86	10.05	10.00	−1.21	−1.14
Node 54 → Epicrates	15	15	7.44	7.44	5.88	5.87	1.56	1.57
Node 54 → Eunectes	15	15	6.24	6.27	5.02	5.00	1.23	1.24
Node 55 → Morelia	15	15	9.84	9.56	7.52	7.36	2.32	2.20
Node 55 → node 56	5	5	−2.00	−3.20	−1.03	−1.62	−5.34	−1.44
Node 56 → Pythonmolurus	10	10	12.35	12.47	9.09	9.15	3.26	3.32
Node 56 → Pythonregius	10	10	4.45	4.57	4.69	4.70	−0.24	−0.20

The changes in precloacal, postcloacal, and total vertebral numbers are reported as a rate. The evolutionary change in vertebral number was inferred for each branch using the species data in Table 9.1 and the node estimates in Table 9.2. Because branch length varies considerable across each tree, the change along each branch was standardized by dividing it by its branch length to yield an index of change in number per million years (MY).

of "odds ratios" summarizing change in regional proportions were also calculated: the difference in body-to-tail at the end of each branch compared with the same ratio at the beginning (ΔTtB) and the ratio of change-in-body to change-in-tail (ΔT:ΔB). The first of these, ΔTtB, was calculated for each branch by first calculating the TtB odds ratio for the node at the beginning and end of each branch as described above. The end value was then subtracted from the beginning and divided through by the length of the branch in millions of years. The ΔTtB ratio thus represents change in the number of tail vertebrae per body vertebrae per million years per branch. The second ratio, ΔT:ΔB, is the ratio of change in the number of tail vertebrae per million years per branch to each body vertebra per million years per branch. This was calculated by dividing the change in tail vertebrae per million years per branch (Table 9.3) into the change in body vertebrae per million years per branch. Because the two ratios are easily derived from the data in Tables 9.2 and 9.3, the numbers are not reported separately. The two ratios were compared with changes in total number of vertebrae to assess whether the addition and deletion of axial segments is normally associated with change in the body region or in the tail region or is equally distributed between the regions.

RESULTS AND DISCUSSION

The number of vertebrae in snakes was variable among species, both in total and region by region (Table 9.1). The total number ranged from 149 in *Storeria* to 361 in *Morelia*, a twofold difference. The mean was 257, the SD was 56.8, and the coefficient of variation (CV) was 22.1. Among-species variation in the tail was found to be greater than that in the body. The number of body vertebrae ranged from 114 in *Storeria* to 286 in *Morelia*. The mean was 203, SD was 41.5, and CV was 20.4. The number in the tail ranged from 15 in *Typhlops* to 142 in *Oxybelis* with a mean of 54, SD of 31.6, and CV of 58.6. The large CV in the number of tail vertebrae indicates that most changes in total vertebral number are probably concentrated in that region. Also, the fact that the species with the longest and shortest tails are not the species with the longest and shortest bodies indicates that body and tail change independently. If the two were universally correlated, then the same species would be at the extreme ends of the ranges of both regions. The variation in regional numbers confirms that the number of segments can be modified, both in individual regions and in toto. Although this suggests that dissociation between body and tail may occur, it does not rule out an allometric relationship between evolutionary changes in the two regions.

There is considerable variation in the ratio of tail to body (TtB) among snakes (Table 9.1). The smallest TtB ratio is 0.08:1 in fossorial *Typhlops*. In other words, *Typhlops* has the smallest number of tail vertebrae in proportion to its body of any of the snakes we examined. The largest TtB is 0.74:1 in the arboreal snake *Ahaetulla*, giving it the largest number of tail vertebrae in proportion to its body of any snake examined. *Ahaetulla* thus has more than nine times as many tail vertebrae per body segment than does *Typhlops*. The mean TtB was 0.27:1, its SD was 0.16:1, and its CV was 58.3. This is again suggestive of dissociation but does not rule out the possibility that the proportion of tail to body is allometric.

The numbers of body and tail vertebrae are both positively correlated with the total number in the skeleton (Fig. 9.6, A and B). This is expected because the total is, by definition, the sum of body and tail. As one or both of the components increase, so does the total. This autocorrelation means that the slopes of the two regressions in Fig. 9.6, A and B, sum to 1.0 because, together, the two regions account for the total number of vertebrae. This means that the slopes do not tell us anything of interest about the evolution of regional proportions. The R^2 values do, however. The total number of vertebrae better predicts the number of body vertebrae than it does the number in the tail. Total number explains 70% of the variance in the body but only 49% of variance in the tail. This suggests that, regardless of how many segments long a snake is, the number of vertebrae in the tail is more variable than the number in the body. There is not a constant relationship between total number of vertebrae and the proportion of body to tail, meaning that the proportion between regions can be changed without changing the total number of vertebrae. This could not be accomplished without some dissociation. The lack of correlation between the numbers of body and tail vertebrae substantiates this (Fig. 9.6C).

The relationship between proportion of body relative to tail is easier to see in a regression of the tail-to-body odds ratio (TtB) on the total number of vertebrae (Fig. 9.7). There is a positive and significant ($P = 0.036$) correlation between the ratio of tail to body segments and the total number of vertebrae, but the latter only explains 15% of the variance of the former. The positive correlation between TtB and total vertebrae number suggests that, on average, body and tail evolution are not dissociated; as segment number increases, so does the number of tail vertebrae relative to the number in the body. However, the low predictive value of the regression means that quite a lot of the change in regional proportions is independent of changes in number of axial segments.

Evolution in the number of tail vertebrae relative to both body vertebrae and the total number of segments can be more fruitfully examined by looking directly at phylogenetic changes in these three variables (Table 9.3). Looking at the branch-by-branch correlation between change in the total number of vertebrae per million years (ΔTot) and change in body number per million years (ΔB), we find a very tight linear relationship in which ΔTot explains 46% of the variance in ΔB as optimized on Tree A and 59% when optimized on Tree B (Fig. 9.8, A and B). Conversely, ΔTot is positively correlated with change in tail number (ΔT) but only accounts for 36% of the variance on Tree A and 39% on Tree B (Fig. 9.8, C and D). This means that adding vertebrae to the body almost always increases the total number of vertebrae, but adding them to the tail may either increase the total or happen while the total decreases (i.e., vertebrae are deleted from the body and added to the tail). This implies that long-

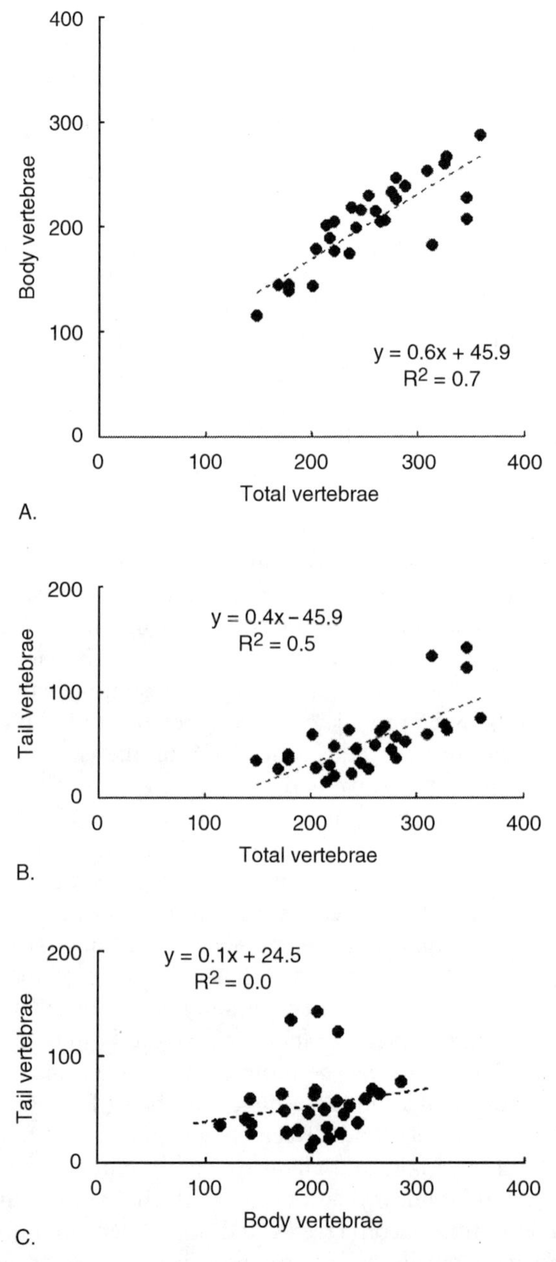

Figure 9.6. Correlation between the total number of vertebrae in the axial skeleton and the number of precloacal (body) vertebrae (A), the total number and the number of postcloacal (tail) vertebrae (B), and the number of pre- and postcloacal vertebrae in 29 snake species (C). A positive relationship between the total number of vertebrae and both precloacal and postcloacal number is always expected because of autocorrelation. The wide spread of outliers along the *y* axis in C indicates that some snake species have regional proportions substantially different from others given the same total number of vertebrae.

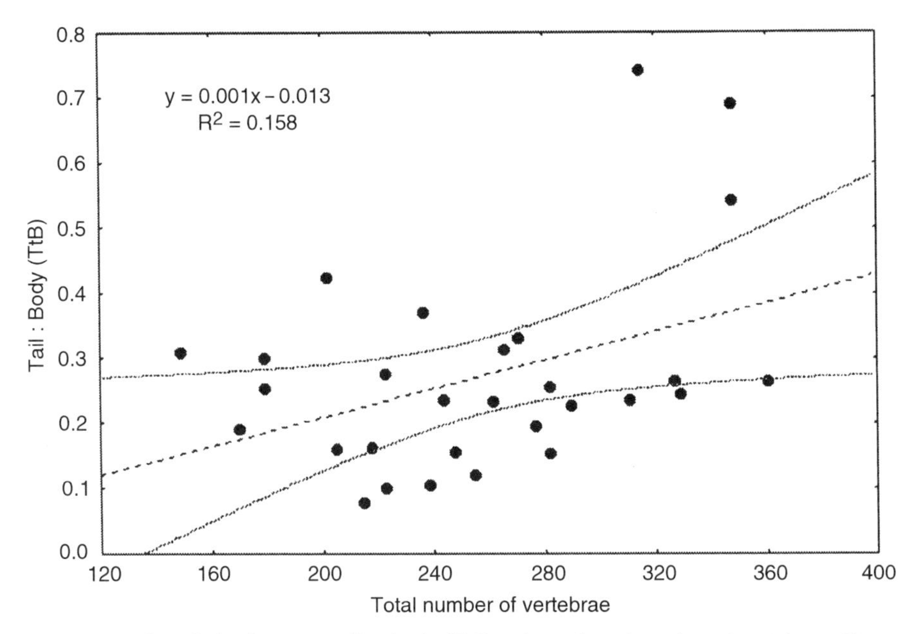

Figure 9.7. Correlation between tail-to-body (TtB) ratio and total number of vertebrae. The *y* axis is the "tail" part of the ratio, as the "body" part always equals one (i.e., Y:1). Linear regression line and 95% prediction intervals for the mean and individuals are indicated. There is a positive correlation between the ratio of tail-to-body and the total number of vertebrae, $R^2 = 0.15$, $P = 0.036$.

bodied snakes evolve by increasing the total number of vertebrae but that long-*tailed* snakes have evolved both by adding to the total number of vertebrae or adding to the tail at the expense of the body. A comparison of ΔT relative to ΔB makes this clearer (Fig. 9.8, E and F). The scatter of points above the regression line and to the left of the line of isometry indicates that, when large numbers of tail vertebrae are added, the number of body vertebrae often simultaneously decreases. However, when body vertebrae are added (points to the right of the line of isometry), the number of tail vertebrae does not usually change. These data strongly suggest that dissociation between body and tail is possible. Note, however, that it is quite common for no change to occur in either the number of tail or the number of body vertebrae, as indicated by the large number of data points clustered at 0,0 (Fig. 9.8, A–F).

The strongest evidence for occasional dissociation between body and tail in the evolution of snakes is shown in Figure 9.9, which illustrates the relationship between ΔTtB and ΔT. The points in the upper left and lower right quadrant represent branches on which the total number of vertebrae has either increased or decreased, but the tail has done the opposite. When the odds ratio on the vertical axis is positive, the number of tail vertebrae relative to the number of body vertebrae has increased. The most common mode of change—indicated

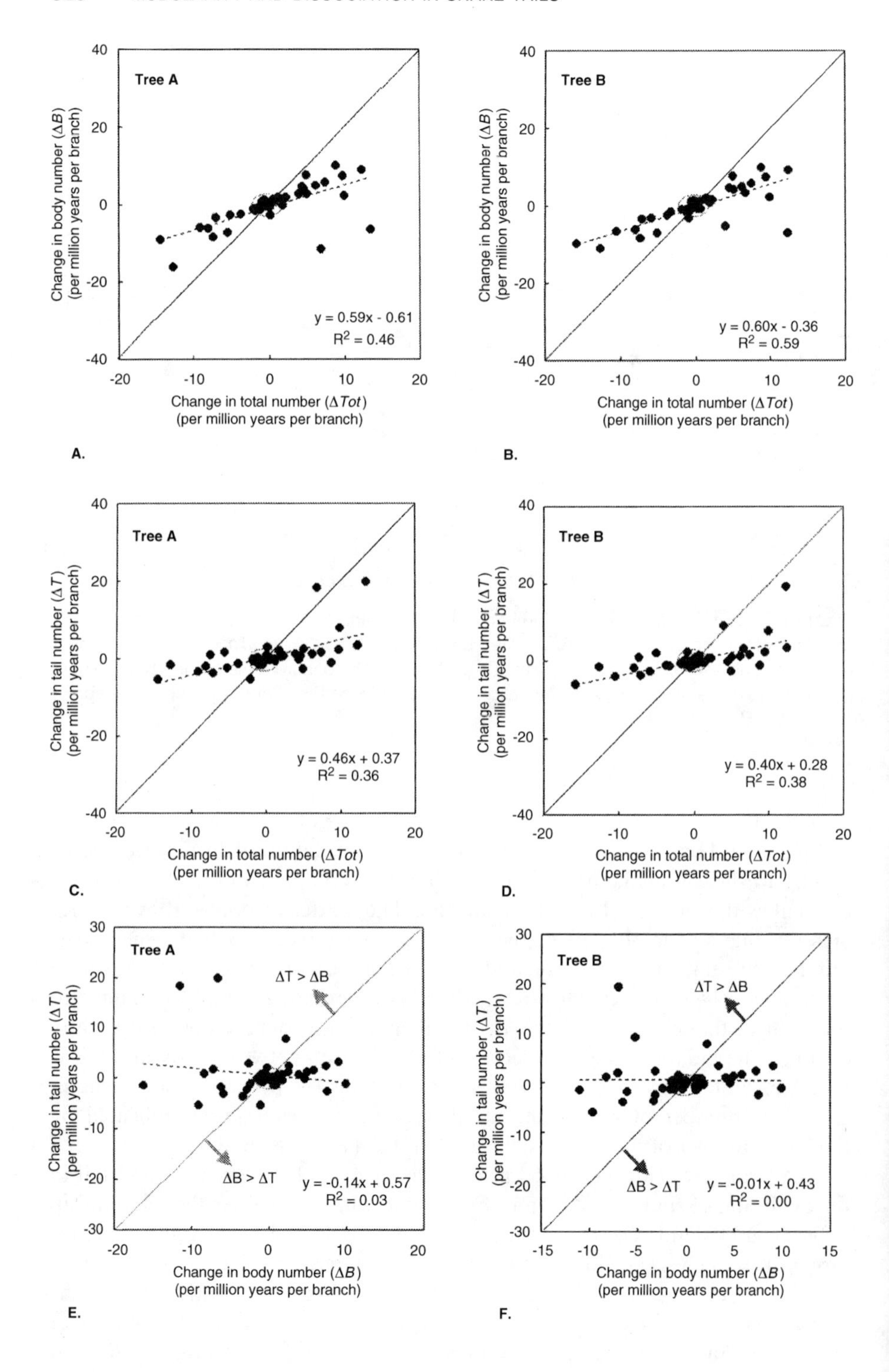
Tree A
Change in body number (ΔB) (per million years per branch)
y = 0.59x - 0.61
R2 = 0.46
Change in total number (ΔTot) (per million years per branch)
A.
Tree B
Change in body number (ΔB) (per million years per branch)
y = 0.60x - 0.36
R2 = 0.59
Change in total number (ΔTot) (per million years per branch)
B.
Tree A
Change in tail number (ΔT) (per million years per branch)
y = 0.46x + 0.37
R2 = 0.36
Change in total number (ΔTot) (per million years per branch)
C.
Tree B
Change in tail number (ΔT) (per million years per branch)
y = 0.40x + 0.28
R2 = 0.38
Change in total number (ΔTot) (per million years per branch)
D.
Tree A
ΔT > ΔB
ΔB > ΔT
y = -0.14x + 0.57
R2 = 0.03
Change in tail number (ΔT) (per million years per branch)
Change in body number (ΔB) (per million years per branch)
E.
Tree B
ΔT > ΔB
ΔB > ΔT
y = -0.01x + 0.43
R2 = 0.00
Change in tail number (ΔT) (per million years per branch)
Change in body number (ΔB) (per million years per branch)
F.

by the circled points at 0,0—is for neither the total number nor the ratio to change. Counting the number of nonzero data points in the various quadrants allows us to estimate the probability of various modes of change. The most likely explanation is that the addition of tail vertebrae corresponds to addition of total body segments or, conversely, deletion of tail vertebrae to deletion of total segments (17 points fall into this category on Tree A and 18 on Tree B). In this mode, the proportion of body-to-tail is changed as a result of change in the overall number of tail vertebrae rather than by a shift in the border between regional boundaries. It is much less likely that the ratio of tail to body segments increases at the same time as the total number of segments decreases (three points fall into this category). In this mode, the border between the boundaries is shifted anteriorly (increasing the ratio of tail to body vertebrae) while the number of segments in the body is decreased. For this to happen, vertebrae must be lost from the body and added to the tail, indicating a dissociation. This mode of change is localized in two clades *Thamnophis* + *Storeria* and in *Agkistrodon*. In Tree B, *Arizona* creeps into this category also. The inverse mode of change, addition of total vertebrae but reduction of the number in the tail, lies in the bottom right quadrant. Three data points fall into this category. This mode occurred within the clade containing *Hydrophis* and *Laticaudata* (perhaps associated with aquatic specialization) and independently in *Python regius*, a long-bodied snake with a relatively short tail. Thus only 6 of 59 branches show evidence for dissociation between body and tail, indicating that it is possible but rare.

CONCLUSIONS

In this chapter, we tested evolutionary transitions in the number of body and tail vertebrae in snakes for patterns of modularity and dissociation. Based on experimental embryology of other amniotes, we hypothesize that, in snakes, these two regions are modules (sensu Raff, 1996). The adult body vertebrae (precloacal vertebrae in snakes) forms from the segmented mesoderm that originates from the primitive streak and node, whereas the postcloacal (tail) skeleton is formed from mesoderm that originates from the tail bud. The physical point of contact between these two processes is at what becomes the cloacal region in the adult (in most other vertebrates, it is more easily recognized as the sacral region). The total number of precloacal segments may then be controlled by anterior growth of the primitive streak and groove, and the

◄

Figure 9.8. Changes in segment number along branches per million years. Each data point represents changes along one of the branches in Tree A and Tree B (Fig. 9.5). A and B: relationship between change in total number of vertebrae and the number of body segments. C and D: relationship between change in total number of vertebrae and number of tail segments. E and F: relationship between change in body vertebrae and tail vertebrae. Solid diagonal lines represent isometric change.

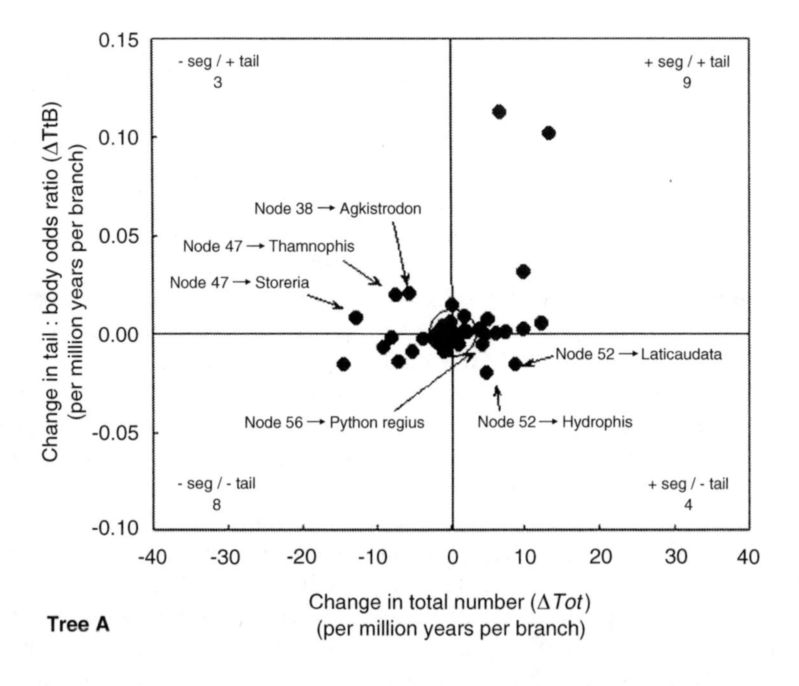

0.15
0.10
0.05
0.00
-0.05
-0.10
- seg / + tail
3
+ seg / + tail
9
- seg / - tail
8
+ seg / - tail
4
Node 38 → Agkistrodon
Node 47 → Thamnophis
Node 47 → Storeria
Node 52 → Laticaudata
Node 56 → Python regius
Node 52 → Hydrophis
-40
-30
-20
-10
0
10
20
30
40
Change in tail : body odds ratio (ΔTtB) (per million years per branch)
Change in total number (ΔTot) (per million years per branch)

Tree A

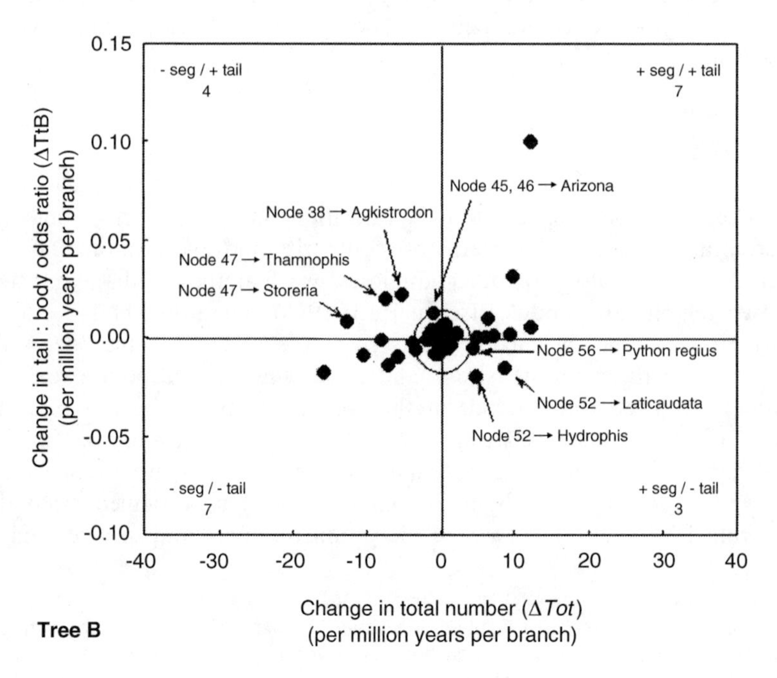

0.15
0.10
0.05
0.00
-0.05
-0.10
- seg / + tail
4
+ seg / + tail
7
- seg / - tail
7
+ seg / - tail
3
Node 45, 46 → Arizona
Node 38 → Agkistrodon
Node 47 → Thamnophis
Node 47 → Storeria
Node 56 → Python regius
Node 52 → Laticaudata
Node 52 → Hydrophis
-40
-30
-20
-10
0
10
20
30
40
Change in tail : body odds ratio (ΔTtB) (per million years per branch)
Change in total number (ΔTot) (per million years per branch)

Tree B

number of tail segments may be controlled by the amount of mesoderm that is generated by the tail bud. There is evidence for links between the two modules, particularly the processes of segmentation and somitogenesis, which traverse both the body and tail regions. Our data indicate that evolutionary changes in the number of body and tail vertebrae are usually coordinated, with segments being either gained or lost simultaneously in both regions. The tail region, however, is more evolutionarily labile than the body because changes in the number of vertebrae happen with greater frequency and magnitude there. The number of vertebrae in the body also changes, however, but more rarely and less radically than in the tail. Our data also indicate that the two regions can be dissociated so that vertebrae are gained in one and lost in the other. Such dissociation is rare, however. Of the 59 branches of the phylogenetic trees that we considered, dissociation only happened for 6 of them. As far as we can localize these, they occured in a clade of terrestrial North American natricines (*Thamnophis* + *Storeria*), in a clade of marine elapids (*Laticaudata* + *Hydrophis*), in the branch leading to *Python regius* (the ball python), and in the branch leading to *Agkistrodon* (the cottonmouth). These four dissociations are the only candidates for heterochronic change among the data and taxa that we considered.

However, as Raff (1996) pointed out, heterochrony requires dissociation, but not all dissociations are heterochronic. Does heterochrony play a role in the evolution of body and tail proportions in snakes? Our data are suggestive, but, we argue, it is difficult to test for heterochrony without knowing the exact developmental mechanism responsible for an observed morphological transition. Given what is known about vertebrate development, some aspects of body-tail evolution may be heterochronic, whereas others are not. The maximum growth of the tail might be a function of the total volume of mesoderm produced by the posterior streak and tail bud. If very little is produced, then the tail might be small or absent. One gene involved in axial elongation is the *T* or *Brachyury* gene (Herrmann, 1991). BRACHYURY is required for movement of cells through the streak, and, in its absence, posterior elongation of the axis is prevented due to lack of mesodermal differentiation (Wilson and Beddington, 1997). Although a role for BRACHYURY in evolutionary changes to the vertebrate body axis has not been established, it is helpful to think of axial elongation in this developmental mechanistic sense. One way of generating more

Figure 9.9. Relationship between changes in tail-to-body ratio and change in total number of vertebrae. Each data point represents changes along one of the branches in Tree A or Tree B (Fig. 9.5). The lower left and upper right quadrants contain points that are not dissociations. In the lower left, both the total number of vertebrae and the number in the tail decrease; in the upper right, both increase. The upper left and lower right contain dissociations. In the upper left, the number of vertebrae in the tail increases, whereas the total number decreases. In the lower right, the number in the tail decreases, whereas the total number increases. In both cases, body and tail must be dissociated for the changes to occur. The data points in the circle are branches on which there was virtually no change in either the total number of vertebrae or the ratio of tail-to-body.

posterior vertebrae would be to increase the number of cells passing through the posterior streak/tail bud. This process would not be heterochronic; it would be a change in volume rather than a change in timing. Of course, the volume of mesoderm may be a function of the rate and duration of cell production and migration. For example, sustained outgrowth of the tail bud beyond the point at which it normally ceases to produce mesoderm would cause the posterior end of the body axis to grow longer, resulting in a greater number of caudal vertebrae. If so, the nonheterochronic process becomes heterochronic again. Furthermore, what was first imagined as a two-variable process—growth of the primitive streak and the tail bud explain both the total number of vertebrae in a snake plus its regional proportions—has now become a multivariate process in which the rate of mesoderm production during gastrulation, the duration of gastrulation, the rate of anterior growth of the primitive groove, the onset of neurulation, and the rate of *hairy1* cycling all contribute to adult snake vertebral number, but identity is determined by superimposition of differential Hox gene expression on these segments. The nature of the relationship between mechanisms that generate somitic mesoderm and those that determine its Hox code has not yet been resolved.

ACKNOWLEDGMENTS

Christopher J. Bell, Susan Evans, Norman McLeod, Mark Wilkinson, and Marvalee Wake all provided stimulating discussion of material in this paper. Drs. Colin McCarthy and Mark Wilkinson (Natural History Museum), Johnathon Campbell and K. C. Rudy (University of Texas at Arlington), Douglas Rossman (Louisiana State Unviersity), Dale Winkler (Southern Methodist University), Barbara Stein (Museum of Vertebrate Zoology), and Christopher J. Bell (University of Texas at Austin) provided access to specimens in their care. The University of London Central Research Fund provided financial support.

REFERENCES

Ashe JS, Marx H (1987): Phylogeny of the viperine snakes (Viperinae): part II. Cladistic analysis and major lineages. Fieldiana Zool 52: 1–29.

Atchley WR, Hall BK (1991): A model for development and evolution of complex morphological structures. Biol Rev 66: 101–157.

Cadle JE (1988): Phylogenetic relationships among advanced snakes, a molecular perspective. Univ Calif Publ Zool 119: 1–77.

Cheverud JM (1982): Phenotypic, genetic, and environmental morphological integration in the cranium. Evolution 36: 499–516.

Coborn J (1994): *The Mini-atlas of Snakes of the World.* Neptune City, NJ: T. F. H. Publications.

Cohn MJ, Tickle C (1999): Developmental basis of limblessness and axial patterning in snakes. Nature 399: 474–479.

Cundall D, Wallach V, Rossman DA (1993): The systematic relationships of the snake genus *Anomochilus*. Zool J Linn Soc 109: 275–299.

Dowling HG, Highton R, Maha GC, Maxon LR (1983): Biochemical evolution of colubrid snake phylogeny. J Zool 201: 309–329.

Gardner JD, Cifelli RL (1998): A primitive snake from the Cretaceous of Utah. In: Unwin DM (ed), *Cretaceous Fossil Vertebrates*. Spec Pap Palaeontol 60: 87–100.

Gasc JP (1976): Snake vertebrae—a mechanism or merely a taxonomist's toy? In Bellairs AA and Cox CB (eds), *Morphology and Biology of Reptiles*. Linn Soc Symp Ser 3: 177–192.

Goff DJ, Tabin CJ (1997): Analysis of Hoxd-13 and Hoxd-11 misexpression in chick limb buds reveals that Hox genes affect both bone condensation and growth. Development 124: 627–636.

Heise PJ, Maxson LR, Dowling HG, Hedges SB (1995): Higher-level snake phylogeny inferred from mitochondrial DNA sequences of 12S rRNA and 16S rRNA genes. Mol Biol Evol 12: 259–265.

Herrmann B (1991): Expression pattern of the Brachyury gene in whole-mount T^{wis}/T^{wis} mutant embryos. Development 113: 913–917.

Hoffstetter R, Gasc J-P (1969): Vertebrae and ribs of modern reptiles. In: Gans C (ed), *Biology of the Reptilia. Morphology*. London: Academic Press, vol. 1, p. 201–310.

Jayne BC (1982): Comparative morphology of the semispinalis-spinalis muscle of snakes and correlations with locomotion and constriction. J Morphol 172: 83–96.

Jayne BC (1988): Muscular mechanisms of snake locomotion: an electromyographic study of lateral undulation of the Florida banded water snake (*Nerodia fasciata*) and the yellow rat snake (*Elaphe obsoleta*). J Morphol 197: 159–181.

Jockusch EL (1997): Geographic variation and phenotypic plasticity of the number of trunk vertebrae in slender salamanders, *Batrachoseps* (Caudata: Plethodontidae). Evolution 51: 1966–1982.

Keogh JS (1998): Molecular phylogeny of elapid snakes and a consideration of their biogeographic history. Biol J Linn Soc 63: 177–203.

King RB, Bittner TD, Queral-Regil A, Cline JH (1999): Sexual dimorphism in neonate and adult snakes. J Zool 247: 19–28.

Kluge AG (1989): A concern for evidence and a phylogenetic hypothesis of relationships among *Epicrates* (Boidae, Serpentes). Syst Zool 37: 7–25.

Kluge AG (1991): Boine snake phylogeny and research cycles. Misc Publ Mus Zool Univ Mich 178: 1–58.

Kluge AG (1993): *Aspidites* and the phylogeny of pythonine snakes. Rec Aust Mus Suppl 19: 1–77.

Knezevic V, DeSanto R, Mackem S (1998): Continuing organizer function during chick tail development. Development 125: 1791–1801.

LaDuke TC (1991a): The fossil snakes of Pit 91, Rancho La Brea, California. Contrib Sci (Los Angel) 424: 1–28.

LaDuke TC (1991b): *Morphometric Variability of the Pre-caudal Vertebrae of Thamnophis sirtalis sirtalis (Serpentes: Colubridae), and Implications for Interpretation of the fossil record.* PhD Thesis. New York: City University of New York.

Leamy LJ, Routman EJ, Cheverud JM (1999): Quantitative trait loci for early- and late-developing skull characters in mice: a test of the genetic independence model of morphological integration. Am Nat 153: 201–214.

LeDouarin NM, Grapin-Botton A, Catala M (1996): Patterning of the neural primordium in the avian embryo. Semin Cell Dev Biol 7: 157–167.

Lindell LK (1994): The evolution of vertebral number and body size in snakes. Funct Ecol 8: 708–719.

Lindell LK (1996): Vertebral number in adders, *Vipera berus*: direct and indirect effects on growth. Biol J Linn Soc 59: 69–85.

Lovejoy CO, Reno PL, McCollum MA, Hamrick MW, Cohn MJ (2000): The evolution of hominoid hands: growth scaling registers with posterior HOXD expression. Am J Phys Anthropol Suppl 30: 214.

Maddison WP (1991): Squared-change parsimony reconstructions of ancestral states for continuous-valued characters on a phylogenetic tree. Syst Zool 40: 304–314.

Martins EP (1995): COMPARE: statistical analysis of comparative data, version 1.0. (Distributed by the author via Internet at http://work.uoregon.edu/~emartins/).

Martins EP, Hansen TF (1997): Phylogenies and the comparative method: a general approach to incorporating phylogenetic information into the analysis of interspecific data. Am Nat 149: 646–667.

Monsoro-Burq A-H, Duprez D, Watanabe Y, Bontoux M, Vincent C, Brickell P, Le Douarin N (1996): The role of bone morphogenetic proteins in vertebral development. Development 122: 3607–3616.

Mundlos S (1999): Cleidocranial dysplasia: clinical and molecular genetics. J Med Genet 36: 177–182.

Olson EC, Miller RJ (1959): *Morphological Integration*. Chicago, IL: University of Chicago Press.

Palmeirim I, Henrique D, Ish-Horowicz D, Pourquié O (1997): Avian hairy gene expression identifies a molecular clock linked to vertebrate segmentation and somitogenesis. Cell 91: 639–648.

Palmeirim I, Dubrulle J, Henrique D, IshHorowicz D, Pourquié O (1998): Uncoupling segmentation and somitogenesis in the chick presomitic mesoderm Dev Genet 23: 77–85

Raff RA (1996): *The Shape of Life: Genes, Development, and the Evolution of Animal Form*. Chicago, IL: University of Chicago Press.

Rage J-C (1984): *Encyclopedia of Paleoherpetology. Serpentes.* Stuttgart: Gustav Fischer Verlag, part 11.

Rage J-C, Richter A (1994): A snake from the lower Cretaceous (Barremian) of Spain: the oldest known snake. Neues Jahrb Geol Palaeontol Monh 1994: 561–565.

Rieppel O (1988): A review of the origin of snakes. Evol Biol 22: 37–130.

Shine R (1986): Ecology of low-energy specialists: food habits and reproductive biology of the Arafura filesnake, *Acrochordus arafurae*. Copeia 1986: 424–437.

Shine R (1993): Sexual dimorphism in snakes. In: Seigel RA, Collins JT, *Snakes—Ecology and Behavior*. Chicago, IL: R. R. Donnelley & Sons Company, p. 49–86.

Sokal RR, Rohlf FJ (1995): *Biometry* (3rd ed). New York: W. H. Freeman and Company.

Tam PPL, Tan S-S (1992): The somitogenic potential of cells in the primitive streak and the tail bud of the organogenesis-stage mouse embryo. Development 115: 703–715.

Thireau M (1967): Contribution a l'etude de la morphologie caudale et de l'anatomie vertebrale et costale des genres *Atheris*, *Actractaspis*, et *Causus* (Viperides de l'Ouest african). Bull Mus Natl Hist Nat 2: 454–470.

Wake MH (1980): Morphometrics of the skeleton of *Dermophis mexicanus* (Amphibia: Gymnophiona). Part 1. The vertebrae, with comparisons to other species. J Morphol 165: 117–130.

Wilson V, Beddington R (1997): Expression of T protein in the primitive streak is necessary and sufficient for posterior mesoderm movement and somite differentiation. Dev Biol 192: 45–58.

Zelditch ML (1995): Allometry and developmental integration of body growth in a Piranha, *Pygocentrus nattereri* (Teleostei: Ostariophysi). J Morphol 223: 341–355.

10

NOVEL FEATURES OF TETRAPOD LIMB DEVELOPMENT IN TWO NONTRADITIONAL MODEL SPECIES: A SKINK AND A DIRECT-DEVELOPING FROG

Michael D. Shapiro and Timothy F. Carl

Department of Organismic and Evolutionary Biology, and Museum of Comparative Zoology, Harvard University, Cambridge, Massachusetts

Cellular, molecular, and morphogenetic mechanisms that mediate limb development are regarded as highly conserved among vertebrates (Shubin and Alberch, 1986; Shubin et al., 1997). Current interest in limb development and evolution stems largely from recent discoveries of the underlying cellular and molecular mechanisms of limb patterning and outgrowth, especially as revealed in the two best-known amniote model systems, the chicken and the mouse. Previous work points to the conservation of fundamental patterning and outgrowth mechanisms of the vertebrate limb and animal appendages in general (see Shubin et al., 1997, for review). At other levels of organization, however (e.g., temporal or spatial coordination of these mechanisms or morphogenetic events downstream of these mechanisms), we may potentially observe ontogenetic differences that correlate with changes in adult morphology. Hence, vertebrate limb development is amenable to analysis at multiple levels of organization.

Beyond Heterochrony: The Evolution of Development, Edited by Miriam Leah Zelditch
ISBN 0-471-37973-5

Beginning with the pioneering experimental studies of the early and mid-20th century (e.g., Harrison, 1918; Saunders 1948; Zwilling, 1955; Saunders et al., 1957), the vertebrate limb has emerged as a model system for studying pattern formation in development (Summerbell, 1974; Hinchliffe and Johnson, 1980; Tickle, 1995; Cohn and Tickle, 1996; Johnson and Tabin, 1997). The limb also offers excellent opportunities to address the roles of developmental processes in organismal evolution (Raff, 1996; Shubin et al., 1997). A paucity of studies in vertebrate species other than chicken and mouse, however, precludes a comprehensive assessment of the evolutionary conservation or lability of limb development processes. Despite the recent studies of a few laboratory species such as the zebrafish (*Danio rerio;* Akimenko and Ekker, 1995; Laforest et al., 1998; Schauerte et al., 1998), African clawed frog (*Xenopus laevis*; Christen and Slack, 1997, 1998), and axolotl (*Ambystoma mexicanum*; Gardiner et al., 1995; Torok et al., 1998), this dearth of information is especially problematic for nonamniote vertebrates.

Among amniotes, studies of reptiles are relatively rare. Although lizards exhibit tremendous diversity in limb morphology, only a few investigations describe squamate (lizard and snake) cartilage condensation patterns in the limb (among the most basic assays of limb developmental morphology; Sewertzoff, 1904, 1931; Steiner, 1922; Mathur and Goel, 1976). A few others describe cellular level changes correlated with limb loss, a common evolutionary trend in lizards (see Raynaud, 1985, for review). Both amphibians and reptiles exhibit a number of specialized life history or reproductive modes, which might be expected to affect embryogenesis (Elinson, 1987; Raff and Wray, 1989); hence, a cursory examination of only a few representative tetrapod species may yield only a superficial understanding of evolutionary patterns and processes of limb development. This shortcoming can only be remedied by examining a multitude of nonmodel species that represent a diverse set of morphologies and life histories.

In this chapter, we discuss aspects of early limb ontogeny in two nontraditional model species: the direct-developing frog *Eleutherodactylus coqui* and the Australian skink *Hemiergis peronii.* Not surprisingly, we find many developmental similarities between these organisms and traditional models of limb development. These two species are also broadly similar in adult morphology, sharing limb skeleton elements common to most terrestrial vertebrates. Both have a single proximal element attached to the limb girdles (humerus in the forelimb, femur in the hindlimb), followed distally by paired elements (radius and ulna, tibia and fibula), a set of wrist (carpal) or ankle (tarsal) elements, and a distal series of segmented digits composed of phalanges (Fig. 10.1). These broad morphological similarities, observed in most tetrapods, imply a conservation of developmental processes in the tetrapod clade. However, we also extract information from these two nontraditional model species not available from studying early ontogeny in typical systems.

For example, the direct-developing frog *Eleutherodactylus coqui*, endemic to the rain forests of Puerto Rico (Fig. 10.2), presents a unique opportunity to study the function of a key morphogenic structure in vertebrate limb develop-

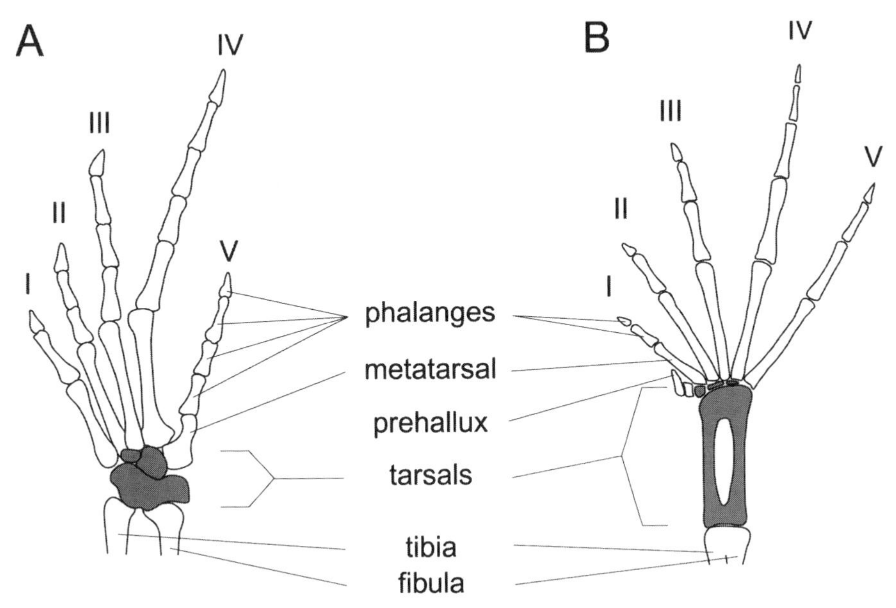

Figure 10.1. Generalized foot (pes) skeletons of a frog and lizard. The tarsals (shaded) lie immediately distal to the tibia and fibula. In the forelimb (not shown), the carpals lie distal to the radius and ulna. Digits are numbered from anterior (I) to posterior (V). A: primitive lizard phalangeal formula (number of bones per digit, beginning with the anterior digit) is 2-3-4-5-4 for the hind limb (2-3-4-5-3 for the forelimb). B: most frogs have a phalangeal formula of 2-2-3-4-3 for the hindlimb (2-2-3-3 for the forelimb). The frog pes features a number of skeletal specializations, including a prehallux (the homology of which is uncertain), fusion of the tibia and fibula, and elongate proximal tarsals. Drawings are not to scale. Frog limb in B redrawn from Duellman and Trueb (1986).

ment. *E. coqui*, unlike most others frogs, does not have a tadpole stage in its life history but instead develops directly from an embryo into an adult form (Elinson, 1990; Hanken et al., 1997). This exceptional life history is associated with interesting developmental modifications, most notably, the lack of an apical ectodermal ridge (AER; Richardson et. al., 1998). The AER is a morphologically distinct ridge of ectoderm located at the distal tip of a developing limb bud. It has been found in most animals in which it has been sought, including mammals, birds, reptiles [but not in the rudimentary hindlimb buds of the python (Cohn and Tickle, 1999)], and the metamorphic frog *Xenopus* (Tschumi, 1957). The AER appears to plays a critical role in proximodistal outgrowth of the developing limb of traditional model species (Saunders, 1948; Zwilling, 1955; Summerbell, 1974; Niswander et al., 1994; Mahmood et al., 1995); nevertheless, an AER is curiously absent from the limb buds of *E. coqui*. Moreover, the removal of the distal ectoderm of the developing limb of *E. coqui* does not have the same effect as removal of this tissue in animals with a distinct AER (Richardson et al., 1998). AER removal during early limb development in the chicken embryo abbreviates limb outgrowth, yielding dramatic proximo-

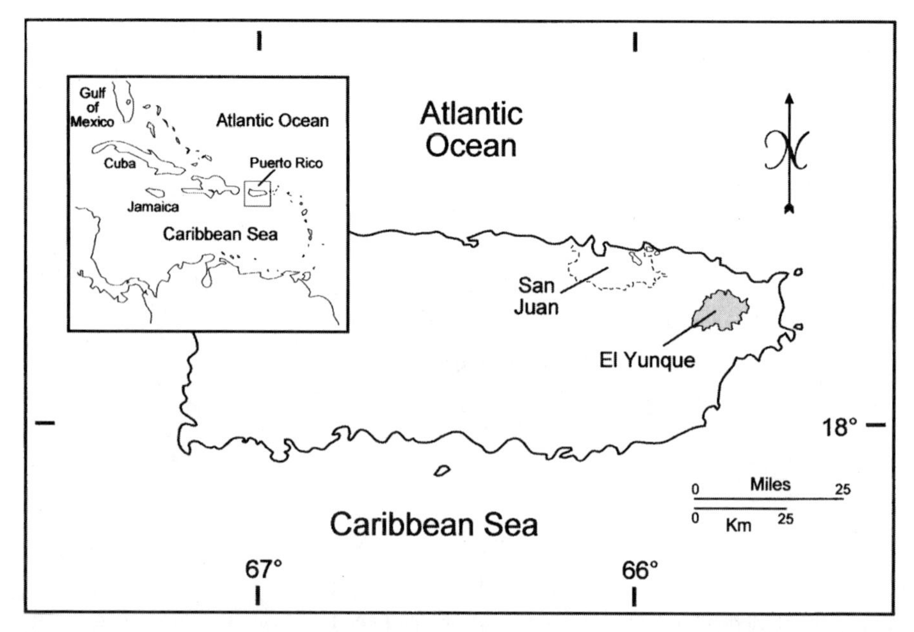

Figure 10.2. Map of Puerto Rico. Puerto Rico is one of the many small islands of the West Indies (inset). *Eleutherodactylus coqui* in this study were collected in the tropical rain forest of El Yunque.

distal truncations of the limb skeleton (Saunders 1948; Summerbell, 1974). In contrast, distal ectoderm removal in *E. coqui* causes only moderate distal limb skeleton malformations.

The skink *Hemiergis peronii* provides a rare chance to study digit loss within a single species. Limb reduction is common among Australian skinks, and the genus *Hemiergis* represents the best example of graded digit loss among humus dwellers (Choquenot and Greer, 1989; Greer, 1989, 1991). In Western Australia, *Hemiergis* contains three live-bearing, semi-fossorial species that inhabit the semi-humid to humid coastal shrublands and woodlands (an additional species inhabits a small, arid, inland region) (Choquenot and Greer, 1989; Greer, 1989; Cogger, 1992; Storr et al., 1999). Within the three closely related (J. Reichert and T. Reeder, unpublished observations) coastal Western Australian representatives, *Hemiergis* features a range of between two and five digits on the forelimbs and hindlimbs. Two distinct populations (with fixed digit numbers) of *H. peronii* are recognized by Choquenot and Greer (1989): a western coastal population with three digits on each limb and a four-digit population that inhabits the southern coast and inland forests (Fig. 10.3).

Together, these skink and direct-developing frog examples permit us to consider two questions not approachable in other model species used to study vertebrate limb development. First, in *Eleutherodactylus coqui*, how does limb development without an AER differ from limb development in a traditional

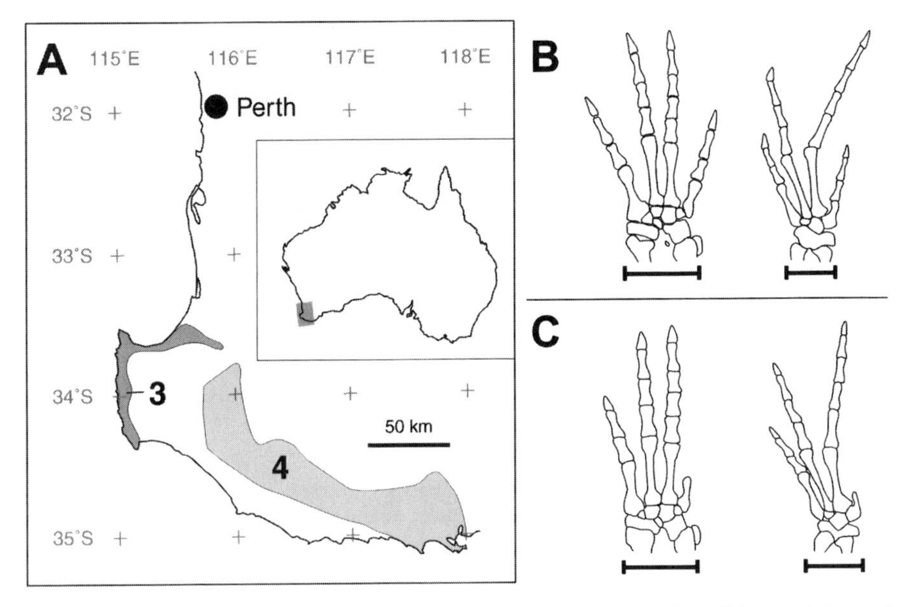

Figure 10.3. Collecting ranges and digit configurations of three- and four-digit populations of *Hemiergis peronii.* A: map of southwestern Western Australia, with collection ranges for three-digit (3; dark gray) and four-digit (4; light gray) populations. A map of continental Australia is inset, and area of detail is shaded. B and C: adult manus and pes morphologies of three-digit (B) and four-digit (C) morphs of *H. peronii.* For each morph, the manus is on the left, and pes is on the right. Both limbs of the three-digit morph have the phalangeal formula X-3-4-5-0, whereas the four-digit formula is X-3-4-5-3. Scale bars in B and C = 1 mm.

model species (the chicken)? Second, in *Hemiergis peronii*, what differences in early skeletal ontogeny yield different numbers of digits in a single species? In addressing these questions, we will discuss our results with respect to heterochrony, broadly defined as evolutionary change in developmental timing. An evolutionary approach to developmental biology has the potential to offer a greater understanding of how differences in developmental processes and sequences produce morphologies—does heterochrony have any predictive or explanatory power in this context?

LIMB DEVELOPMENT IN THE FROG *ELEUTHERODACTYLUS COQUI*

Distal Limb Ectoderm Removal in *Eleutherodactylus coqui* Leads to Moderate (but not Major) Skeletal Malformations

The typical anuran hindlimb consists of five digits with a phalangeal formula of 2-2-3-4-3 (Duellman and Trueb, 1986). In addition, several tarsal elements and a prehallux are present in most frog species; the limb skeleton of *E. coqui* matches this general anuran pattern (Fig. 10.1). As is the case in other frogs,

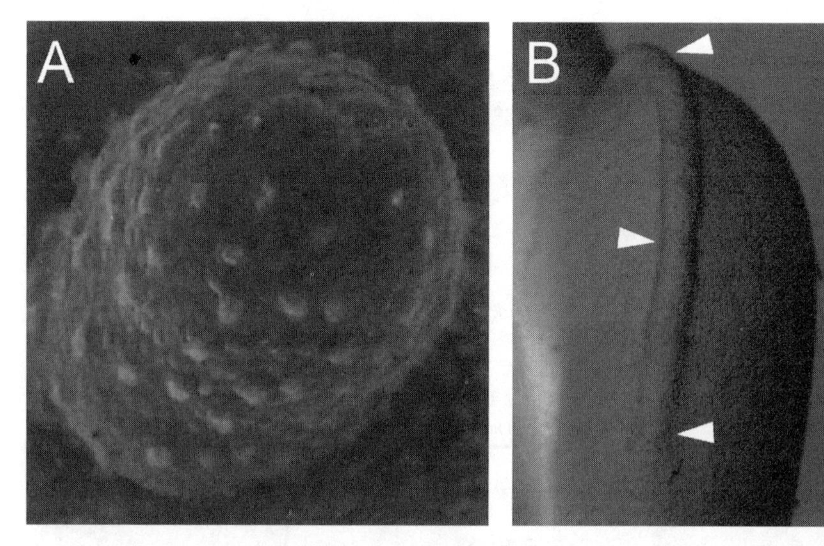

Figure 10.4. Comparison of roughly equivalent embryonic stages of *Eleutherodactylus coqui* and chicken forelimbs. Scanning electron micrographs show TS stage 6 *E. coqui* forelimb (A) and chicken hindlimb (B) in distal and cranial views. Note that the distinct ridge of ectoderm (arrowheads) located along the distal edge of the developing chick limb bud is absent in *E. coqui*. Photograph in B courtesy of M. Richardson.

limb skeleton chondrogenesis in *E. coqui* proceeds in a proximodistal sequence, with the most distal phalanges chondrifying last.

In the chick, the AER is found at the distal end of the developing limb (Fig. 10.4), and removal of this structure induces a progressive loss of distal elements in a stage-dependent manner (Fig. 10.5; Saunders, 1948; Summerbell, 1974; Niswander et al., 1994; Mahmood et al., 1995). To test whether the most distal ectoderm of the limb bud in *E. coqui* mediates limb outgrowth in a way similar to the AER of other organisms, Richardson et al. (1998) removed the ectoderm at the distal tip of the hindlimb buds of early embryos [Townsend and Stewart, 1985 (TS), stages 4–6]. The earliest manipulations were performed at TS stage 4, when limbs are barely visible (Fig. 10.5). At TS stage 6, the latest stage of manipulations, the limbs are prominent (Fig. 10.6). These stages in *E. coqui* roughly correspond to stages 21–27 in the chick embryo (Hamburger and Hamilton, 1951), stages during which removal of the AER causes acute limb truncations.

Of the manipulated stage 4 embryos, 30% of the specimens maintained normal limb morphologies at hatching. The remaining 70% exhibited moderate skeletal defects, including fusion of the proximal ends of metatarsals I and II and the occasional loss of the most distal phalanx of digit I (Fig. 10.5). Roughly 60% of the specimens manipulated at stage 5 exhibited fusion of the proximal ends of metatarsals I and II, whereas 40% completely lost metatarsal I. All stage 5 manipulations resulted in either a complete loss or a reduction in

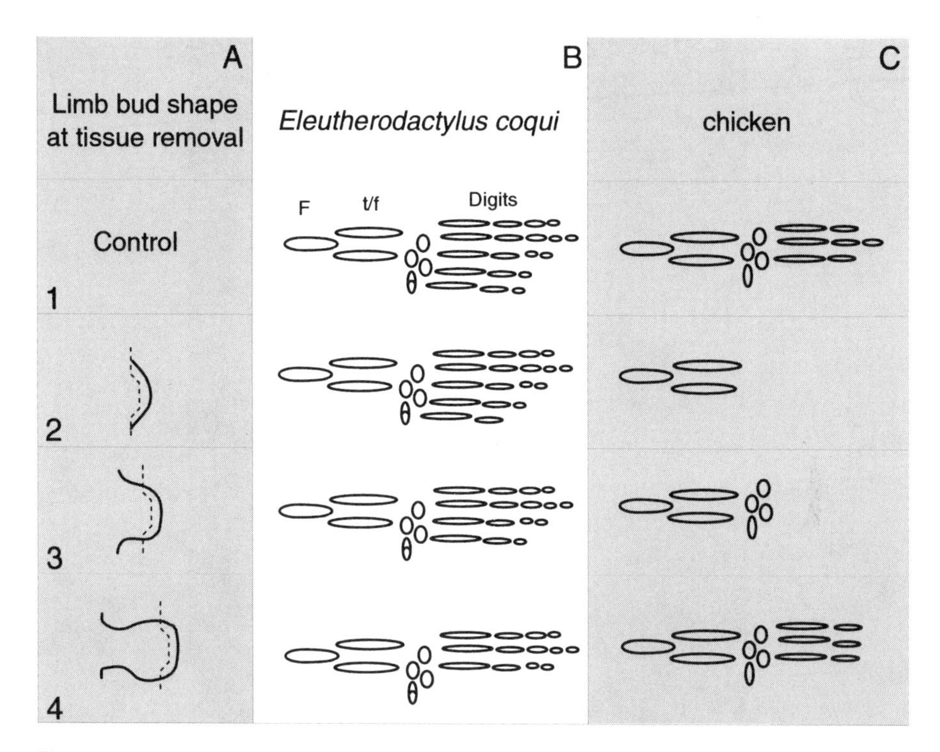

Figure 10.5. Comparison of distal ectoderm removal in *Eleutherodactylus coqui* and chicken at progressively later stages of limb development. Column A: diagrams representing hindlimb buds from which the distal ectoderm (*E. coqui*) or AER (chicken) was removed during development (dashed lines indicate sites of removal). Column B: when the most distal ectoderm is removed from limb buds in *E. coqui*, limb development does not cease. Moreover, manipulations performed early in development (row 2 is TS stage 4) have less of an effect than those done at later stages (row 4 is TS stage 6). Column C: in contrast to *E. coqui*, when the AER is removed in the chicken limb, development ceases, causing loss of distal skeletal elements. Hence, when manipulations are performed early in limb development [row 2 in C is Hamburger and Hamilton, 1951, (HH) stage 21], only the most proximal skeletal elements form; later manipulations induce less severe truncations (row 4 in C is HH stage 27). Chicken results adapted from Saunders (1948) and Summerbell (1974). F, femur; t/f, tibia/fibula.

the number of digit I phalanges. All manipulated stage 6 specimens lost digit I and all phalanges of digits II and III. No specimen, regardless of the stage of ectoderm removal, had any modification to the skeletal elements of digits IV and V (Fig. 10.5).

Tissue Geometry Alterations in *Eleutherodactylus coqui* May Affect Function of the Distal Ectoderm

The loss of an AER and apparent modification of distal ectoderm function are significant changes correlated with major modifications in the life history of

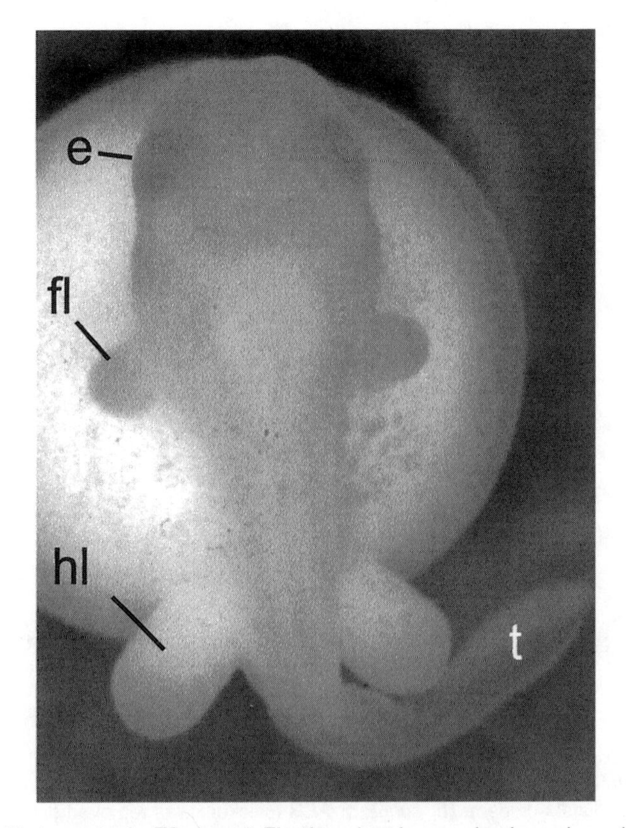

Figure 10.6. Photograph of a TS stage 6 *Eleutherodactylus coqui* embryo shown in dorsal view; anterior is toward the top. This direct-developing frog bypasses a larval stage, developing from embryo to adult without a tadpole. All four limb buds are beginning to flatten dorsoventrally. The prominent tail will regress until it is nearly absent at hatching (TS stage 15). e, eye; fl, forelimb; hl, hindlimb; t, tail.

E. coqui. Most species of frogs have two successive posthatching life history stages: an herbivorous, aquatic larva and a carnivorous, terrestrial adult. These stages are separated by a discrete metamorphosis. Direct-developing species, however, bypass the free-living larval stage and develop directly into adult forms (Elinson, 1990; Hanken et al., 1997). Many adult features (such as the limbs) that form only after hatching in metamorphic anurans instead form during embryogenesis in direct developers (Elinson, 1994). Consequently, direct-developing anurans develop in a manner more similar to amniotes than typical metamorphosing frogs.

Unlike AER removal in early chicken embryos, removal of the distal ectoderm in *E. coqui* is not sufficient to cease limb outgrowth but does induce malformation of distal limb skeletal elements (Richardson et al., 1998). Digits I and II were most severely affected by ectoderm removal; the most common modification was the fusion of digits I and II at the level of the metatarsal and the loss of digit I and II phalanges. Surprisingly, ectoderm removal during early limb bud outgrowth had less of an effect on final limb morphology than manipulations at later stages of limb bud development, the opposite of the effect observed in chickens. Two scenarios may explain these observations: 1) a loss of limb tissue during distal ectoderm removal resulted in the reduction of digit quantity, or 2) both the morphology and function of the distal ectoderm in *E. coqui* are different from the AER of other tetrapods.

Regarding the first possibility, removal of tissue early in limb bud development may reduce the total net volume of tissue resources available to the developing limb. Thus, elements that form earlier may use up the pool of available cells, reducing or eliminating tissues available to late-forming digits. In this scenario, the amount of tissue in a developing anuran limb would be directly proportional to the number of digits. Notably, digits IV and V were never affected as a result of distal ectoderm manipulations in *E. coqui*. As in other anurans (Hinchliffe and Johnson, 1980), metacarpal IV forms first in *E. coqui*, followed by simultaneous appearance of metacarpals V and III, and finally metacarpals II and I, respectively. Hence, when tissue is removed from an *E. coqui* hindlimb, one possible outcome is the loss of elements in digits I and II, the last digits to form. Indeed, Alberch and Gale (1983, 1985) observed a similar phenomenon in salamander experiments: mitotic inhibition of limb bud tissue resulted in a reduced number of digits, presumably by limiting tissue resources available for digit condensations. Although Alberch and Gale's salamander results are similar to those found in *E. coqui*, loss of digits in their experiments was achieved by "removal" (by limiting proliferation) of limb mesenchyme, not ectoderm. Although slight amounts of limb mesenchyme may have been removed during ectoderm manipulations in *E. coqui*, great care was taken to guard against this possibility (Richardson et al., 1998).

A more intriguing explanation points to a modification of both the structure and function of the distal ectoderm in *E. coqui* relative to other tetrapods. This scenario challenges the widely cited model that proximodistal limb growth and differentiation result from interactions between a morphological AER and its associated mesenchyme (see Dudley and Tabin, 2000, for review); however, these interactions have been documented in only a few species. Niswander et al. (1994) hypothesized that the physical ridge is important because flattening the AER may help to concentrate ectodermal molecular signals across the distal limb bud, but this hypothesis of a morphological prerequisite for the AER is not supported by *E. coqui*. Salamanders, too, lack an AER, and they also develop recognizable tetrapod limbs; hence, limb development in the absence of a morphological AER is not limited to *E. coqui*. Indeed, a morphologically distinct AER clearly is not a prerequisite for limb development in all vertebrates.

Experimental manipulations suggest that the distal limb ectoderm of *E. coqui* may differ from the AERs of amniotes in another fundamental way. Late manipulations in *E. coqui* ultimately have a greater impact on skeletal patterning than early ones. As noted above, this result is opposite to the effect observed in chick AER removals, which decrease in severity at later stages. The chick AER does not regenerate after surgical removal at any stage; instead, excision arrests proximodistal growth and patterning of the limb. However, results obtained by Richardson et al. (1998) imply that the distal ectoderm of *E. coqui* may have regenerative capacity. If removed early, the distal ectoderm may regenerate and resume its role in limb patterning, thus potentially highlighting another major difference between limb development in *E. coqui* and traditional model species. This interesting possibility has yet to be verified and is a topic ripe for future study.

LIMB DEVEOPMENT IN THE SKINK *HEMIERGIS PERONII*

Autopod Skeletal Condensation in *Hemiergis peronii*

Early stages of *H. peronii* limb development closely resemble ontogenetic sequences of pentadactyl lizards (Sewertzoff, 1904, 1931; Steiner, 1922; Mathur and Goel, 1976) and amniotes in general (Burke and Alberch, 1985; Shubin and Alberch, 1986) (Fig. 10.7). Metacarpal and metatarsal IV—the first digital elements to appear in the fore- and hindlimb, respectively—invariably condense along the primary axis in amniotes; proximal to the autopod, this axis also includes the ulna and humerus (forelimb) or fibula and femur (hindlimb) (Burke and Alberch, 1985; Shubin and Alberch, 1986; Müller and Alberch, 1990). The axis of digital development then curves preaxially and sequentially yields digits III and II. Digit V, a probable neomorph that does not originate from a bifurcation event from the digital axis (Shubin and Alberch, 1986), appears temporally after digit III but before digit II in *H. peronii*. No trace of digit I ever appears in either limb.

In general, phalanges are added to each digit in the order of appearance of the metacarpals and metatarsals. Hence, the first phalanx to appear in each limb is the proximal phalanx of digit IV (IV-1). The addition of phalanx IV-1 to both sets of limbs marks the final common stage in the development of the two morphs (but see below). At this stage of common skeletal morphology, embryos of the two *H. peronii* morphs are nearly identical externally as well: both show the same degree of development of the lower jaw, eyes, endolymphatic sacs, and (closed) branchial slits (Fig. 10.8; stage 33 of Dufaure and Hubert, 1961). The limbs are also similar in external view and appear to contain the early condensations of four digits in both fore- and hindlimbs.

Subsequent to this shared stage, the four-digit morph adds phalanges across the second through fifth digits to bring the phalangeal formula to X-1-1-2-1 (X denotes a missing digit, whereas 0 indicates a metacarpal/metatarsal but no phalanges) (Fig. 10.9). Differential staining intensities suggest that the proximal phalanges of digits II and V are the last to be added, with IV-2 added slightly

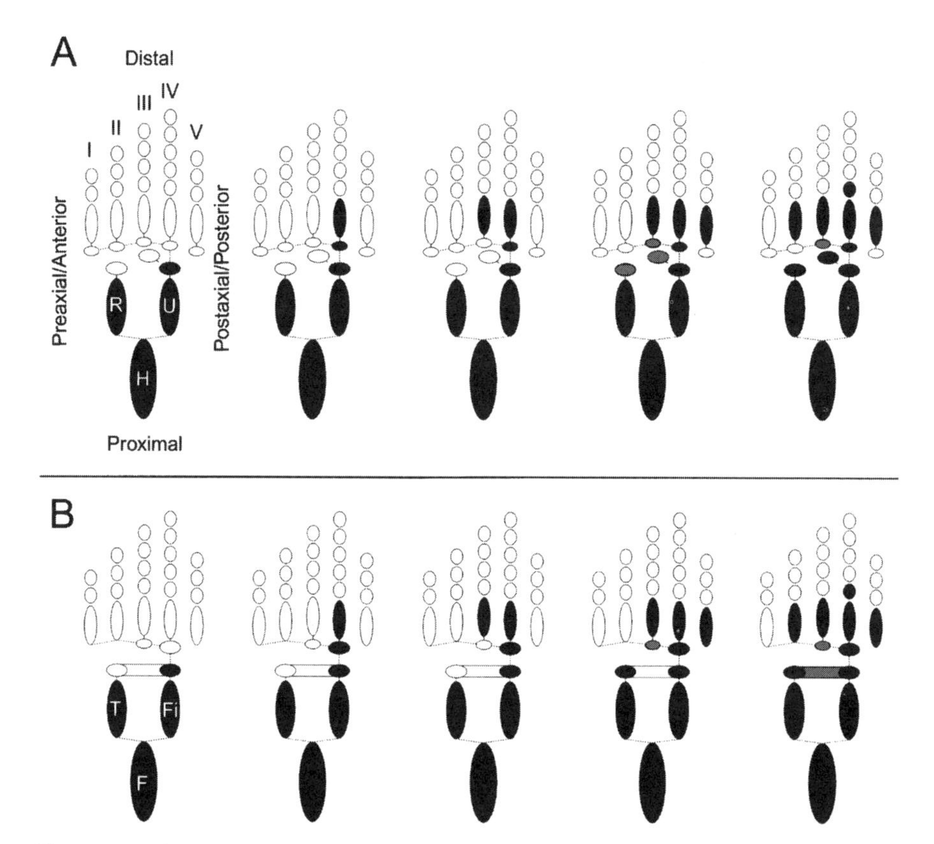

Figure 10.7. Schematic representation of early limb skeleton condensations in both morphs of *Hemiergis peronii*. Cartilaginous elements present in both morphs at a given stage appear as blackened shapes, whereas carpal and tarsal elements exhibiting variability appear in gray. A: forelimb condensations first appear proximally and include the humerus (H), radius (R), and ulna (U). The first through fifth digits (I–V) appear in the order IV > III > V > II; no elements of digit I ever appear in either morph. B: hindlimb elements appear in a similar pattern, beginning with the femur (F), tibia (T), and fibula (Fi).

earlier. This configuration implies a hypothetical intermediate stage in which only the two oldest phalanges, IV-1 and III-1, are present, yielding a phalangeal formula of X-0-1-1-0. Differential staining in three-digit embryos of a similar stage suggests an identical intermediate configuration (although the configuration is not represented by actual specimens of this morph either), and hence these two morphs may share at least one additional common phalangeal configuration (X-0-1-1-0) than is indicated by actual specimens.

In four-digit *H. peronii* embryos, additional phalanges subsequently condense from proximal to distal until the adult configuration of X-3-4-5-3 is attained. Note that digits do not develop one at a time; instead, phalanges

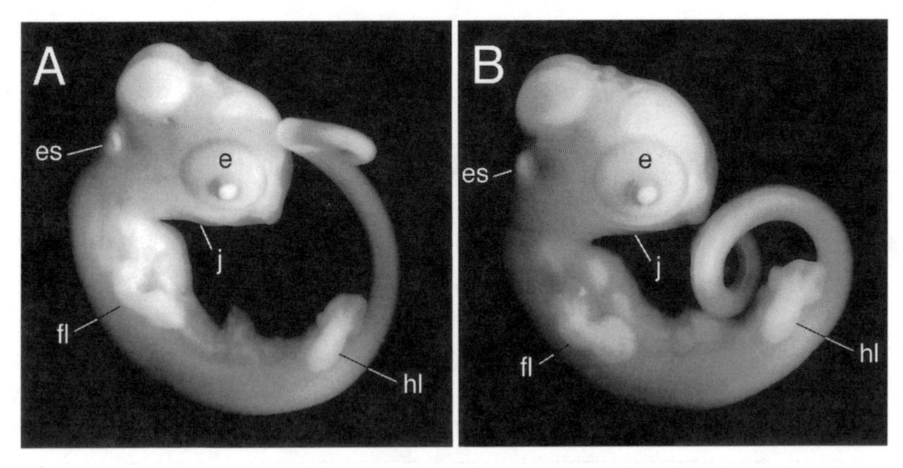

Figure 10.8. Stage 33 embryos of *Hemiergis peronii*. Embryos of the three-digit (A) and four-digit (B) populations have similar morphologies at this stage. e, eye; es, endolyphatic sac; fl, forelimb; hl, hindlimb; j, lower jaw.

condense in parallel across all four digits. A similar condensation pattern is observed in the three-digit morph, although digit V is excluded from the skeletal developmental program. Following the last common configuration shared by both morphs, phalanges in three-digit *H. peronii* are added only to the second through fourth digits, yielding a formula of X-1-1-2-0. As with the four-digit morph, phalanges then condense proximal to distal, although across only three digits, until the adult configuration of X-3-4-5-0 is reached.

The Three-Digit *Hemiergis peronii* Limb Is not Simply a Truncated Four-Digit Limb

The limb skeleton developmental sequences of the three- and four-digit morphs of *H. peronii* are essentially indistinguishable until stage 33+. After that point, the four-digit morph begins adding phalanges to digit V, whereas the three-digit morph does not. Despite the difference in digit number, however, phalanges are added proximal to distal across all remaining digits in the embryos of both morphs. This mode of digit assembly, common to all known amniote limb development sequences, limits the mechanisms with which limb reduction can be described. If digits formed one at a time, for example, early truncations of an ancestral limb development program would produce fewer (but still complete) digits. In this hypothetical situation, the last digits to form in the ancestral program would be lost, but the remaining digits would maintain their ancestral morphologies with full complements of phalanges. If digit V in *H. peronii* formed after digits II–IV were fully formed, for example, the loss of digit V in

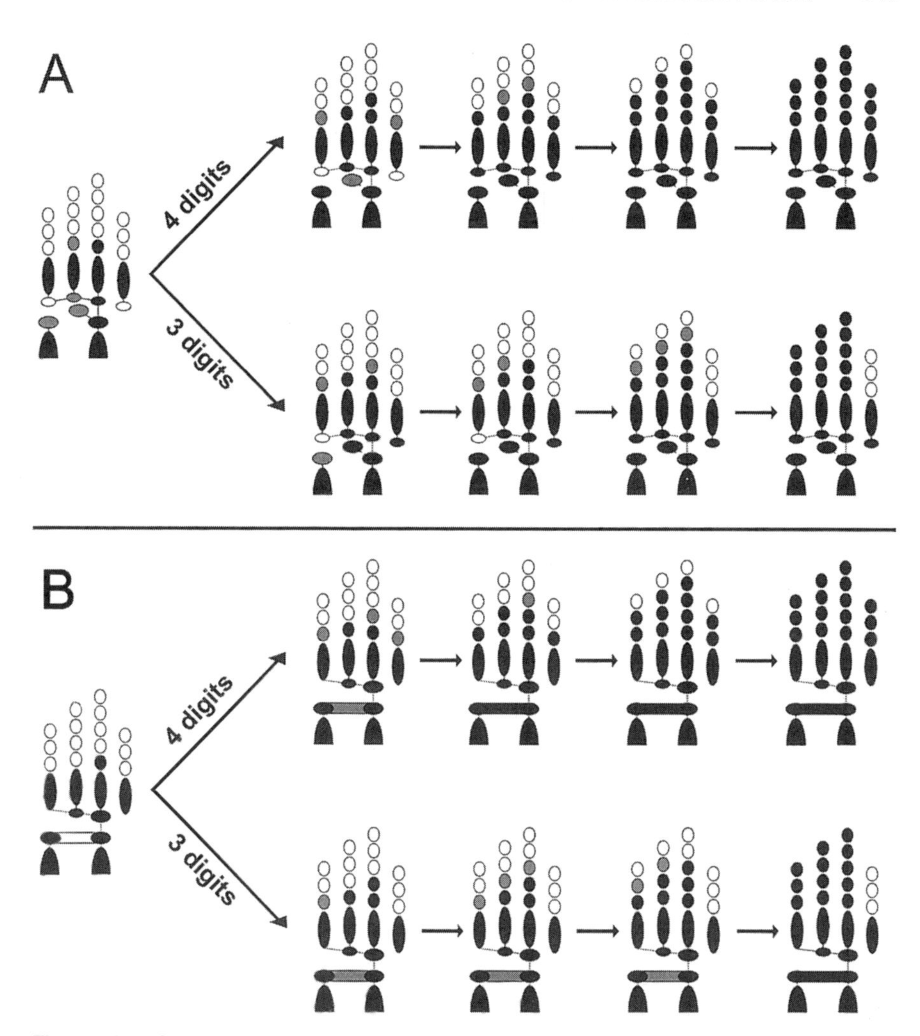

Figure 10.9. Comparisons of condensation patterns in three- and four-digit morphs of *H. peronii*. Cartilaginous elements are indicated in black, precartilaginous in gray. A: in the forelimb, both morphs share a common configuration with a phalangeal formula of X-0-1-1-0 before developmental trajectories diverge. In the four-digit morph (top row), phalanges are added proximally to distally across all four digits. The adult formula of X-2-3-4-3 is indicated at top right. In the three-digit morph, phalanges are only added across digits II–IV; only a metacarpal remains in the digit V position. The adult configuration of the three-digit morph is X-3-4-5-0. B: hindlimb condensation sequences follow a similar pattern, and adult phalangeal formulae mirror those of the forelimbs for each morph.

the three-digit morph could be explained by a truncation of the four-digit developmental program.

However, the absence of digit V phalanges in the three-digit morph cannot be explained by a simple truncation of another developmental program. If the four-digit developmental program were to be truncated at the latest stage when digit V has no phalanges, the resulting phalangeal formula would be X-0-1-1-0, not the three-digit adult formula of X-3-4-5-0. The reduced digital configuration of the three-digit morph is simply not represented in the developmental sequence of the four-digit morph; therefore, the former cannot represent a truncated version of the latter. In fact, at no known intermediate stage in the development of any pentadactyl lizard—including the closely related pentadactyl *H. initialis*—is the adult configuration of either the three- or four-digit *H. peronii* represented (Shapiro, unpublished observations). Shubin and Alberch (1986) describe a parallel example in the salamander *Proteus*, in which three-fingered and two-toed adult limbs do not correspond to a single developmental stage of another salamander. A similar situation can also be observed in the speciose scincid *Lerista*, which features a morphocline of digit quantities from zero to five on both limbs (Greer, 1987, 1990, 1991). As is the case in *Hemiergis*, the reduced adult digit morphologies in *Lerista* do not correspond to an intermediate developmental stage in any known ontogenetic sequence (however, no *Lerista* limb development sequences have been described) (Greer, 1991).

In contrast, other instances of limb reduction in lizards likely do represent truncations of appendicular skeletal ontogenies. For example, the adult morphologies of some species of the teiid *Bachia* are paralleled by intermediate developmental stages of other lizards (Presch, 1975) (Fig. 10.10). Unlike *Hemiergis* and *Lerista*, adult *Bachia* forelimbs match up with early ontogenetic stages of other squamates (although carpal and tarsal homologies are difficult to establish in cases of extreme reduction and miniaturization). Hence, the mode of digit loss observed in *Hemiergis* and *Lerista* is not generalized for all cases of limb reduction in lizards.

Figure 10.10. Adult forelimb morphologies of *Bachia* spp. (A, C, E, G) closely resemble embryonic configurations of other lizards (B, D, F, H). Anterior is to the left and distal is up in all panels. Blackened shapes in diagrams of embryonic morphologies represent cartilaginous, precartilaginous, and mesenchymatous condensations. Dashed lines in B, D, F, and H indicate probable developmental connectivity between condensations. A: forelimb of *B. trisanale*. From proximal to distal, the elements of the limb likely include the humerus, radius and ulna, ulnare (possibly fused with the radiale), fourth distal tarsal, and fourth metacarpal. B: these elements should be present in all amniotes with at least one digit, according to models of limb morphogenesis (Burke and Alberch, 1985; Shubin and Alberch, 1986). C, D: the adult forelimb of *B. monodacylus* and an embryonic limb configuration of *Calotes versicolor* (Mathur and Goel, 1976) both have four metacarpals distal to the carpals. E, F: the adult forelimb of *B. dorbignyi*, with three metacarpals and a single phalanx over digit IV, corresponds closely to an embryonic configuration in *Hemiergis*; however, *Hemiergis* has an additional metacarpal (metacarpal 2, shaded) when the first phalanx appears. G, H: the adult forelimb of *B. heteropa*, with two phalanges over digits I–IV and a single phalanx over digit V, resembles an embryonic configuration of *H. initialis* with a reversed anteroposterior axis (Shapiro, unpublished observations). The *H. initials* diagram in H is reversed so that digit V is on the far left and digit I is on the far right. *Bachia* figures redrawn from Presch (1975). Scale bar in G = 1 mm for all *Bachia* drawings.

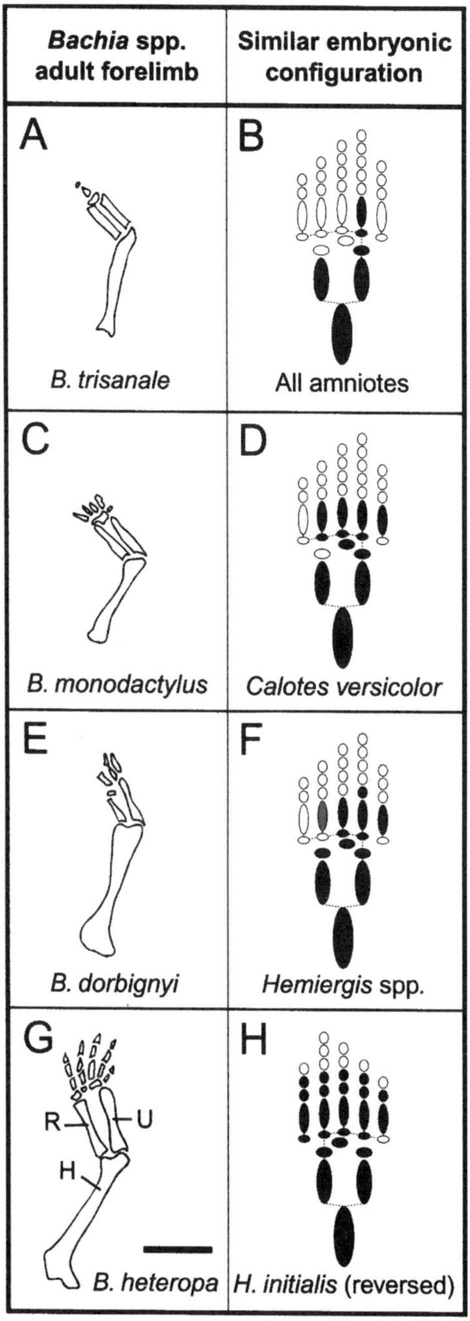

***Bachia* spp. adult forelimb**	**Similar embryonic configuration**
A *B. trisanale*	B All amniotes
C *B. monodactylus*	D *Calotes versicolor*
E *B. dorbignyi*	F *Hemiergis* spp.
G R U H *B. heteropa*	H *H. initialis* (reversed)

EVOLUTIONARY DIGIT LOSS AND THE TERMINOLOGY OF HETEROCHRONY

Two seemingly opposing interpretations of the same case of skink limb reduction highlight the diversity of thought regarding the relationship between ontogenetic timing changes and morphology. Greer (1987, 1990, 1991) studied in detail the reduced adult limbs of *Lerista* and other lizards and notes that lizards with reduced numbers of digits often have complete (or nearly complete) complements of phalanges in their remaining fingers and toes. He suggests that, since no pentadactyl lizards (or other amniotes for that matter) have intermediate stages of limb development that correspond to these adult forms, many reduced limb morphologies likely do not result from terminal deletions of ancestral developmental programs (Greer, 1991).

Müller (1991), too, cites the example of limb reduction in *Lerista* in a discussion of heterochrony and limb development but interprets developmental timing mechanisms differently from Greer. Müller notes that the phylogenetic loss of digits in skinks occurs in the exact reverse sequence of digit (i.e., metacarpal or metatarsal) appearance in the alligator (Müller and Alberch, 1990); therefore, truncations of ancestral developmental programs, manifested as failure of chondrogenic condensation segmentation, are responsible for digit loss. Müller concludes that, "These truncations and the processes of nonsegmentation of chondrogenic arrays suggest that a primary mechanism of limb transformation is paedomorphic heterochrony" (p. 402).

In the traditional terminology of heterochrony (e.g., Gould, 1977; Alberch et al., 1979), Greer's observations of adult skeletal pattern do not implicate paedomorphosis (or more specifically progenesis) as a mechanism of limb reduction (although Greer does not mention these heterochrony terms specifically). The derived conditions of reduced-limbed skinks do not parallel developmental intermediates of the ancestral condition, and thus they do not represent truncated ontogenies at the level of the whole limb. Müller's argument, however, implies heterochronic events at the level of bifurcations and segmentations of the digital arch: digits may be lost because branching processes or events in an ancestral ontogenetic sequence are truncated or eliminated, resulting in the loss of digits. It is important to note, however, that these foreshortened events are not terminal in the ontogenetic sequence at the level of the limb skeleton developmental program. In both lizards and crocodiles (and other amniotes), metacarpals and metatarsals branch off the digital arch well before *all* phalanges condense.

The *Lerista* example highlights that the evaluation of evolutionary changes in developmental timing depends largely on the level of organization examined (also see Raff and Wray, 1989). At the level of the entire limb skeletal development program, for example, digit loss in *Hemiergis* cannot be regarded as a truncation: the overall timing of the likely ancestral sequence is not modified. At the level of specific events within this sequence, however, truncations or deletions yield an absence of metacarpals and phalanges. For instance, a lack of anterior digital arch bifurcation leads to an absence of digit I in both morphs of *H. peronii*, and "non-segmentation" of digit V in the three-digit morph produces a metacarpal without phalanges.

We question the utility of identifying these examples at the level of the digits as heterochronic events. Although one is tempted to argue that they represent truncations of the developmental plans of specific digits, those digits are integral parts—not terminal events—of a larger unit. By ascribing heterochrony to any developmental process, or in this case a nested subset of a process, essentially all evolutionary phenotypic changes can be attributed to heterochrony. Consequently, "heterochrony" becomes synonymous with phenotypic change and thus loses any meaning of its own (Wake, 1996; Zelditch and Fink, 1996; Rice, 1997). It is beyond the scope of this chapter to offer a new or improved definition of heterochrony; rather, when changes in ontogentic trajectories are discussed, we advocate explicit discussion of the level of organization involved.

In the case of *Hemiergis*, for example, analyses of the underlying mechanism(s) of digit reduction can focus on several different levels. In addition to the above discussion of cartilage condensation programs and events, we could choose to examine cell proliferation. Although we have not studied this process in *Hemiergis*, previous experimental work on amphibians and lizards has shown that mitotic inhibition of early limb buds is sufficient to cause digit loss (e.g., Alberch and Gale, 1983, 1985; Alberch, 1985; Raynaud and Clergue-Gazeau, 1986; Raynaud, 1991). In these cases, cell proliferation is slowed so that, during morphogenesis of the autopod, not enough limb mesenchyme is present to produce a full complement of digits. If cell proliferation is slowed or terminated early in *Hemiergis* limb ontogeny such that digit development is influenced, is it sufficient to simply implicate "heterochrony" or "paedomorphosis" as the mechanism of digit loss? As argued above, these terms are not adequate to describe morphological changes at the level of limb skeleton condensation. Furthermore, what if the spatial and/or temporal expression patterns of genes regulating cell proliferation differ between the three- and four-digit morphs of *H. peronii*? Simply stating that digits are lost by paedomorphosis, or that descendant morphologies are not truncations of ancestral ones, does not adequately address the myriad of possible changes in the evolution of ontogeny. Oster and Alberch (1982), for instance, view development as a "hierarchically organized sequence of processes" (p. 452); this view of ontogeny permits changes at the level of genomic interactions, gene expression, cell interactions and behavior, tissue interactions, and the series of coordinating mechanisms that comprise a developmental program.

Attributing all morphological change to heterochrony is further complicated because developmental alterations at any of these levels of organization may not be exclusively related to timing and rate changes. Heterotopy—a change in tissue geometry—for example, is recognized as a potential mechanism of digit loss (e.g., Alberch, 1985). We also cannot exclude changes in developmental processes themselves—or changes in the interactions between these processes—as mechanisms for ontogenetic and morphological diversity. For example, although unlikely in the case of *Hemiergis*, morphological novelty in digit number could result if the *processes themselves* that specify digits are changed, as opposed to a rearrangement (or truncation) of a *sequence* of conserved events or processes.

Similarly, in *E. coqui*, the terminology of heterochrony does not adequately describe the loss of an important developmental structure in the limb. The dramatic shift in timing of limb formation that accompanied the evolution of direct development in *E. coqui* appears to have involved relatively few modifications to ancestral limb development mechanisms, especially those that mediate limb outgrowth and patterning. Instead of betraying dramatic differences from either amniotes or metamorphosing anurans, the limb development program in *E. coqui* appears to have evolved largely as an intact module (sensu Raff, 1996), producing a relatively conserved adult morphology. However, the developmental changes that effected the gross alteration in timing of limb formation from posthatching to embryogenesis, as well as possible changes in the location and identity of limb precursor cell populations that accompanied the shift from larval to embryonic development, remain largely unknown (Elinson, 1994).

One consequence of these shifts may be the loss of a morphological AER. The distal ectoderm that exists in its place likely plays a similar role in proximodistal outgrowth and patterning of the limb, but we discuss differences in the structure and function of this tissue in *E. coqui*. In this case, the loss of an AER appears to be a change in tissue geometry (heterotopy) rather than a shift in developmental timing (heterochrony) at the level of limb bud outgrowth.

FUTURE DIRECTIONS

Assays of gene expression patterns are commonplace in molecular studies of development in traditional model species but are often overlooked by evolutionary morphologists. Similarly, molecular developmental biologists often do not realize the evolutionary implications of their gene expression studies by considering their results in a comparative or phylogenetic context. We do not see fundamental differences in the value of data gathered by either approach; these two perspectives differ in the scale and level of organization examined, but they can be complementary. A synthesis of developmental biology and comparative morphology facilitates an integrative approach to evolutionary morphology in which differences in structure potentially can be correlated with changes in gene expression patterns.

Gene expression studies in *E. coqui* and *H. peronii* may prove instrumental in identifying developmental mechanisms of evolutionary change in morphology. For example, the highly conserved Distalless homeobox-containing gene family is expressed in distal embryonic structures of a diverse group of organisms, including the AER of vertebrates (Dollé et al., 1992; Ferrari et al., 1995;

Bendall and Abate-Shen, 2000). The specific functions of Distalless family genes are largely unknown, but their expression patterns may provide valuable molecular markers that define developmental domains of functionally distal structures. In addition, members of the fibroblast growth factor (FGF) gene family are known to be critical for AER maintenance and limb bud proliferation in mouse and chicken embryos (e.g., Fallon et al., 1994; Niswander et al., 1994; Cohn et al., 1995; Vogel et al., 1996; Dudley and Tabin, 2000). Hence, comparative studies of both Distalless and FGF expression patterns in *E. coqui* and other frogs with an AER may help identify the functional regions of the distal ectoderm and, possibly, the molecular mechanisms associated with the loss of an AER.

Further insight may also be gained though comparative studies of tissue geometry and cell movement at stages before the prospective distal ectoderm is specified. Examining the terminal stages of specification (i.e., once molecular AER markers are expressed) may provide limited clues about the developmental processes that actually influence changes in distal ectoderm morphology. However, studies of cell and tissue interactions at the earliest stages of limb development may lead us to a better understanding of subsequent morphogenetic events (or lack thereof).

Morphological studies of *H. peronii* limb development could also be augmented to include expression assays of genes that regulate the anteroposterior limb axis, digit specification, and cell proliferation. Expression of the gene Sonic Hedgehog (*Shh*), for example, is localized within the early limb bud in the zone of polarizing activity (Riddle et al., 1993; Pearse and Tabin, 1998). *Shh* gene products play a role in limb mesenchyme proliferation and anteroposterior limb patterning, including digit specification (Riddle et al., 1993, Tickle, 1996; Drossopoulou et al., 2000; Dahn and Fallon, 2000). By reducing mesenchyme available for digital condensations or altering anteroposterior patterning, differences in spatial or temporal expression of *Shh* may underlie the loss of phalanges in digit V of the three-digit morph.

SUMMARY AND CONCLUSIONS

Heterochrony likely represents an important class of evolutionary changes in development. However, evolutionary changes in developmental timing may not account for all evolutionary changes in morphology. Changes in tissue geometry and changes in developmental processes themselves (rather than simple chronological permutations of conserved processes) likely are also important sources of ontogenetic diversity. As other authors have noted, by attempting to ascribe all phenotypic alterations to heterochrony, we risk synonomizing heterochrony and morphological change. Morphological novelty can arise from several categories of ontogenetic processes; carefully distinguishing among these possibilities permits more useful generalizations about categories of evolutionary changes in development. If everything is classified as heterochrony, this category becomes so heterogeneous that we lose predictive or explanatory power.

In this chapter, we presented two examples of evolutionary changes in limb development that may not be directly attributable to heterochrony. In *Hemiergis peronii*, an Australian skink, adult limbs with three digits do not result from a simple truncation of a four-digit (or any other) developmental program. Rather than yield fewer numbers of complete digits, truncations of the four-digit program at intermediate developmental stages would yield incomplete digits. Comparative analyses of three- and four-digit developmental sequences revealed little, if anything, about heterochrony as an evolutionary *mechanism* of limb reduction; the observed morphologies in *H. peronii* digit loss are neither predicted nor explained by heterochrony. The *H. peronii* example also highlights the importance of specifying the level of analysis in evolutionary developmental studies. Although we did not observe a truncation of the entire limb skeleton development program in the three-digit morph, truncations are implied at other levels of analysis (e.g., segmentation of digit V). However, we question the utility of ascribing heterochronic changes to nested subsets of developmental programs. If changes in developmental timing are observed, it is critical to explicitly communicate which events or processes are affected.

The loss of an AER in the limb of the direct-developing frog *Eleutherodactylus coqui* likely has little to do with changes in developmental timing. Rather, changes in tissue geometry (heterotopy) probably underlie the loss of this structure, which is present in most other tetrapods. AER loss in *E. coqui* does not preclude normal limb development because the distal limb ectoderm likely retains AER signaling function; a similar situation likely applies in salamanders. Hence, a morphological AER is not required for limb development in all tetrapods. Furthermore, experimental manipulations of *E. coqui* distal ectoderm suggest a possible, unique regenerative capacity not observed in traditional model species of tetrapod limb development. The distal ectoderm of *E. coqui* differs from model species (and most other frogs) in its morphology and developmental plasticity but not in its developmental timing.

Traditional model systems of vertebrate development are indispensable for elucidating the basic patterns and mechanisms of limb ontogeny, but they do not always afford us the opportunity to study the evolution of development. Comparisons of distantly related organisms often lack the resolution to reflect how modest changes in development can result in novel phenotypes. To dissect the ontogentic mechanisms of evolutionary morphological change, we must examine a multitude of diverse species, thereby increasing the resolution at which we examine a developmental process.

As an analogy, an image on a computer screen with a resolution of only a few pixels per inch is sufficient for gaining a broad understanding of a particular photograph, but one needs increased resolution to discern the details of the entire picture. Understanding every possible detail of a single pixel does little to help resolve the larger image. Similarly, resolving all possible details of any one species' limb development program does not necessarily provide a clear understanding of vertebrate limb development in general or how subtle (or not so subtle) changes in that program can yield evolutionary changes in morphology. Only when additional species (each one contributing to the pixel density in our analogy) are studied will a detailed picture of limb development emerge. In this chapter, we try to increase the "resolution" of vertebrate limb development using two nontraditional model species and find that, although some aspects are readily comparable to traditional model species, others are not.

Ultimately, we seek to document how novel phenotypes are produced during early ontogeny. Studies of related organisms in a phylogenetic context provide greater resolution of evolutionary changes in developmental mechanisms than do studies of single or unrelated species. The significance of an evolutionary approach to development, including studies of nontraditional organisms, is that it promises a richer understanding of both anatomical pattern and developmental process in vertebrate morphology.

ACKNOWLEDGMENTS

We thank Miriam Zelditch for the opportunity to contribute to this exciting project and for her comments on an earlier draft of the manuscript. We also thank James Hanken and Nadia Rosenthal for their guidance and generous support and Dr. Hanken for the opportunity to work on *E. coqui*. We gratefully acknowledge the many kind people at El Verde Tropical Field Station in Puerto Rico; Ken Aplin and the staff of the Western Australian Museum; Brad Maryan, Robert Browne-Cooper, and the Western Australian Society of Amateur Herpetologists; and numerous other intrepid collectors for their assistance in the jungles of Puerto Rico and the bush of Western Australia. Michael Richardson contributed his experimental expertise on the *E. coqui* limb work and the photograph for Figure 10.4B, Jennifer Reichert and Tod Reeder kindly shared their unpublished *Hemiergis* phylogeny, and Kathryn Kavanagh and Neil Shubin offered helpful comments on earlier drafts of the manuscript. This work was supported by (separate) grants to M. D. Shapiro and T. F. Carl from the National Science Foundation, the Society for Integrative and Comparative Zoology, and Sigma Xi. The Putnam Expeditionary Fund of the MCZ (to M. D. Shapiro) and the University of Colorado at Boulder (to T. F. Carl) granted additional support.

REFERENCES

Akimenko MA, Ekker M (1995): Anterior duplication of the Sonic hedgehog expression pattern in the pectoral fin buds of zebrafish treated with retinoic acid. Dev Biol 170: 243–247.

Alberch P (1985): Problems with interpretation of developmental sequences. Syst Zool 34: 46–58.

Alberch P, Gale EA (1983): Size dependence during the development of the amphibian foot. Colchicine-induced digital loss and reduction. J Embryol Exp Morphol 76: 177–197.

Alberch P, Gale EA (1985): A developmental analysis of an evolutionary trend: digital reduction in amphibians. Evolution 39: 8–23.

Alberch P, Gould SJ, Oster GF, Wake DB (1979): Size and shape in ontogeny and phylogeny. Paleobiology 5: 296–317.

Bendall AJ, Abate-Shen C (2000): Roles for Msx and Dlx homeoproteins in vertebrate development. Gene 247: 17–31.

Burke AC, Alberch P (1985): The development and homology of the chelonian carpus and tarsus. J Morphol 186: 119–131.

Choquenot D, Greer AE (1989): Intrapopulational and interspecific variation in digital limb bones and presacral vertebrae of the genus *Hemiergis* (Lacertilia, Scincidae). J Herpetol 23: 274–281.

Christen B, Slack JM (1997): FGF-8 is associated with anteroposterior patterning and limb regeneration in *Xenopus*. Dev Biol 192: 455–466.

Christen B, Slack JM (1998): All limbs are not the same. Nature 395: 230–231.

Cogger HG (1992): *Reptiles and Amphibians of Australia* (5th ed). Ithaca, NY: Cornell University Press.

Cohn MJ, Izpisúa-Belmonte J-C, Abud H, Heath JK, Tickle C (1995): FGF-2 application can induce additonal limb bud formation from the flank of chick embryos. Cell 80: 739–746.

Cohn MJ, Tickle C (1996): Limbs: a model for pattern formation within the vertebrate body plan. Trends Genet 12: 253–257.

Cohn MJ, Tickle C (1999): Developmental basis of limblessness and axial patterning in snakes. Nature 399: 474–479.

Dahn RD, Fallon JF (2000): Interdigital regulation of digit identity and homeotic transformation by modulated BMP signaling. Science 289: 438–441.

Dollé P, Price M, Duboule D (1992): Expression of the murine Dlx-1 homeobox gene during facial, ocular and limb development. Differentiation 49: 93–99.

Drossopoulou G, Lewis DE, Sanz-Ezquerro JJ, Nikbakht N, McMahon AP, Hofmann C, Tickle C (2000): A model for anteroposterior patterning of the vertebrate limb based on sequential long- and short-range Shh signalling and Bmp signalling. Development 127: 1337–1348.

Dudley AT, Tabin CJ (2000): Constructive antagonism in limb development. Curr Opin Genet Dev 10: 387–392.

Duellman WE, Trueb L (1986): *Biology of Amphibians.* Baltimore, MD: The Johns Hopkins University Press.

Dufaure JP, Hubert J (1961): Table de développment du lézard vivipare: *Lacerta* (*Zootoca*) *vivipara* Jacquin. Arch Anat Microsc Morphol Exp 50: 309–327.

Elinson RP (1987): Changes in developmental patterns: embryos of amphibians with large eggs. In: Raff RA, Raff EC (eds), *Development as an Evolutionary Process.* New York: Alan R. Liss, Inc, p. 1–21.

Elinson RP (1990): Direct development in frogs: wiping the recapitulationist slate clean. Semin Dev Biol 1: 263–270.

Elinson RP (1994): Leg development in a frog without a tadpole (*Eleutherodactylus coqui*). J Exp Zool 270: 202–210.

Fallon J, López A, Ros M, Savage M, Olwin B, Simandl B (1994): FGF-2, apical ectodermal ridge growth signal for chick limb development. Science 264: 104–107.

Ferrari D, Sumoy L, Gannon J, Sun H, Brown AM, Upholt, WB, Kosher RA (1995): The expression pattern of the Distal-less homeobox-containing gene Dlx-5 in the developing chick limb bud suggests its involvement in apical ectodermal ridge activity, pattern formation, and cartilage differentiation. Mech Dev 52: 257–264.

Gardiner DM, Blumberg B, Komine Y, Bryant SV (1995): Regulation of HoxA expression in developing and regenerating axolotl limbs. Development 121: 1731–1741.

Gould SJ (1977): *Ontogeny and Phylogeny.* Cambridge, MA: Harvard University Press.

Greer AE (1987): Limb reduction in the lizard genus *Lerista.* 1. Variation in the number of phalanges and presacral vertebrae. J Herpetol 21: 267–276.

Greer AE (1989): *The Biology and Evolution of Australian Lizards.* Chipping Norton, NSW: Surrey Beatty & Sons Pty Limited.

Greer AE (1990): Limb reduction in the scincid lizard genus *Lerista.* 2. Variation in the bone complements of the front and rear limbs and the number of postsacral vertebrae. J Herpetol 24: 142–150.

Greer AE (1991): Limb reduction in squamates: identification of the lineages and discussion of the trends. J Herpetol 25: 166–173.

Hamburger V, Hamilton HL (1951): A series of normal stages in the development of the chick embryo. J Morphol 88: 49–92.

Hanken J, Jennings DH, Olsson L (1997): Mechanistic basis of life-history evolution in anuran amphibians: direct development. Am Zool 37: 160–171.

Harrison RG (1918): Experiments on the development of the forelimb of *Ambystoma*, a self-differentiating equipotential system. J Exp Zool 25: 413–461.

Hinchliffe JR, Johnson DR (1980): The development of the vertebrate limb. Oxford: Clarendon Press.

Johnson RL, Tabin CJ (1997): Molecular models for vertebrate limb development. Cell 90: 979–90.

Laforest L, Brown CW, Poleo G, Geraudie J, Tada M, Ekker M, Akimenko MA (1998): Involvement of the *Sonic Hedgehog*, *patched 1* and *bmp2* genes in patterning of the zebrafish dermal fin rays. Development 125: 4175–4184.

Mahmood R, Bresnick J, Hornbruch A, Mahony C, Morton N, Colquhoun K, Martin P, Lumsden A, Dickson C, Mason I (1995): A role for FGF-8 in the initiation and maintenance of vertebrate limb bud outgrowth. Curr Biol 5: 797–806.

Mathur JK, Goel SC (1976): Patterns of chondrogenesis and calcification in the developing limb of the lizard, *Calotes versicolor*. J Morphol 149: 401–420.

Müller GB (1991): Evolutionary transformation of limb pattern: heterochrony and secondary fusion. In: Hinchliffe JR, Hurle JM, Summerbell D (eds), *Developmental Patterning of the Vertebrate Limb.* New York: Plenum Press, p. 395–405.

Müller GB, Alberch P (1990): Ontogeny of the limb skeleton in *Alligator mississippiensis*: developmental invariance and change in the evolution of archosaur limbs. J Morphol 203: 151–164.

Niswander L, Tickle C, Vogel A, Martin G (1994): Function of FGF-4 in limb development. Mol Reprod Dev 39: 83–89.

Oster G, Alberch P (1982): Evolution and bifurcation of developmental programs. Evolution 36: 444–459.

Pearse RV, Tabin CJ (1998): The molecular ZPA. J Exp Zool 282: 677–690.

Presch W (1975): The evolution of limb reduction in the teiid lizard genus *Bachia.* Bull So Calif Acad Sci 74: 113–121.

Raff RA (1996): *The Shape of Life: Genes, Development, and the Evolution of Animal Form.* Chicago, IL: The University of Chicago Press.

Raff RA, Wray GA (1989): Heterochrony: developmental mechanisms and evolutionary results. J Evol Biol 2: 409–434.

Raynaud A (1985): Development of limbs and embryonic limb reduction. In: Gans C, Billet F (eds), *Biology of the Reptilia, Development B.* New York: John Wiley and Sons, p. 59–148.

Raynaud A (1991): Modifications de la structure des mains et des pieds des embyons de lézard vert (*Lacerta viridis* Laur.) sous l'effect de la cytosine-arabinofuranoside. Ann Sci Nat, Zool 12: 11–38.

Raynaud A, Clergue-Gazeau M (1986): Identification des doigts réduits ou manquants des les pattes des embryons de Lézard vert (*Lacerta viridis*) traités par la cystosine-arabinofuranoside. Comparison avec les réductions digitales naturelles des espèces de reptiles serpentiformes. Arch Biol (Bruxelles) 97: 279–299.

Rice S (1997): The analysis of ontogenetic trajectories: when a change in size or shape is not a heterochrony. Proc Natl Acad Sci USA 94: 907–912.

Richardson MK, Carl TF, Hanken J, Elinson RP, Cope C, Bagley P (1998): Limb development and evolution: a frog embryo with no apical ectodermal ridge (AER). J Anat 192: 379–390.

Riddle RD, Johnson RL, Laufer E, Tabin C (1993): Sonic hedgehog mediates the polarizing activity of the ZPA. Cell 75: 1401–1416.

Saunders JW (1948): The proximo-distal sequence of origin of the parts of the chick wing and the role of the ectoderm. J Exp Zool 108: 363–404.

Saunders JW, Cairns JM, Gasleling MT (1957): The role of the apical ridge of ectoderm in the differntiation of the morphological structure and inductive specificity of limb parts of the chick. J Morphol 101: 57–88.

Schauerte HE, van Eeden FJ, Fricke C, Odenthal J, Strahle U, Haffter P (1998): Sonic hedgehog is not required for the induction of medial floor plate cells in the zebrafish. Development 125: 2983–2993

Sewertzoff AN (1904): Die Entwickelung der pentadaktylen Extremitat der Wirbeltiere. Anat Anz 25: 472–494.

Sewertzoff AN (1931): Studien uber die Reduktion der Organe der Wirbeltiere. Zool Jahrb Abt Anat Ontog Tiere 53: 611–700.

Shubin NH, Alberch P (1986): A morphogenetic approach to the origin and basic organization of the tetrapod limb. Evol Biol 20: 319–387.

Shubin N, Tabin C, Carroll S (1997): Fossils, genes and the evolution of animal limbs. Nature 388: 639–648.

Steiner H (1922): Die ontogenetische und phylogenetische Entwicklung des Vogelflugelskelettes. Acta Zool 3: 307–360.

Storr GM, Smith LA, Johnstone RE (1999): *Lizards of Western Australia. I. Skinks* (2nd ed). Perth: Western Australian Museum.

Summerbell D (1974): A quantitative analysis of the effect of excision of the AER from the chick limb-bud. J Embryol Exp Morphol 32: 651–660.

Tickle C (1995): Vertebrate limb development. Curr Opin Genet Dev 5: 478–484.

Tickle C (1996): Genetics and limb development. Dev Genet 19: 1–8.

Torok MA, Gardiner DM, Shubin NH, Bryant SV (1998): Expression of HoxD genes in developing and regenerating axolotl limbs. Dev Biol 200: 225–233.

Townsend DS, Stewart MM (1985): Direct development in *Eleutherodactylus coqui* (Anura: Leptodactylidae): a staging table. Copeia. 423–436.

Tschumi PA (1957): The growth of the hind limb bud of *Xenopus laevis* and its dependence upon the epidermis. J Anat 91: 149–172.

Vogel A, Rodriguez C, Izpisúa-Belmonte J-C (1996): Involvement of FGF-8 in initiation, outgrowth and patterning of the vertebrate limb. Development 122: 1737–1750.

Wake DB (1996): Evolutionary developmental biology—prospects for an evolutionary synthesis at the developmental level. In: Ghiselin MT, Pinna G (eds), *New Perspectives on the History of Life.* San Francisco, CA: California Academy of Sciences, p. 97–107.

Zelditch ML, Fink WL (1996): Heterochrony and heterotopy: Stability and innovation in the evolution of form. Paleobiology 22: 241–254.

Zwilling E (1955): Ectoderm-mesoderm relationship in the development of the chick embryo limb bud. J Exp Zool 128: 423–441.

INDEX